INTRODUCTION TO PROBABILITY AND STATISTICS

INTRODUCTION TO PROBABILITY AND STATISTICS

FROM A BAYESIAN VIEWPOINT

PART 2
INFERENCE

BY

D. V. LINDLEY

Head of the Department of Statistics
University College London

CAMBRIDGE
AT THE UNIVERSITY PRESS
1970

CAMBRIDGE UNIVERSITY PRESS
Cambridge, New York, Melbourne, Madrid, Cape Town, Singapore, São Paulo, Delhi

Cambridge University Press
The Edinburgh Building, Cambridge CB2 8RU, UK

Published in the United States of America by Cambridge University Press, New York

www.cambridge.org
Information on this title: www.cambridge.org/9780521055635

First published 1965
Reprinted 1970
Re-issued in this digitally printed version 2008

A catalogue record for this publication is available from the British Library

ISBN 978-0-521-05563-5 hardback
ISBN 978-0-521-29866-7 paperback

To

M. P. MESHENBERG

in gratitude

CONTENTS

PREFACE

The content of the two parts of this book is the minimum that, in my view, any mathematician ought to know about random phenomena—probability and statistics. The first part deals with probability, the deductive aspect of randomness. The second part is devoted to statistics, the inferential side of our subject.

The book is intended for students of mathematics at a university. The mathematical prerequisite is a sound knowledge of calculus, plus familiarity with the algebra of vectors and matrices. The temptation to assume a knowledge of measure theory and general integration has been resisted and, for example, the concept of a Borel field is not used. The treatment would have been better had these ideas been used, but against this, the number of students able to study random phenomena by means of the book would have been substantially reduced. In any case the intent is only to provide an introduction to the subject, and at that level the measure theory concepts do not appreciably assist the understanding. A statistical specialist should, of course, continue his study further; but only, in my view, at a postgraduate level with the prerequisite of an honours degree in pure mathematics, when he will necessarily know the appropriate measure theory.

A similar approach has been adopted in the level of the proofs offered. Where a rigorous proof is available at this level, I have tried to give it. Otherwise the proof has been omitted (for example, the convergence theorem for characteristic functions) or a proof that omits certain points of refinement has been given, with a clear indication of the presence of gaps (for example, the limiting properties of maximum likelihood). Probability and statistics are branches of applied mathematics—in the proper sense of that term, and not in the narrow meaning that is common, where it means only applications to physics. This being so, some slight indulgence in the nature of the rigour is perhaps permissible. The applied nature of the subject means that the student using this book needs to supplement it with some experience of practical data handling. No attempt has been made to provide such experience in the present book, because it would have made the book too large, and in any case other books that do provide it are readily available. The student should be trained in the use of various computers and be given exercises in the handling of data. In this way he will obtain the necessary understanding of the practical stimuli that have led to the mathematics, and the use of the mathematical results in understanding the numerical data. These two aspects of the subject, the mathematical and the practical, are complementary, and both are necessary for a full understanding of our subject. The fact that only one aspect is fully discussed here ought not to lead to neglect of the other.

The book is divided into eight chapters, and each chapter into six sections. Equations and theorems are numbered in the decimal notation: thus equation 3.5.1 refers to equation 1 of section 5 of chapter 3. Within §3.5 it would be referred to simply as equation (1). Each section begins with a formal list of definitions, with statements and proofs of theorems. This is followed by discussion of these, examples and other illustrative material. In the discussion an attempt has been made to go beyond the usual limits of a formal treatise and to place the ideas in their proper contexts; and to emphasize ideas that are of wide use as distinct from those of only immediate value. At the end of each chapter there is a large set of exercises, some of which are easy, but many of which are difficult. Most of these have been taken from examination papers, and I am grateful for permission from the Universities of London, Cambridge, Aberdeen, Wales, Manchester and Leicester to use the questions in this way. (In order to fit into the Bayesian framework some minor alterations of language have had to be made in these questions. But otherwise they have been left as originally set.)

The second part of the book, the last four chapters, 5 to 8, is devoted to statistics or inference. The first three chapters of the first part are a necessary prerequisite. Much of this part has been written in draft twice: once in an orthodox way with the use only of frequency probabilities; once in terms of probability as a degree of belief. The former treatment seemed to have so many unsatisfactory features, and to be so difficult to present to students because of the mental juggling that is necessary in order to understand the concepts, that it was abandoned. This is not the place to criticize in detail the defects of the purely frequentist approach. Some comments have been offered in the text (§5.6, for example). Here we merely cite as an example the concept of a confidence interval in the usual sense. Technically the confidence level is the long-run coverage of the true value by the interval. In practice this is rarely understood, and is typically regarded as a degree of belief. In the approach adopted here it is so regarded, both within the formal mathematics, and practically. We use the adjective *Bayesian* to describe an approach which is based on repeated uses of Bayes's theorem.

In chapter 5 inference problems for the normal distribution are discussed. The use of Bayes's theorem to modify prior beliefs into posterior beliefs by means of the data is explained, and the important idea of vague prior knowledge discussed. These ideas are extended in chapter 6 to several normal distributions leading as far as elementary analysis of variance. In chapter 7 inferences for other distributions besides the normal are discussed: in particular goodness-of-fit tests and maximum likelihood ideas are introduced. Chapter 8 deals with least squares, particularly with tests and estimation for linear hypotheses. The intention has been to provide a sound basis consisting of the most important inferential concepts. On this basis a student should be able to apply these ideas to more specialised topics in statistics: for example, analysis of more complicated experimental designs and sampling schemes.

The main difficulty in adopting, in a text-book, a new approach to a subject (as the Bayesian is currently new to statistics) lies in adapting the new ideas to current practice. For example, hypothesis testing looms large in standard statistical practice, yet scarcely appears as such in the Bayesian literature. An unbiased estimate is hardly needed in connexion with degrees of belief. A second difficulty lies in the fact that there is no accepted Bayesian school. The approach is too recent for the mould to have set. (This has the advantage that the student can be free to think for himself.) What I have done in this book is to develop a method which uses degrees of belief and Bayes's theorem, but which includes most of the important orthodox statistical ideas within it. My Bayesian friends contend that I have gone too far in this: they are probably right. But, to give an example, I have included an account of significance testing within the Bayesian framework that agrees excellently, in practice, with the orthodox formulation. Most of modern statistics is perfectly sound in practice; it is done for the wrong reason. Intuition has saved the statistician from error. My contention is that the Bayesian method justifies what he has been doing and develops new methods that the orthodox approach lacks. The current shift in emphasis from significance testing to interval estimation within orthodox statistics makes sense to a Bayesian because the interval provides a better description of the posterior distribution.

In interpreting classical ideas in the Bayesian framework I have used the classical terminology. Thus I have used the phrase *confidence interval* for an interval of the posterior distribution. The first time it is introduced it is called a Bayesian confidence interval, but later the first adjective is dropped. I hope this will not cause trouble. I could have used another term, such as posterior interval, but the original term is apposite and, in almost all applications, the two intervals, Bayesian and orthodox, agree, either exactly or to a good approximation. It therefore seemed foolish to introduce a second term for something which, in practice, is scarcely distinguishable from the original.

There is nothing on decision theory, apart from a brief explanation of what it is in §5.6. My task has been merely to discuss the way in which data influence beliefs, in the form of the posterior distribution, and not to explain how the beliefs can be used in decision making. One has to stop somewhere. But it is undoubtedly true that the main flowering of the Bayesian method over the next few years will be in decision theory. The ideas in this book should be useful in this development, and, in any case, the same experimental results are typically used in many different decision-making situations so that the posterior distribution is a common element to them all.

I am extremely grateful to J. W. Pratt, H. V. Roberts, M. Stone, D. J. Bartholomew; and particularly to D. R. Cox and A. M. Walker who made valuable comments on an early version of the manuscript and to D. A. East who gave substantially of his time at various stages and generously helped with the proof-reading. Mrs M. V. Bloor and Miss C. A. Davies made life easier by their efficient and accurate typing. I am most grateful to the University Press for the excellence of their printing.

D.V.L.

Aberystwyth
April 1964

5

INFERENCES FOR NORMAL DISTRIBUTIONS

In this chapter we begin the discussion of the topic that will occupy the rest of the book: the problem of inference, or how degrees of belief are altered by data. We start with the situation where the random variables that form the data have normal distributions. The reader may like to re-read § 1.6, excluding the part that deals with the justification of the axioms, before starting the present chapter.

5.1. Bayes's theorem and the normal distribution

A *random sample of size n* from a distribution is defined as a set of n independent random variables each of which has this distribution (cf. §§ 1.3, 3.3). If for each real number, θ, belonging to a set (say, the set of positive numbers or the set of all real numbers), $f(x|\theta)$ is the density of a random variable, then θ is called a *parameter* of the *family* of distributions defined by the densities $\{f(x|\theta\}$ (cf. the parameter, p, of the binomial distribution, § 2.1). We consider taking a random sample from a distribution with density $f(x|\theta)$ where θ is fixed but unknown and the function f is known. Let H denote our state of knowledge before the sample is taken. Then θ will have a distribution dependent on H; this will be a distribution of probability in the sense of degree of belief, and we denote its density by $\pi(\theta|H)$. As far as possible π will be used for a density of beliefs, p will be used for a density in the frequency sense, the sense that has been used in applications in chapters 2–4. If the random sample is $\mathbf{x} = (x_1, x_2, \ldots, x_n)$ then the density of it will be, because the x_i are independent,

$$\prod_{i=1}^{n} f(x_i|\theta) = p(\mathbf{x}|\theta, H), \quad \text{say.} \tag{1}$$

(The symbol H should strictly also appear after θ on the left-hand side.) The density of beliefs about θ will be changed by the sample according to Bayes's theorem (theorem 1.4.6 and its generalization, equation 3.2.9) into $\pi(\theta|\mathbf{x}, H)$ given by

$$\pi(\theta|\mathbf{x}, H) \propto p(\mathbf{x}|\theta, H)\,\pi(\theta|H) \tag{2}$$

according to the density form of the theorem (equation 3.2.9). The constant of proportionality omitted from (2) is

$$\left\{\int p(\mathbf{x}|\theta, H)\,\pi(\theta|H)\,d\theta\right\}^{-1} = \left\{\pi(\mathbf{x}|H)\right\}^{-1} \tag{3}$$

say, and does not involve θ. H will often be omitted from these and similar equations in agreement with the convention that an event which is always part of the conditioning event is omitted (§1.2). It will accord with the nomenclature of §1.6 if $\pi(\theta|H)$ is called the *prior* density of θ; $p(\mathbf{x}|\theta, H)$, as a function of θ, is called the *likelihood*; and $\pi(\theta|\mathbf{x}, H)$ is called the *posterior* density of θ. We first consider the case of a single observation where $\mathbf{x} = x$ and $f(x|\theta)$ is the normal density.

Theorem 1. *Let x be $N(\theta, \sigma^2)$, where σ^2 is known, and the prior density of θ be $N(\mu_0, \sigma_0^2)$. Then the posterior density of θ is $N(\mu_1, \sigma_1^2)$, where*

$$\mu_1 = \frac{x/\sigma^2 + \mu_0/\sigma_0^2}{1/\sigma^2 + 1/\sigma_0^2}, \quad \sigma_1^{-2} = \sigma^{-2} + \sigma_0^{-2}. \tag{4}$$

(Effectively this is a result for a random sample of size one from $N(\theta, \sigma^2)$.) The likelihood is (omitting H)

$$p(x|\theta) = (2\pi\sigma^2)^{-\frac{1}{2}}\exp\left[-(x-\theta)^2/2\sigma^2\right] \tag{5}$$

and the prior density is

$$\pi(\theta) = (2\pi\sigma_0^2)^{-\frac{1}{2}}\exp\left[-(\theta-\mu_0)^2/2\sigma_0^2\right] \tag{6}$$

so that, omitting any multipliers which do not involve θ and may therefore be absorbed into the constant of proportionality, the posterior density becomes

$$\begin{aligned}\pi(\theta|x) &\propto \exp\left\{-\frac{(x-\theta)^2}{2\sigma^2} - \frac{(\theta-\mu_0)^2}{2\sigma_0^2}\right\}\\ &\propto \exp\{-\tfrac{1}{2}\theta^2(1/\sigma^2 + 1/\sigma_0^2) + \theta(x/\sigma^2 + \mu_0/\sigma_0^2)\}\\ &= \exp\{-\tfrac{1}{2}\theta^2/\sigma_1^2 + \theta\mu_1/\sigma_1^2\}\\ &\propto \exp\{-\tfrac{1}{2}(\theta-\mu_1)^2/\sigma_1^2\},\end{aligned} \tag{7}$$

where, in the first and third stages of the argument, terms not involving θ have been respectively omitted and introduced. The missing constant of proportionality can easily be found from the requirement that $\pi(\theta|x)$ must be a density and therefore integrate to one. It is obviously $(2\pi\sigma_1^2)^{-\frac{1}{2}}$ and so the theorem is proved. (Notice that it is really not necessary to consider the constant at all: it must be such that the integral of $\pi(\theta|x) = 1$, and a constant times (7) is a normal distribution.)

Corollary. *Let* $\mathbf{x} = (x_1, x_2, \ldots, x_n)$ *be a random sample of size* n *from* $N(\theta, \sigma^2)$, *where* σ^2 *is known and the prior density of* θ *is* $N(\mu_0, \sigma_0^2)$. *Then the posterior density of* θ *is* $N(\mu_n, \sigma_n^2)$, *where*

$$\mu_n = \frac{n\bar{x}/\sigma^2 + \mu_0/\sigma_0^2}{n/\sigma^2 + 1/\sigma_0^2}, \quad \sigma_n^{-2} = n\sigma^{-2} + \sigma_0^{-2}, \tag{8}$$

and $\bar{x} = n^{-1} \sum_{i=1}^{n} x_i$.

The likelihood is (equation (1))

$$\begin{aligned} p(\mathbf{x}|\theta) &= (2\pi\sigma^2)^{-n/2} \exp\left[-\sum_{i=1}^{n} (x_i - \theta)^2/2\sigma^2\right] \\ &\propto \exp\left[-\tfrac{1}{2}\theta^2(n/\sigma^2) + \theta\bar{x}(n/\sigma^2)\right] \\ &\propto \exp\left[-\tfrac{1}{2}(\bar{x}-\theta)^2\,(n/\sigma^2)\right], \end{aligned} \tag{9}$$

where again terms not involving θ have been omitted and then introduced. Equation (9) is the same as (5) with $\bar{x}$ for x and n/σ^2 for σ^2, apart from a constant. Hence the corollary follows since (8) is the same as (4), again with $\bar{x}$ for x and n/σ^2 for σ^2.

Random sampling

We have mentioned random samples before (§§1.3, 3.3). They usually arise in one of two situations: either samples are being taken from a large (or infinite) population or repetitions are being made of a measurement of an unknown quantity. In the former situation, if the members of the sample are all drawn according to the rule that each member of the population has the same chance of being in the sample as any other, and the presence of one member in the sample does not affect the chance of any other member being in the sample, then the random variables, x_i, corresponding to each sample member will have a common distribution and be independent, the two conditions for a random sample.† In the second situation the repetitions are made under similar circumstances and one measurement does not influence any other, again ensuring that the two conditions for a random sample are satisfied. The purpose of the repetition in the two cases is the same: to increase one's knowledge, in the first case of the population and in the second case of the unknown quantity—the latter knowledge usually being expressed by saying that the random error of the determination is reduced. In this section we want to see in more detail than previously how the extent of this increase in knowledge can be expressed quantitatively in a special case. To do so it is necessary to express one's knowledge quantitatively; this can be done using probability as a degree of belief (§ 1.6). Thus our task is to investigate, in a special case, the changes in degrees of belief, due to random sampling. Of course, methods other than random sampling are often used in practice (see, for example, Cochran (1953)) but even with other methods the results for random sampling can be applied with modifications and therefore are basic to any sampling study. Only random sampling will be discussed in this book.

Likelihood and parameters

The changes in knowledge take place according to Bayes's theorem, which, in words, says that the posterior probability is proportional to the product of the likelihood and the prior probability. Before considering the theorem and its consequences let us take the three components of the theorem in turn, beginning with the likelihood. The likelihood is equivalently the probability density of the random variables forming the sample and will have the form (1): the product arising from the independence and the multiplication law (equation 3.2.10) and each term involving the same density because of the common distribution. Hence, consideration of the likelihood reduces to consideration of the density of a single member of

† Some writers use the term 'random sample from a population' to mean one taken without replacement (§ 1.3). In which case our results only apply approximately, though the approximation will be good if the sample is small relative to the population.

the sample. This density is purely a frequency idea, empirically it could be obtained through a histogram (§2.4), but is typically unknown to us. Indeed if it were known then there would be little point in the random sampling: for example, if the measurements were made without bias then the mean value of the distribution would be the quantity being measured, so knowledge of the density implies knowledge of the quantity. But when we say 'unknown', all that is meant is 'not completely known', we almost always know something about it; for example that the density increases steadily with the measurement up to a maximum and then decreases steadily—it is unimodal—or that the density is small outside a limited range—it being very unlikely that the random variable is outside this range. Such knowledge, all part of the 'unknown', consists of degrees of belief about the structure of the density and will be expressed through the prior distribution. It would be of great help if these beliefs could be expressed as a density of a finite number of real variables when the tools developed in the earlier chapters could be used. Otherwise it would be necessary to talk about densities, representing degrees of belief, of functions, namely frequency densities, for which adequate tools are not available. It is therefore usual to suppose that the density of x may be written in the form $f(x \mid \theta_1, \theta_2, \ldots, \theta_s)$ depending on a number, s, of real values θ_i called parameters; where the function f is known but the parameters are unknown and therefore have to be described by means of a prior distribution. Since we know how to discuss distributions of s real numbers this can be done; for example, by means of their joint density. It is clear that a very wide class of densities can be obtained with a fixed functional form and varying parameters; such a class is called a *family* and later we shall meet a particularly useful class called the *exponential family* (§5.5). In this section we consider only the case of a single parameter, which is restrictive but still important.

Sometimes f is determined by the structure of the problem: for example, suppose that for each member of a random sample from a population we only observe whether an event A has, or has not, happened, and count the number of times it happens, x, say. Then x has a binomial distribution (§2.1) and the only parameter is $\theta = p$, the probability of A on a single trial. Hence the density is known, as binomial, apart from the value of an unknown parameter: the knowledge of the parameter will have to be expressed through a prior distribution. In other situations such reasons do not exist and we have to appeal to other considerations. In the present section the function f is supposed to be the density of a normal distribution with known variance, σ^2, say, and unknown mean. These are the two parameters of the normal distribution (§2.5). The mean has previously been denoted by μ but we shall now use θ to indicate that it is unknown and reserve μ to denote the true, but unknown, value of the mean. Notice that this true value stays constant throughout the random sampling. The assumption of normality might be reasonable in those cases where past, similar experience has shown that the normal distribution occurs (§3.6). For example, suppose that repeated measurements of a quantity are being made with an instrument of a type which has been in use for many years. Experience with the type might be such that it was known to yield normal distributions and therefore that the same might be true of this particular instrument. If, in addition, the particular instrument had been extensively used in the past, it may have been found to yield results of known, constant accuracy (expressed through the variance or standard deviation). In these circumstances every set of measurements of a single quantity with the instrument could be supposed to have a normal distribution of known variance, only the mean changing with the quantity being measured: if the instrument was free from bias, the mean would be the required value of the quantity. Statistically we say that the scientist is *estimating* the mean of a normal distribution.† This situation could easily occur in routine measurements carried out in the course of inspecting the quality of articles coming off a production line. Often the normal distribution with known variance is assumed with little or no grounds for the normality assumption, simply because it is very easy to handle. That is why it is used here for the first example of quantitative inference.

† Estimation is discussed in §5.2.

Prior distribution

The form of the prior distribution will be discussed in more detail in the next section. Here we consider only the meaning of a prior density of θ. We saw, in §1.6, what a prior probability meant: to say that a hypothesis H has prior probability p means that it is considered that a fair bet of H against not-H would be at odds of p to $(1-p)$. We also saw that a density is a function which, when integrated (or summed), gives a probability (§2.2). Hence a prior density means a function which, when integrated, gives the odds at which a fair bet should be made. If $\pi(\theta)$ is a prior density then $\int_0^\infty \pi(\theta)\,d\theta$ is the prior probability that θ is positive, and a fair bet that θ was positive would be at odds of $\int_0^\infty \pi(\theta)\,d\theta$ to $\int_{-\infty}^0 \pi(\theta)\,d\theta$ on. In particular, to suppose, as has been done in the statement of the theorem, that θ has prior density $N(\mu_0, \sigma_0^2)$ means, among other things, that

(i) θ is believed to be almost certainly within the interval $(\mu_0-3\sigma_0, \mu_0+3\sigma_0)$ and most likely within $(\mu_0-2\sigma_0, \mu_0+2\sigma_0)$ (compare the discussion of the normal distribution in §2.5. We are arbitrarily and conventionally interpreting 'most likely' to mean that the odds against lying outside the interval are 19 to 1).

(ii) θ is just as likely to be near $\mu_0+\lambda\sigma$ as it is to be near $\mu_0-\lambda\sigma$, for any λ, and in particular is equally likely to be greater than μ_0 as less than μ_0.

(iii) Within any interval $(\mu_0-\lambda\sigma_0, \mu_0+\lambda\sigma_0)$ the central values are most probable and the further θ is from the mean, the less likely are values near θ.

Posterior distribution and precision

Often these three reasons are held to be sufficient for assuming a normal prior density. But an additional reason is the theorem, which shows that, with a normal likelihood, the posterior distribution is also normal. The extreme simplicity of the result makes it useful in practice, though it should not be used as an excuse for assuming a normal prior distribution when that assumption conflicts with the actual beliefs.

The posterior distribution is, like the prior distribution, one of probability as a degree of belief and because of the normality enables statements like (i)–(iii) above to be made in the light of the datum, the single value of x, but with different values of the mean and variance. Let us first consider how these are related to the corresponding values of the prior density and the likelihood; taking the variance first because it is the simpler. We shall call the inverse of the variance, the *precision*. The nomenclature is not standard but is useful and is partly justified by the fact that the larger the variance the greater the spread of the distribution and the larger the intervals in (i) above and therefore the smaller the precision. The second equation in (4) therefore reads:

posterior precision equals the datum precision
plus the prior precision (10)

(this, of course, for normal distributions of datum and prior knowledge and a sample of size 1). The datum precision is the inverse of the random error in the terminology of §3.3. It follows therefore that the posterior precision is necessarily greater than the prior precision and that it can be increased either by an increase in the datum precision (that is by a decrease in the variance of the measurement, or the random error) or by an increase in the prior precision. These statements are all quantitative expressions of rather vaguer ideas that we all possess: their great merit is the numerical form that they assume in the statistician's language. It is part of the statistician's task to measure precision. Notice again that it is the inverse of the variance that occurs naturally here, and not the standard deviation which is used in statements (i)–(iii) above. This agrees with earlier remarks (§§2.4, 3.3) that the variance is easier to work with than the more meaningful standard deviation which can always be obtained by a final square root operation.

The first equation in (4) can also be conveniently written in words provided the idea of a weighted mean is used. A *weighted mean* of two values a_1 and a_2 with *weights* w_1 and w_2 is defined as $(w_1a_1+w_2a_2)/(w_1+w_2)$. With equal weights, $w_1 = w_2$, this is the ordinary arithmetic mean. As w_1 increases relative to w_2 the weighted mean moves towards a_1. Only the ratio of weights is relevant and the definition obviously extends to any number of values. In this terminology

the posterior mean equals the weighted mean of the datum value and the prior mean, weighted with their precisions. (11)

Information about θ comes from two sources, the datum and the prior knowledge. Equation (11) says how these should be combined. The more precise the datum the greater is the weight attached to it; the more precise the prior knowledge the greater is the weight attached to it. Again this is a quantitative expression of common ideas.

Small prior precision

With equations (10) and (11), and the knowledge that the posterior density is normal, revised statements like (i)–(iii) can be made with μ_1 and σ_1 replacing μ_0 and σ_0. The most important effect of the datum is that the intervals in these statements will necessarily be narrower, since $\sigma_1 < \sigma_0$; or, expressed differently, the precision will be greater. A most important special case is where the prior precision is very low, or σ_0 is very large. In the limit as $\sigma_0 \to \infty$ (10) and (11) reduce to saying that the posterior precision and mean are equal to the datum precision and value. Furthermore, both posterior and datum distributions are normal. Consequently there are two results which are quite distinct but which are often confused:

(*a*) the datum, x, is normally distributed about a mean μ with variance σ^2;

(*b*) the parameter, θ, is normally distributed about a mean x with variance σ^2.

The first is a statement of frequency probability, the second a statement of (posterior) beliefs. The first is a distribution of x, the second a distribution of θ. So they are truly different. But it is very easy to slip from the statement that x lies within three standard deviations of μ (from (*a*)) to the statement that θ lies within three standard deviations of x (from (*b*)—cf. (i) above). Scientists (and statisticians) quite often do this and we see that it is quite all right for them to do so provided the prior precision is low in comparison with the datum precision and they are dealing with normal distributions.

Precision of random samples

The corollary establishes similar results for a normal random sample of size n instead of for a single value. It can also usefully be expressed in words by saying:

> a random sample of size n from a normal distribution is equivalent to a single value, equal to the mean of the sample, with n times the precision of a single value. (12)

(An important proviso is that normal distributions are assumed throughout.) The result follows since, as explained in the proof of the corollary, (8) is the same as (4) with $\bar{x}$ for x and n/σ^2 for σ^{-2}. The result is related to theorem 3.3.3 that, under the same circumstances, $\mathscr{D}^2(\bar{x}) = \sigma^2/n$, but it goes beyond it because it says that the mean, $\bar{x}$, is equivalent to the whole of the sample. The earlier result merely made a statement about $\bar{x}$, for example that it was a more precise determination than a single observation; the present result says that, with normal distributions it is the most precise determination. This equivalence between $\bar{x}$ and the sample may perhaps be most clearly expressed by considering two scientists both with a random sample of n measurements. Scientist 1 uses the procedure of the corollary. Scientist 2 is careless and only retains the number and mean of his measurements: he then has a single value $\bar{x}$, with mean θ and variance σ^2/n (§3.3), and a normal distribution (§3.5), and can use the theorem. The two scientists end up with the same posterior distribution, provided they had the same prior distribution, so that scientist 2's discarding of the results, except for their number and their mean, has lost him nothing under the assumptions stated. One of a statistician's main tasks used to be called the *reduction of data*, replacing a lot of numbers by a few without losing information, and we see now how this can be done in the special case of a normal distribution of known variance: n values can be replaced by two, n and $\bar{x}$. But remember that this does assume normality, a very important proviso.

Notice that the proof of the corollary does not use any of the distributional theory of §§3.3 and 3.5. It follows a direct and simple calculation in which everything about the sample, except $\bar{x}$ and n, is absorbed into the irrelevant constant of proportionality.

Beliefs about the sample

The constant is not always irrelevant. The general expression is given in equation (3). $\pi(x|H)$, which will now be written $\pi(x)$, can be thought of as the distribution of x obtained as a marginal distribution from the joint distribution of x and θ; the joint distribution being defined by means of the conditional distribution of x for fixed θ and the prior distribution of θ. $\pi(x)$ can be obtained by evaluating (3), but is most easily obtained by using the results on the bivariate normal distribution in §3.2. In the notation of that section: if y, for fixed x, is $N(\alpha+\beta x, \sigma^2)$ and x is $N(\mu_1, \sigma_1^2)$ then y is $N(\mu_2, \sigma_2^2)$ with $\mu_2 = \alpha+\beta\mu_1$, $\sigma_2^2 = \sigma^2+\beta^2\sigma_1^2$. Here, in the notation of this section, x, for fixed θ, is $N(\theta, \sigma^2)$ and θ is $N(\mu_0, \sigma_0^2)$; consequently x is $N(\mu_0, \sigma^2+\sigma_0^2)$. The meaning to be attached to this distribution requires some care. Suppose that, before making the observation x, and therefore when one's knowledge was described by H, one had been asked one's beliefs about what value the observation would have. There are two sources of doubt present about x; first, the doubt that arises from x having a distribution even if θ is known, and secondly the doubt about the value of θ itself. The former is described by $p(x|\theta)$ and the latter by $\pi(\theta)$. They may be compounded in the usual way, as above, and yield $\pi(x) = \int p(x|\theta)\pi(\theta)d\theta$. To illustrate the meaning of $\pi(x)$ we may say that if x_0 is the median of $\pi(x)$ then, before experimentation, a bet at evens would be offered that x would be below x_0. It is a degree of belief in the outcome of the experiment. Notice that $p(x|\theta)$ can also be thought of as a degree of belief in the outcome of the experiment, but when the parameter is known to have the value θ, as distinct from $\pi(x)$ which expresses the belief given H. That $p(x|\theta)$ is both a frequency and belief probability causes no confusion since, as explained in §1.6, when both types of statement are possible the two probabilities concerned are equal. For example, if the parameter were known to have the value θ and x_1 were the median of $p(x|\theta)$, then a bet at evens would be offered that x would be below x_1 because, on a frequency basis, x would be below x_1 one half of the time.

Example

In the preparation of an insulating material, measurements are made of the conductivity using an instrument of known standard deviation which we can suppose, by suitable choice of units, to be one. Prior knowledge of the production process suggests that most likely the conductivity will lie between 15 and 17 (cf. (i) above) and therefore it seems reasonable to suppose a prior distribution of conductivity that, in these units† is $N(16, \frac{1}{4})$; that is, $\mu_0 = 16$, $\sigma_0 = \frac{1}{2}$. Ten readings are made giving values 16·11, 17·37, 16·35, 15·16, 18·82, 18·12, 15·82, 16·34, 16·64, 15·01, with a mean of 16·57. Hence, in the notation of the corollary, $n = 10$, $\sigma = 1$, $\bar{x} = 16{\cdot}57$ and from (8)

$$\mu_{10} = \frac{10 \times 16{\cdot}57 + 4 \times 16}{10+4} = 16{\cdot}41$$

and $$\sigma_{10}^{-2} = 10+4 = 14, \quad \sigma_{10} = 1/\sqrt{14} = 0{\cdot}27.$$

Hence the posterior distribution is $N(16{\cdot}41, (0{\cdot}27)^2)$. On the basis of this it can be said that the mean conductivity of the material most likely lies between 15·87 and 16·95, the most probable value being 16·41. We shall see in the next section the formal language that the statistician uses. Notice that the prior mean is 16, the sample mean is 16·57, and the posterior mean at 16·41 occupies a position between these two but nearer the latter than the former because the sample mean has precision (n/σ^2) of 10 and the prior precision (σ_0^{-2}) is only 4. The posterior precision, at 14, is of course the sum of the two. If the prior knowledge is very imprecise we could allow σ_0 to tend to infinity and attach no weight to it. The posterior mean is then the sample mean, 16·57, but its precision has decreased to 10.

It is instructive to consider what would have been the result

† Notice that in the notation $N(\mu, \sigma^2)$, the second argument is the variance (here 1/4) and not the standard deviation (1/2).

had the prior distribution been $N(10, \frac{1}{4})$, with a much lower mean. The corollary can still be used but the sample and the prior knowledge are incompatible: before sampling the mean was almost certainly less than 11·5 $(\mu_0 + 3\sigma_0)$ yet the sample values are all above 15. It is therefore absurd to take a weighted mean. The experimenter is in the position of obtaining readings around 16 when he had expected readings in the interval (6·64, 13·36) (that is $\mu_0 \pm 3\{\sigma^2 + \sigma_0^2\}^{-\frac{1}{2}}$, from $\pi(x)$). Clearly he would say that somewhere a mistake has been made either in the prior assessment, or in obtaining the data, or even in the arithmetic. All probabilities are conditional and these are conditional on a mistake not having been made; this is part of H. One should bear such points in mind in making any statistical analysis and not proceed only by the text-book rules.

Robustness

A general remark, that will apply to all the methods to be developed in the remainder of this book, is that any inference, any statement of beliefs, is conditional not only on the data but also on the assumptions made about the likelihood. Thus here, the posterior normal distribution depends on the normality assumptions made about the data. It might, or might not, be affected if the data had not a normal distribution. We say that an inference is *robust* if it is not seriously affected by small changes in the assumptions on which it is based. The question of robustness will not be investigated in this book but it is believed that most, if not all, of the inference methods given are reasonably robust.

5.2. Vague prior knowledge and interval estimates for the normal mean

Theorem 1. *A random sample, $\mathbf{x} = (x_1, x_2, \ldots, x_n)$, of size n is taken from $N(\theta, \sigma^2)$, where σ^2 is known. Suppose there exist positive constants; α, ϵ, M and c (small values of α and ϵ are of interest†), such that in the interval I_α defined by*

$$\bar{x} - \lambda_\alpha \sigma/\sqrt{n} \leqslant \theta \leqslant \bar{x} + \lambda_\alpha \sigma/\sqrt{n}, \quad (1)$$

† Strictly the constants depend on $\mathbf{x}$ but the dependence will not be indicated in the notation.

where $2\Phi(-\lambda_\alpha) = \alpha$, the prior density of θ lies between $c(1-\epsilon)$ and $c(1+\epsilon)$: and outside I_α it is bounded by Mc. Then the posterior density $\pi(\theta|\mathbf{x})$ satisfies the inequalities

$$\frac{(1-\epsilon)}{(1+\epsilon)(1-\alpha)+M\alpha}\left(\frac{n}{2\pi\sigma^2}\right)^{\frac{1}{2}}\exp\left\{-\frac{n(\bar{x}-\theta)^2}{2\sigma^2}\right\} \leqslant \pi(\theta|\mathbf{x})$$

$$\leqslant \frac{(1+\epsilon)}{(1-\epsilon)(1-\alpha)}\left(\frac{n}{2\pi\sigma^2}\right)^{\frac{1}{2}}\exp\left\{-\frac{n(\bar{x}-\theta)^2}{2\sigma^2}\right\} \quad (2)$$

inside I_α, and

$$0 \leqslant \pi(\theta|\mathbf{x}) \leqslant \frac{M}{(1-\epsilon)(1-\alpha)}\left(\frac{n}{2\pi\sigma^2}\right)^{\frac{1}{2}} e^{-\frac{1}{2}\lambda_\alpha^2} \quad (3)$$

outside I_α.

The likelihood of the sample is given by equation 5.1.9 which, on inserting a convenient constant, is

$$p(\mathbf{x}|\theta) = \left(\frac{n}{2\pi\sigma^2}\right)^{\frac{1}{2}}\exp\left\{-\frac{n(\bar{x}-\theta)^2}{2\sigma^2}\right\}.$$

Hence by Bayes's theorem (equation 5.1.2), within I_α

$$Ac(1-\epsilon)\left(\frac{n}{2\pi\sigma^2}\right)^{\frac{1}{2}}\exp\left\{-\frac{n(\bar{x}-\theta)^2}{2\sigma^2}\right\} \leqslant \pi(\theta|\mathbf{x})$$

$$\leqslant Ac(1+\epsilon)\left(\frac{n}{2\pi\sigma^2}\right)^{\frac{1}{2}}\exp\left\{-\frac{n(\bar{x}-\theta)^2}{2\sigma^2}\right\}, \quad (4)$$

and outside I_α

$$0 \leqslant \pi(\theta|\mathbf{x}) \leqslant AMc\left(\frac{n}{2\pi\sigma^2}\right)^{\frac{1}{2}}\exp\left\{-\frac{n(\bar{x}-\theta)^2}{2\sigma^2}\right\}, \quad (5)$$

where A is a constant equal to $\pi(\mathbf{x})^{-1}$, equation 5.1.3. The right-hand inequality in (4) gives

$$\int_{I_\alpha}\pi(\theta|\mathbf{x})\,d\theta \leqslant Ac(1+\epsilon)\int_{I_\alpha}\left(\frac{n}{2\pi\sigma^2}\right)^{\frac{1}{2}}\exp\left\{-\frac{n(\bar{x}-\theta)^2}{2\sigma^2}\right\}d\theta$$

$$= Ac(1+\epsilon)\int_{-\lambda_\alpha}^{\lambda_\alpha}\frac{1}{\sqrt{(2\pi)}}e^{-\frac{1}{2}t^2}dt, \quad \text{where} \quad t = \sqrt{n}(\bar{x}-\theta)/\sigma,$$

$$= Ac(1+\epsilon)\,[\Phi(\lambda_\alpha)-\Phi(-\lambda_\alpha)] = Ac(1+\epsilon)(1-\alpha)$$

since $\Phi(-\lambda_\alpha) = 1-\Phi(\lambda_\alpha)$. Similarly, the same integral exceeds $Ac(1-\epsilon)(1-\alpha)$ and, if J_α is the outside of I_α,

$$0 \leqslant \int_{J_\alpha}\pi(\theta|\mathbf{x})\,d\theta \leqslant AMc\alpha.$$

Combining these results we have, since $\int_{I_\alpha+J_\alpha} \pi(\theta|\mathbf{x})\,d\theta = 1$,

$$Ac(1-\epsilon)(1-\alpha) \leqslant 1 \leqslant Ac[(1+\epsilon)(1-\alpha)+M\alpha],$$

and hence

$$\frac{1}{(1+\epsilon)(1-\alpha)+M\alpha} \leqslant Ac \leqslant \frac{1}{(1-\epsilon)(1-\alpha)}. \tag{6}$$

Inserting (6) in (4) immediately gives (2); a similar insertion in (5) gives (3) on remarking that the maximum value of the exponential in J_α occurs at the end-points $\theta = \bar{x} \pm \lambda_\alpha \sigma/\sqrt{n}$ where it has the value $e^{-\frac{1}{2}\lambda_\alpha^2}$.

If ϵ and α are small, so that $e^{-\frac{1}{2}\lambda_\alpha^2}$ is also small, the results say that the posterior distribution of θ is approximately $N(\bar{x}, \sigma^2/n)$.

Definition. If $\pi(\theta|\mathbf{x})$ is any posterior distribution of θ after observing $\mathbf{x}$ and $I_\beta(\mathbf{x})$ is any interval of θ, depending on $\mathbf{x}$ and β, $0 \leqslant \beta \leqslant 1$, such that

$$\int_{I_\beta(\mathbf{x})} \pi(\theta|\mathbf{x})\,d\theta = \beta, \tag{7}$$

then $I_\beta(\mathbf{x})$ is called a 100β % (Bayesian) *confidence interval for* θ (given $\mathbf{x}$). The words in brackets are usually omitted. $I_\beta(\mathbf{x})$ is often called a (Bayesian) *interval estimate* of θ. β is called the (Bayesian) *confidence coefficient*, or (Bayesian) *confidence level*. The definition is not restricted to the case of any particular prior distribution.

Discussion of the theorem

The importance of the theorem lies in the fact that it enables a good approximation to be made to the posterior distribution when sampling from a normal distribution of known variance, without being too precise about the prior distribution. The idea behind the theorem and its proof can be applied to distributions other than the normal, and is an important statistical tool. With the normal sample the likelihood function is given, apart from a constant of proportionality, by equation 5.1.9. If a constant $(n/2\pi\sigma^2)^{\frac{1}{2}}$ is inserted it is proportional to

$$\left(\frac{n}{2\pi\sigma^2}\right)^{\frac{1}{2}} \exp\left\{-\frac{n(\bar{x}-\theta)^2}{2\sigma^2}\right\}, \tag{8}$$

which, as a function of θ, is a normal density function of mean $\bar{x}$ and standard deviation $\sigma/\sqrt{n}$. We know from the properties of the normal density that (8) decreases rapidly as θ departs from $\bar{x}$ and so does the indefinite integral of (8) (the normal distribution function) if† $\theta < \bar{x}$. If θ had the density given by (8) we could say that θ almost certainly lay within three standard deviations ($3\sigma/\sqrt{n}$) of the mean $\bar{x}$ (cf. equation 2.5.13 and 5.1(i)): generally we could say that the probability that θ lay within λ_α standard deviations of $\bar{x}$ (that is, within I_α) is $1-2\Phi(-\lambda_\alpha) = 1-\alpha$, say. But, in fact, θ has not the posterior density given by (8), its density is obtained by multiplying (8) by the prior density and dividing by an appropriate constant. Nevertheless, in I_α, which is the only part of the range of θ where the likelihood is not very small, the prior density from the theorem is itself almost constant. Consequently the true posterior density of θ is, in I_α, almost a constant times (8): this is equation (4). Now what happens outside I_α, in J_α? There the likelihood contribution, (8), is very small. Hence unless the prior density is very large in J_α their product, the posterior density, must be very small, apart again from this multiplying constant. So, with the boundedness condition on $\pi(\theta)$ in J_α, we obtain (5). It only remains to determine the multiplying constant to make the integral one. This is done by evaluating separately the integrals over I_α and J_α. The result is (6). If ϵ, α and $M\alpha$ are small Ac is almost equal to 1 and the limits in (2) differ but little from the density (8), that is $N(\bar{x}, \sigma^2/n)$. The upper bound given in (3) is also small provided α is small, because then $e^{-\frac{1}{2}\lambda_\alpha^2}$ will be small. Hence the posterior distribution is approximately $N(\bar{x}, \sigma^2/n)$.

Example

Consider a numerical example. Suppose $\lambda_\alpha = 1{\cdot}96$, or about 2, so that $\alpha = 0{\cdot}05$. The interval I_α then extends two standard deviations ($2\sigma/\sqrt{n}$) either side of $\bar{x}$. Consider the values of θ within this interval: it may be judged that prior to taking the sample no one value of θ in I_α was more probable than any other so that $\pi(\theta)$ is constant within I_α and we can put $\epsilon = 0$. Consider the values of θ outside I_α: it may be judged

† If $\theta > \bar{x}$, the indefinite integral rapidly approaches one.

that prior to taking the sample no value of θ there is more than twice as probable as any value of θ in I_α; that is $M = 2$. Then (2) says that the true density certainly lies between multiples $(1-\alpha+M\alpha)^{-1}$ and $(1-\alpha)^{-1}$ of the normal density within I_α; that is within multiples $(1\cdot05)^{-1}$ and $(0\cdot95)^{-1}$, or within about 5 % of the normal density. Thus the posterior probability that θ lies within I_α is within 5 % of 0·95, the value for the normal density: this posterior probability is at least 0·90. If λ_α is increased to 3·29 so that $\alpha = 0\cdot001$ then, again taking $M = 2$, the true density lies within 0·1 % of the normal density, and the posterior probability that θ lies within two standard deviations of $\bar{x}$ differs by only 0·1 % from 0·95. These statements, with a quite accurate probability interpretation, can be made without too precise assumptions about the prior density.

Interpretation of the prior distribution

The restrictions on the prior density correspond to a certain amount of vagueness about the value of θ before sampling. Within the effective range of the likelihood no value of θ is thought to be substantially more probable than any other and values outside this range are not much more probable. This is certainly not the attitude of someone who has strong prior ideas about the value of θ, as in the example of §5.1 where the prior distribution was $N(16, \frac{1}{4})$ and $\sigma/\sqrt{n}$ was $1/\sqrt{10}$. In the modification of this example in which σ_0 was allowed to tend to infinity, the prior distribution does satisfy, for large σ_0, the conditions of the theorem, and the posterior distribution is $N(\bar{x}, \sigma^2/n)$. The scientists' practice of passing from statement (*a*) to (*b*) in §5.1 is allowable provided the prior distribution has the degree of vagueness prescribed by the theorem.

Large samples

The theorem also has a useful interpretation as a limiting result. It will surely be true for a wide class of prior distributions that the conditions of the theorem will be satisfied for sufficiently large n. As n increases, the width of the interval I_α, namely $2\lambda_\alpha\sigma/\sqrt{n}$, tends to zero, and therefore the condition that $\pi(\theta)$ be almost constant in I_α becomes less of a restriction. This is a particular case of a general result that as the sample size increases the sample values, the data, influence the posterior distribution more than the prior distribution. We have already had one example of this in equation 5.1.10: the datum precision is n/σ^2, increasing with n, whilst the prior precision remains constant at σ_0^{-2}. Indeed we know from the laws of large numbers (§3.6) that $\bar{x}$ converges (weakly and strongly) to μ, the true value of θ, as $n \to \infty$ so that prior information necessarily becomes irrelevant: at least with one proviso. If $\pi(\theta) = 0$ for any θ then no amount of experimentation will make $\pi(\theta|\mathbf{x})$ other than 0 (a direct consequence of Bayes's theorem). Hence $\pi(\theta) = 0$ represents an extreme pig-headed view that will not be influenced by any evidence. The proviso is therefore that $\pi(\theta) \neq 0$, for any possible parameter value. (If θ is, for example, a binomial parameter then only values $0 \leqslant \theta \leqslant 1$ are possible.) Notice that, as $n \to \infty$, the posterior distribution tends to be concentrated around one particular value, $\bar{x}$, and the variance about this value tends to zero. A distribution which has this form means that one is almost certain that θ is very near to $\bar{x}$ and, in the limit $n \to \infty$, that one knows the value of θ. This acquisition of complete knowledge in the limit as $n \to \infty$ is a general feature of Bayesian inferences.

Uniform prior distribution

In subsequent sections we shall often find it convenient to use a particular prior distribution: namely, one with constant density for all θ, the *uniform distribution* (cf. §3.5). The reason for this is that it is a reasonable approximation to a distribution satisfying the conditions of the theorem, and is particularly easy to handle. It should not be treated too literally as a distribution which says that any value of θ is as likely as any other, but rather as an approximation to one which satisfies the conditions of the theorem; namely, that over the effective range of the likelihood any value of θ is about as likely as any other, and outside the range no value has much higher probability. If θ has infinite range (as with a normal mean) then the uniform distribution cannot be defined in the usual way; there is no $\pi(\theta) = c$ such that

$$\int_{-\infty}^{\infty} c\,d\theta = 1.$$

Instead it must be defined as a *conditional density*: if F is any set of θ of finite length, then the distribution, conditional on θ belonging to F, has density $\pi(\theta|F) = m(F)^{-1}$, where $m(F)$ is the length of F, so that

$$\int_F \pi(\theta|F)\,d\theta = m(F)^{-1}\int_F d\theta = 1.$$

In this way we can talk about the uniform distribution on the real line. As an illustration of the simplicity obtained using the uniform distribution consider the case of normal sampling (§5.1) with $\pi(\theta)$ constant. The likelihood is given by equation 5.1.9 and the posterior density must be the same, apart from the constant of proportionality, which can include the (constant) prior density. The constant of proportionality is obviously $(n/2\pi\sigma^2)^{\frac{1}{2}}$ and the posterior distribution is $N(\bar{x}, \sigma^2/n)$.

Sample information

The uniform distribution, representing vagueness, can often be used even when one's prior distribution is quite precise, for a different reason. Much effort has been devoted by scientists and statisticians to the task of deriving statements that can be made on the basis of the sample alone without prior knowledge (see §5.6 for more details). For example, they have tried to answer the question, what does a sample have to say about θ? Our approach does not attempt to do this: all we claim to do is to show how a sample can change beliefs about θ. What change it effects will depend not only on the sample but also on the prior beliefs. Modern work suggests that the question just posed has no answer. The approach adopted in this book would suggest that what the questioner means is, what does a sample say about θ when the prior knowledge of θ is slight; or when the scientist is not influenced by strong prior opinions about θ. What, in other words, does the sample have to say about θ when the sample provides the bulk of the information about θ? This can be answered using the theorem and the uniform prior distribution, so that even when one has some appreciable prior knowledge of θ one may like to express the posterior beliefs about θ without reference to them. This has the additional advantage of making the results meaningful to a wider range of people, namely those with vague prior beliefs, and the results have a superficial appearance of more objectivity than if a subjective prior distribution had been used.

Problems with substantial prior information

Although the uniform prior distribution will be used in most of the book, for the reason that most interest centres on the situation where we wish to express the contribution of the sample to our knowledge, there are problems wherein the contributions from prior knowledge and the likelihood are comparable. They may sometimes be treated by the methods used for small prior knowledge, in the following way. If, in respect of observations $\mathbf{x}$ now being considered, the prior knowledge has been obtained from past observations $\mathbf{y}$, known to the experimenter, the relevant statement of prior knowledge when discussing $\mathbf{x}$ will be the posterior distribution with respect to the past observations $\mathbf{y}$. Three distributions of degrees of belief are therefore available: (1) before the observations $\mathbf{y}$; (2) after $\mathbf{y}$, but before $\mathbf{x}$; (3) after $\mathbf{x}$ and $\mathbf{y}$. Although the original problem may have involved the passage from (2) to (3), and hence a statement of appreciable prior knowledge due to $\mathbf{y}$, it is possible to consider instead the passage from (1) to (3), incorporating both the observations $\mathbf{x}$ and $\mathbf{y}$, and for this the prior knowledge is weak and may therefore be treated by the methods of this book. An example is given in the alternative proof of theorem 6.6.1. The method is always applicable if $\mathbf{x}$ and $\mathbf{y}$ come from the same exponential family (§5.5) and the distributions (1) to (3) are all members of the conjugate family.

There remain problems in which there is an appreciable amount of prior knowledge but it is not possible to be precise about the observations from which it has been obtained. These cannot be treated directly by the methods of this book, though minor modifications to the methods are usually available. To see the form of this modification we must await the development of further results. For the moment we merely recall the fact that, accepting the arguments summarized in §1.6, any state of knowledge or uncertainty can be expressed numerically and therefore in the form of a probability distribution. The form of this distribution can usually be ascertained by asking and answering questions like, 'Do you think the parameter lies below such and such a value?', in the way discussed in §5.1. If the likelihood belongs to the exponential family and the prior distribution so obtained can be adequately† described by a member of the family conjugate to this family (§5.5) then again the methods appropriate to vague knowledge may be used. For we may suppose that the prior distribution of the conjugate family has been obtained by, possibly imaginary, observations $\mathbf{y}$ from the same exponential family starting from weak prior knowledge. An example is given in example 2 of §6.6 and another is given in §7.2.

Non-normal distributions

It is possible to generalize Theorem 1 to distributions other than the normal. The basic idea of the theorem is that if the prior density is sensibly constant over that range of θ for which the likelihood function is appreciable and not too large over that range of θ for which the likelihood function is small, then the posterior density is approximately equal to the likelihood function (apart from a possible constant). The result thus extends to distributions other than the normal. In extensions, the uniform distribution will often be used to simplify the analysis. The principle of using it has been called by Savage, the *principle of precise measurement*. (Cf. the discussion as $n \to \infty$ above.)

The theorem is often held to be important for another reason but the argument is not satisfactory. In §3.6 we discussed the Central Limit Theorem 3.6.1 and saw that, provided $\mathscr{D}^2(x_i)$ exists, $\bar{x}$ will have, as n increases, an approximately normal distribution, $N(\theta, \sigma^2/n)$: or, more exactly, $n^{\frac{1}{2}}(\bar{x}-\theta)/\sigma$ will have a distribution which tends to $N(0, 1)$. Consequently if $\mathbf{x} = (x_1, x_2, \ldots, x_n)$ is a random sample from any distribution with finite variance, the density‡ of $\bar{x}$ is approximately given by

† More research needs to be carried out on what is meant by 'adequately' here. It is usually difficult to describe how the final inference is affected by changes in the prior distribution.

‡ The central limit theorem concerns the distribution function, not the density function: but a density is a function which, when integrated, gives the distribu-

(5.1.9) again and hence the posterior density, for any sufficiently large random sample, is $N(\bar{x}, \sigma^2/n)$. But the unsatisfactory feature of this reasoning is that the equivalent, for the non-normal sample, of the three lines before (5.1.9) has been omitted. In other words (5.1.9) is not the likelihood of $\mathbf{x}$, but of $\bar{x}$, and hence what is true is that for a sufficiently large random sample from any distribution of finite variance $\pi(\theta \mid \bar{x})$ (but not necessarily $\pi(\theta \mid \mathbf{x})$) is approximately $N(\bar{x}, \sigma^2/n)$. It might happen that inferences based on $\bar{x}$ are substantially different from inferences based on the whole of the sample available. We saw in the last section that with normal samples $\bar{x}$ and n are equivalent to $\mathbf{x}$, but this is not true for samples from other distributions. Indeed, in §3.6, it was shown that if the sample is from a Cauchy distribution then $\bar{x}$ has the same distribution as x_1, and is therefore only as good as a single observation. Obviously $\mathbf{x}$ contains much more information than just x_1, say; though the above reasoning would not be used since the Cauchy distribution does not satisfy the conditions for the Central Limit Theorem. The situation with the Cauchy distribution is extreme. It is probably true that in many situations $\pi(\theta \mid \bar{x})$ will not differ substantially from $\pi(\theta \mid \mathbf{x})$ and it may not be worth the extra labour of finding the latter.

Confidence intervals

The definition of confidence interval given above is not made for mathematical use but for convenience in practical applications. Undoubtedly the proper way to describe an inference is by the relevant distribution of degrees of belief, usually the posterior distribution. But, partly for historical reasons, partly because people find the idea of a distribution a little elaborate and hard to understand (compare the use of features of distributions in §2.4), inferences have not, in practice, been described this way. What is usually required is an answer to a question such as 'in what interval does θ most likely lie?'. To answer this the concept of a confidence interval has been introduced. A value

tion function and in this sense the integration of the normal density gives the correct approximation. The central limit theorem does not say that the density of $\bar{x}$ tends to the normal density, though usually this is true and conditions for it are known.

of β commonly used is 0·95 and then an interval is quoted such that the posterior probability of θ lying in the interval is 0·95. For example, in the situation of the theorem

$$(\bar{x}-1\cdot96\sigma/\sqrt{n},\ \bar{x}+1\cdot96\sigma/\sqrt{n})$$

is an approximate 95 % confidence interval for θ: one would be prepared to bet 19 to 1 against θ lying outside the interval. A higher value of β gives a wider interval and a statement of higher probability: $\beta = 0\cdot99$ is often used, giving $(\bar{x}-2\cdot58\sigma/\sqrt{n},\ \bar{x}+2\cdot58\sigma/\sqrt{n})$ in the normal case. Values of β near one are those most commonly used but there is sometimes an advantage in using $\beta = 0\cdot50$ so that θ is as likely to lie inside the interval as outside it: indeed this used to be common practice. In the normal case the result is $(\bar{x}-0\cdot67\sigma/\sqrt{n},\ \bar{x}+0\cdot67\sigma/\sqrt{n})$ and $0\cdot67\sigma/\sqrt{n}$ is called the *probable error* of $\bar{x}$. Modern practice uses $\sigma/\sqrt{n}$ and calls it the *standard error*. Thus the sample mean, plus or minus two standard errors, gives a 95 % confidence interval for the normal mean.

An important defect of a confidence interval is that it does not say whether any values of θ within the interval are more probable than any others. In the normal case, for example, values at the centre, $\theta = \bar{x}$, are about seven times more probable than values at the ends, $\theta = \bar{x} \pm 1\cdot96\sigma/\sqrt{n}$, in the case of a 95 % interval ($\phi(0) = 0\cdot3989$, $\phi(2) = 0\cdot0540$ and their ratio is 7·39: for the notation see §2.5). The difficulty is often avoided in the case of the normal mean by quoting the interval in the form $\bar{x} \pm 1\cdot96\sigma/\sqrt{n}$; thus, in the numerical example of the last section with $\sigma_0 \to \infty$, the mean conductivity would be described as $16\cdot57 \pm 0\cdot63$. This indicates that the most probable value is 16·57, but that values up to a distance 0·63 away are not improbable. Sometimes the most probable value alone is quoted but this is bad practice as no idea of the precision (in the sense of the last section) is provided. Such a single value is an example of a *point estimate* (as distinct from an interval estimate). Point estimates have their place, either in conjunction with a standard error, or in decision theory, but will not be discussed in detail in this book. Their place in most problems will be taken by sufficient statistics (§5.5).

Estimation

Something, however, must be said about estimates and *estimation* because the terms are so commonly used in statistical writing. Current statistical thinking divides problems of inference rather loosely into problems of estimation and problems of *tests of hypotheses*. The latter will be discussed in §5.6. It is difficult to draw a hard and fast distinction between the two types of problem, but a broad dividing line is obtained by saying that hypothesis testing is concerned with inferences about a fixed value, or set of values, of θ (for example, is $\theta = 7$ a reasonable value, or is it reasonable that $6 \leqslant \theta \leqslant 8$) whereas estimation problems have no such fixed value in mind and, for example, may conclude with a statement that θ lies between 6 and 8 (as with a confidence interval); the values 6 and 8 having no prior significance. The distinction may be illustrated by inferences appropriate to the two situations:

(*a*) Is the resistance of a new alloy less than that of aluminium?

(*b*) What is the resistance of this new alloy?

The former demands a statement relative to the resistance of aluminium: the latter requires no such consideration. We shall define significance tests of hypotheses in §5.6. We shall occasionally use the term, 'an estimate of a parameter', when we refer to a value which seems a fairly likely (often the most likely if it is the maximum likelihood estimate, §7.1) value for the parameter. As just mentioned such an estimate should have associated with it some idea of its precision. A rigorous definition of a least-squares estimate is given in §8.3.

Choice of a confidence interval

Confidence intervals are not unique: there are several intervals containing an assigned amount of the posterior probability. Thus, in the case of the normal mean the infinite interval $\theta > \bar{x} - 1{\cdot}64\sigma/\sqrt{n}$ is a 95 % confidence interval since $\Phi(-1{\cdot}64) = 0{\cdot}05$. In this book a confidence interval will usually be chosen such that the density is larger at any point in the interval than it is at any point outside the interval; points inside are more probable than points outside. This rules out the infinite interval just quoted since $\bar{x}-1\cdot70\sigma/\sqrt{n}$ has higher density, for example, than $\bar{x}+1\cdot80\sigma/\sqrt{n}$. The reason for the choice is that the interval should contain the more probable values and exclude the improbable ones. It is easy to see that this rule gives a unique interval (apart from arbitrariness if there are values of equal probability). It can also be shown that the length of the interval so chosen is typically as small as possible amongst all confidence intervals of prescribed confidence coefficient. This is intuitively obvious: thinking of a probability density in terms of the mass density of a rod (§2.2) the part of the rod having given mass (confidence coefficient) in the least length is obtained by using the heavier parts (maximum density). A rigorous proof can be provided by the reader. Notice that the rule of including the more probable values is not invariant if one changes from θ to some function of θ, $\phi(\theta)$, because in so doing the density changes by a factor $|d\phi/d\theta|$ (theorem 3.5.1) so that the relative values of the densities at different values of θ (and hence ϕ) change and a high density in terms of θ may have a low one in terms of ϕ. Usually, however, there is some reason for using θ instead of ϕ. For example, here it would be unnatural to use anything other than θ, the mean.

Several parameters

The idea of a confidence interval can be generalized to a *confidence set*. If $S_\beta(\mathbf{x})$ is any set of values of θ (not necessarily an interval) with

$$\int_{S_\beta(\mathbf{x})} \pi(\theta \mid \mathbf{x})\, d\theta = \beta, \tag{9}$$

then $S_\beta(\mathbf{x})$ is a confidence set, with confidence coefficient β.

The definition of confidence sets enables one to make confidence statements about several parameters, though this is rarely done. It is only necessary to consider the joint posterior distribution of several parameters and to replace (9) by a multiple integral.

The definition of confidence interval given here is not the usual one and hence the qualification, Bayesian. The usual one will be given later (§5.6) together with our reasons for altering the definition. In most problems the intervals produced according to our definition will be identical with those produced by the usual definition, and from a statistician's practice one would not be able to tell which definition was being used.

5.3. Interval estimates for the normal variance

In this section, as in the last, the data are a random sample from a normal distribution but instead of the mean being unknown it is the variance whose prior and posterior distributions interest us, the mean having a known value. If a random variable x, in this context often denoted by χ^2, has a density

$$\exp\{-\tfrac{1}{2}x\}x^{m-1}/2^m(m-1)!$$
$$= \exp\{-\tfrac{1}{2}\chi^2\}(\chi^2)^{\frac{1}{2}\nu-1}/2^{\frac{1}{2}\nu}(\tfrac{1}{2}\nu-1)! \qquad (1)$$

for $x \geqslant 0$, and zero for $x < 0$, it is said to have a *χ^2-distribution with ν degrees of freedom*, where $\nu = 2m > 0$.

Theorem 1. *Let $\mathbf{x} = (x_1, x_2, \ldots x_n)$ be a random sample of size n from $N(\mu, \theta)$, where μ is known, and the prior density of $\nu_0\sigma_0^2/\theta$ be χ^2 with ν_0 degrees of freedom; then the posterior density of $(\nu_0\sigma_0^2+S^2)/\theta$ is χ^2 with $\nu_0+n = \nu_1$, say, degrees of freedom, where*

$$S^2 = \sum_{i=1}^{n}(x_i-\mu)^2.$$

If the random variable $x = \nu_0\sigma_0^2/\theta$ has prior density given by (1) it follows from theorem 3.5.1 that $\theta = \nu_0\sigma_0^2/x$ has prior density

$$\exp\left\{-\frac{\nu_0\sigma_0^2}{2\theta}\right\}\left(\frac{\nu_0\sigma_0^2}{\theta}\right)^{\frac{1}{2}\nu_0-1}\frac{\nu_0\sigma_0^2}{\theta^2}\Big/2^{\frac{1}{2}\nu_0}(\tfrac{1}{2}\nu_0-1)!$$
$$\propto \exp\left\{-\frac{\nu_0\sigma_0^2}{2\theta}\right\}\theta^{-\frac{1}{2}\nu_0-1}, \qquad (2)$$

since $$dx = -\nu_0\sigma_0^2 d\theta/\theta^2.$$

The likelihood of the sample is

$$p(\mathbf{x}\,|\,\theta) = (2\pi\theta)^{-\frac{1}{2}n}\exp\left\{-\sum_{i=1}^{n}(x_i-\mu)^2/2\theta\right\} \propto e^{-S^2/2\theta}\theta^{-\frac{1}{2}n}. \qquad (3)$$

Hence using Bayes's theorem with (2) and (3) as the values of prior density and likelihood, the posterior density is proportional to

$$e^{-(\nu_0\sigma_0^2+S^2)/2\theta}\,\theta^{-\frac{1}{2}(n+\nu_0)-1}. \qquad (4)$$

It only remains to note that (2) and (4) are the same expressions with $\nu_0\sigma_0^2$ and ν_0 of the former replaced by $\nu_0\sigma_0^2+S^2$ and $\nu_0+n = \nu_1$ in the latter. Since (2) is obtained from χ^2, so is (4) and the theorem follows.

We record, for future reference, the following result.

Theorem 2.

$$\int_0^\infty e^{-A/\theta}\,\theta^{-m}\,d\theta = (m-2)!/A^{m-1} \quad (A > 0, m > 1).$$

The substitution $x = A/\theta$, with $dx = -A\,d\theta/\theta^2$ gives (§2.3)

$$\int_0^\infty e^{-x}x^{m-2}\,dx/A^{m-1} = (m-2)!/A^{m-1}.$$

Example

The situation envisaged in this section where a random sample is available from a normal distribution of known mean but unknown variance rarely occurs in practice: but the main result (theorem 1) is similar in form to, but simpler than, a more important practical result (theorem 5.4.2) and it may help the understanding to take the simpler result first. It can occur when a new piece of testing apparatus is being used for the first time: for example, suppose someone comes along with a new instrument for measuring the conductivity, in the example of §5.1, which he claims has smaller standard deviation than the instrument currently in use (of standard deviation unity). A natural thing to do is to measure several times, n say, the conductivity of a piece of material of known conductivity, μ. If the instrument is known to be free from bias and is assumed to yield a normal distribution, each x_i is $N(\mu, \theta)$ with unknown precision θ^{-1}: the problem is to make inferences about θ (or equivalently θ^{-1}). The snag here is the phrase 'known to be free from bias'. This is rather an unusual situation; normally μ is also unknown, methods for dealing with that problem will be discussed in §5.4.

The χ^2-distribution

The χ^2-distribution was known to Helmert in 1876, but its importance in statistics dates from its introduction in 1900 by

Karl Pearson in a problem to be considered later (§7.4). It is not a complete stranger to us, for suppose y has a $\Gamma(n, \lambda)$-distribution (§2.3) and let $x = 2\lambda y$; then since $dx = 2\lambda dy$ it follows from theorem 3.5.1 that x has the density (1). Hence if y is $\Gamma(n, \lambda)$, $2\lambda y$ is χ^2 with $2n$ degrees of freedom: conversely, if x is χ^2 with ν degrees of freedom then $x/2\lambda$ is $\Gamma(\frac{1}{2}\nu, \lambda)$. The reasons for using the χ^2-distribution will appear when we examine statement (*b*) below. Essentially they are that it is a convenient representation of many states of belief about θ and that it leads to analytically simple results. There is no obligation to use χ^2: it is merely convenient to do so. The reason for the name 'degrees of freedom' for ν will appear later (§6.1).

For $\nu > 2$ the density, (1), of the χ^2-distribution increases from zero at $\chi^2 = 0$ to a maximum at $\chi^2 = \nu - 2$ and then diminishes, tending to zero as $\chi^2 \to \infty$. For $\nu = 2$ the density diminishes from a finite, non-zero value at $\chi^2 = 0$ to zero as $\chi^2 \to \infty$. For $0 < \nu < 2$ the density tends to infinity as $\chi^2 \to 0$, to zero as $\chi^2 \to \infty$ and decreases steadily with χ^2. The mean of the distribution is ν and the variance 2ν. All these results follow from similar properties of the Γ-distribution (§2.3, and equation 2.4.9). For large ν the distribution of χ^2 is approximately normal. A proof of this is provided by relating χ^2 to Γ and using the facts that the sum of independent Γ-distributions with the same parameter has also a Γ-distribution with index equal to the sum of the indices (§3.5), and the Central Limit Theorem (3.6.1) (see §3.6).

Prior distribution

Let us now consider what it means to say that the prior density of $\nu_0\sigma_0^2/\theta$ is χ^2 with ν_0 degrees of freedom, so that the density of θ (the unknown variance of the normal distribution) is given by (2). In this discussion we omit the suffix 0 for simplicity. For all $\nu > 0$ the latter density increases from 0 at $\theta = 0$ to a maximum at $\theta = \sigma^2\nu/(\nu+2)$ and then tends to zero as $\theta \to \infty$. The density has therefore the same general shape for all degrees of freedom as the χ^2-distribution itself has for degrees of freedom in excess of two. To take it as prior density for θ is equivalent to saying that values that are very large or very small are improbable, the most probable value is $\sigma^2\nu/(\nu+2)$ and, because the decrease from the maximum as θ increases is less than the corresponding decrease as θ diminishes, the values above the maximum are more probable than those a similar distance below the maximum. The expectation and variance of θ are most easily found by remarking that since $\nu\sigma^2/\theta$ is χ^2 with ν degrees of freedom $x = \nu\sigma^2/2\theta$ is $\Gamma(\frac{1}{2}\nu, 1)$ with density $e^{-x}x^{\frac{1}{2}\nu-1}/(\frac{1}{2}\nu-1)!$: hence

$$\mathscr{E}(\theta) = E(\nu\sigma^2/2x) = \tfrac{1}{2}\nu\sigma^2\int e^{-x}x^{\frac{1}{2}\nu-2}dx/(\tfrac{1}{2}\nu-1)!$$

$$= \sigma^2\nu/(\nu-2) \tag{5}$$

and

$$\mathscr{E}(\theta^2) = \tfrac{1}{4}\nu^2\sigma^4\int e^{-x}x^{\frac{1}{2}\nu-3}dx/(\tfrac{1}{2}\nu-1)!$$

$$= \sigma^4\nu^2/(\nu-2)(\nu-4),$$

so that

$$\mathscr{D}^2(\theta) = \mathscr{E}(\theta^2)-\mathscr{E}^2(\theta) = 2\sigma^4\nu^2/(\nu-2)^2(\nu-4). \tag{6}$$

These results are only valid if $\nu > 4$, otherwise the variance is undefined (or infinite). If ν is large the values are approximately σ^2 and $2\sigma^4/\nu$. Hence the two numbers at our disposal, σ^2 and ν, enable us to alter the mean and variance of the prior distribution: σ^2 is approximately the mean (and also the most probable value) and $\sqrt{(2/\nu)}$ is approximately the coefficient of variation. Large values of ν correspond to rather precise knowledge of the value of θ prior to the experiment. The two quantities, σ^2 and ν, therefore allow considerable variation in the choice of prior distribution within this class of densities.

We note, for future reference (§7.1), that the prior distribution of θ, like χ^2, tends to normality as $\nu \to \infty$. To prove this consider $z = (\theta-\sigma^2)/(2\sigma^4/\nu)^{\frac{1}{2}}$ with approximately zero mean and unit standard deviation. From (2), z has a density whose logarithm is

$$-\frac{\nu\sigma^2}{2(z\sigma^2\sqrt{(2/\nu)}+\sigma^2)}-(\tfrac{1}{2}\nu+1)\ln(z\sigma^2\sqrt{(2/\nu)}+\sigma^2)$$

$$= -\tfrac{1}{2}\nu(1+z\sqrt{(2/\nu)})^{-1}-(\tfrac{1}{2}\nu+1)\ln(1+z\sqrt{(2/\nu)})$$

omitting terms not involving z. Expansion of this in powers of $\nu^{-\frac{1}{2}}$ gives $-\frac{1}{2}z^2+O(\nu^{-\frac{1}{2}})$ which proves the result.

Simpler results are obtained by considering $\nu\sigma^2/\theta$, the parameter in terms of which the theorem is couched. From the mean and variance of χ^2 quoted above $\mathscr{E}(\nu\sigma^2/\theta) = \nu$ and $\mathscr{D}^2(\nu\sigma^2/\theta) = 2\nu$ so that $\mathscr{E}(\theta^{-1}) = \sigma^{-2}$ and $\mathscr{D}^2(\theta^{-1}) = 2\sigma^{-4}/\nu$. θ^{-1} is what we called the precision in §5.1. The mean precision is therefore σ^{-2} (hence the notation) and the coefficient of variation of the precision is (now exactly) $\sqrt{(2/\nu)}$.

Likelihood

Next consider the likelihood (equation (3)). The remarkable thing about this is that it only depends on $S^2 = \sum_{i=1}^{n} (x_i - \mu)^2$ and n. In other words, the scientist can discard all his data provided only that he retains the sum of squares about the known mean and the size of the sample. The situation is comparable to that in §5.1 where only the mean $\bar{x}$ and n were needed to provide all the information about the mean: here S^2 and n provide all the information about the variance. This is a strong reason for evaluating S^2 and not some other statistic (see §2.4) such as $\sum_{i=1}^{n} |x_i - \mu|$ in order to estimate the variance: but notice the result assumes the distribution from which the sample is taken to be normal (compare the discussion of $\bar{x}$ and the Central Limit Theorem in §5.2).

Posterior distribution

The theorem says that the posterior distribution is of the same form as the prior distribution but with $\nu_0\sigma_0^2$ replaced by $\nu_0\sigma_0^2 + S^2$ and ν_0 by $\nu_0 + n = \nu_1$. The interpretation of the posterior distribution is therefore the same as for the prior distribution with these numerical changes. The result is most easily understood by introducing the quantity

$$s^2 = S^2/n = \sum_{i=1}^{n} (x_i - \mu)^2/n.$$

The random variables $(x_i - \mu)^2$ are independent and identically distributed with mean $\mathscr{E}[(x_i - \mu)^2] = \sigma^2$, the true value of the variance of the sample values. Hence since s^2 is a mean of n such values it tends to σ^2, as $n \to \infty$, by the strong law of large numbers (theorem 3.6.3). Consequently s^2 is an estimate of σ^2 from the sample. Now prior to sampling the most probable value of θ was $\sigma_0^2\nu_0/(\nu_0+2)$ and its mean was $\sigma_0^2\nu_0/(\nu_0-2)$, $\nu_0 > 2$, so that σ_0^2, between these two values, to avoid complications with odd 2's, is an estimate of σ^2 from prior knowledge. The posterior value corresponding to σ_0^2 is $(\nu_0\sigma_0^2+ns^2)/(\nu_0+n)$, which is a weighted mean of prior knowledge (σ_0^2) and sample knowledge (s^2) with weights ν_0 and n. The weights are appropriate because we saw that large values of ν_0 correspond to rather precise knowledge of θ before the experiment and large values of n correspond naturally to a lot of knowledge from the sample. Hence the result for the variance is very similar to that for the mean (equation 5.1.8); in both cases evidence from the sample is combined with the evidence before sampling, using a weighted mean. The posterior density of θ has mean and variance

$$\mathscr{E}(\theta\,|\,\mathbf{x}) = (\nu_0\sigma_0^2+S^2)/(\nu_0+n-2), \tag{7}$$

$$\mathscr{D}^2(\theta\,|\,\mathbf{x}) = 2(\nu_0\sigma_0^2+S^2)^2/(\nu_0+n-2)^2\,(\nu_0+n-4) \tag{8}$$

from (5) and (6). These expressions are valid provided $\nu_0+n-4 > 0$. The approximate results are that the mean is $(\nu_0\sigma_0^2+ns^2)/\nu_1 = \sigma_1^2$, say, and the coefficient of variation is $\surd(2/\nu_1)$. The coefficient of variation is thus reduced by sampling, from $\surd(2/\nu_0)$, corresponding to an increase in our knowledge of θ due to sampling. As $n \to \infty$ the variance of θ tends to zero and the mean behaves like $S^2/n = s^2$, tending to σ^2. So that with increasing sample size we eventually gain almost complete knowledge about the true value of θ. Similar results are available in terms of the precision θ^{-1}.

Vague prior information

In the normal mean situation (§5.1) it was explained that special attention is paid to the case where the prior information is very imprecise: in the notation of that section, $\sigma_0 \to \infty$. Also, in §5.2, it was shown that a wide class of prior distributions could lead to results equivalent to large σ_0 and that a convenient prior distribution would be a uniform distribution of the unknown parameter, there the mean. In these circumstances the weighted mean of μ_0 and $\bar{x}$ depends less on μ_0 and more on $\bar{x}$ and in the limit as $\sigma_0 \to \infty$, θ is $N(\bar{x}, \sigma^2/n)$. Closely related results apply in the normal variance situation. Very imprecise prior information clearly corresponds to $\nu_0 \to 0$; for example, the coefficient of variation of θ^{-1}, $\sqrt{(2/\nu_0)}$, $\to \infty$. Then the weighted mean depends less on the prior value σ_0^2 and more on s^2, and in the limit as $\nu_0 \to 0$ we have the simple result that S^2/θ is χ^2 with n degrees of freedom. We shall not give the equivalent result to theorem 5.2.1 for the variance situation, but there exists a wide class of prior distributions, satisfying conditions similar to those in that theorem, which leads to this last result as an approximation. A convenient prior distribution is obtained by letting $\nu_0 \to 0$ in (2). That expression may be written, with a slight rearrangement, as

$$\exp\left\{-\frac{\nu_0\sigma_0^2}{2\theta}\right\}\left(\frac{\nu_0\sigma_0^2}{2\theta}\right)^{\frac{1}{2}\nu_0}\frac{1}{(\frac{1}{2}\nu_0-1)!}\frac{1}{\theta}$$

and $(\frac{1}{2}\nu_0-1)!$ times this tends, as $\nu_0 \to 0$, to θ^{-1}. Hence the prior distribution suggested has density proportional to θ^{-1}. This is not a usual form of density since, like the uniform distribution of the mean, it cannot be standardized to integrate to one. But like the uniform distribution it can be treated as a conditional density (§5.2). With this prior distribution and the likelihood given by (3), the posterior distribution is obviously, apart from the constant of proportionality, $e^{-S^2/2\theta}\theta^{-\frac{1}{2}n-1}$, which is (4) with $\nu_0 = 0$. Hence the usual form of inference made in the situation of the theorem, that is with imprecise prior knowledge, is

(*b*) the parameter θ is such that S^2/θ is distributed in a χ^2-distribution with n degrees of freedom.

This statement should be compared with (5.1(*b*)) to which corresponds a parallel statement (5.1(*a*)) with, however, quite a different meaning. It is interesting to note that there is a similar parallel statement corresponding to (*b*). To obtain this we recall a remark made in §3.5 and used above that the sum of two (and therefore of any number) of independent Γ-variables with the same parameter has also a Γ-distribution with that parameter and index equal to the sum of the indices. Also, from example 3.5.1 we know that $(x_i-\mu)^2$ is $\Gamma(\frac{1}{2}, 1/2\sigma^2)$, so that $S^2 = \Sigma(x_i-\mu)^2$ is $\Gamma(n/2, 1/2\sigma^2)$, and hence, by the relationship between the Γ- and χ^2-distributions, we have

(*a*) the datum S^2 is such that S^2/σ^2 is distributed in a χ^2-distribution with n degrees of freedom. Warnings about confusing (*a*) and (*b*) similar to those mentioned in §5.1 apply here.

The conditional density θ^{-1} is related to the uniform distribution over the whole real line in the following way. Let $\phi = \ln\theta$. Then the density of ϕ is constant, since $d\phi/d\theta = \theta^{-1}$ (theorem 3.5.1) and as $\ln\theta$ extends from $-\infty$ to $+\infty$, the logarithm of the variance has a uniform distribution over the whole real line. The distribution arises in yet another way. The density of any x_i is $(2\pi\theta)^{-\frac{1}{2}}\exp\{-(x_i-\mu)^2/2\theta\}$. Effect transformations of x_i and θ as follows: $z_i = \ln[(x_i-\mu)^2]$, $\phi = \ln\theta$. The density of z_i is proportional to $e^{\frac{1}{2}(z_i-\phi)}\exp\{-\frac{1}{2}e^{(z_i-\phi)}\}$, a function of $z_i-\phi$ only. Similarly, in the case of the normal mean the density of x_i (without any transformation) is proportional to $\exp\{-\frac{1}{2}(x_i-\theta)^2\}$, a function of $x_i-\theta$ only. If a random variable x has a density $p(x|\theta)$, depending on a single parameter θ, which is a function only of $x-\theta$ then θ is said to be a *location parameter* for x (cf. §2.4). It means that as θ changes the density remains unaltered in shape and merely moves along the x-axis. Our remarks show that the normal mean and the normal log-variance (or equally log-standard deviation) are both location parameters. It therefore seems in good agreement with our choice of a uniform distribution of the mean in §5.2 to use a uniform distribution of log-variance here. Both in the case of the normal mean and the normal variance, it is because the parameter is transformable to one of location that there exist related pairs of statements (*a*) and (*b*). If

$$p(x|\theta) = f(x-\theta)$$

an operation on θ is an operation on $x-\theta$ or equivalently x: thus,

$$\int_{t-1}^{t} f(t-\theta)\,d\theta = \int_0^1 f(u)\,du = \int_w^{w+1} f(x-w)\,dx$$

so that statements made about θ (type (*b*)) correspond to statements about the datum (type (*a*)).

Confidence intervals

It remains to discuss how confidence interval statements may be made about θ on the basis of the posterior distribution. We shall suppose the posterior distribution to be such that S^2/θ is χ^2 with n degrees of freedom ($\nu_0 \to 0$), the case of general ν_0 follows similarly. To make confidence interval statements we need the distribution function of the χ^2-distribution for integral values of ν, the degrees of freedom. This is extensively tabulated; see, for example, Lindley and Miller (1961). To simplify the exposition we introduce some standard terminology. If $F(x)$ is a distribution function, the value $\bar{x}_\alpha$, supposed unique, such that $F(\bar{x}_\alpha) = 1-\alpha$ is called the *upper* 100α % *point* of the distribution: the value $\underline{x}_\alpha$ such that $F(\underline{x}_\alpha) = \alpha$ is called the *lower* 100α % *point*. If x is a random variable with $F(x)$ as its distribution function $p(x \leqslant \underline{x}_\alpha) = p(x > \bar{x}_\alpha) = \alpha$. The numerical values for the normal distribution function given in equations 2.5.13 can be rephrased by saying 'the upper $2\frac{1}{2}$ % point of the standardized normal distribution is 1·96', etc. Let $\bar{\chi}^2_\alpha(\nu)$ [$\underline{\chi}^2_\alpha(\nu)$] be the upper [lower] 100α % points of the χ^2-distribution with ν degrees of freedom. These are tabulated by Lindley and Miller† for $\alpha = 0{\cdot}05, 0{\cdot}025, 0{\cdot}01$ and $0{\cdot}005$. A confidence interval for θ, with confidence coefficient β, can then be found from the result

$$\pi(S^2/\theta > \underline{\chi}^2_\alpha(n) \,|\, \mathbf{x}) = 1-\alpha = \beta,$$

which gives
$$\pi(\theta < S^2/\underline{\chi}^2_\alpha(n) \,|\, \mathbf{x}) = \beta, \tag{9}$$

on rephrasing the description of the event being considered. Here, in a convention which will be adhered to, $\beta = 1-\alpha$, α is typically small and β near one. This (semi-infinite) interval may be enough for some practical purposes for it says, with large β, that one is fairly certain that θ is less than $S^2/\underline{\chi}^2_\alpha(n)$. In the example of conductivity measurements it may be enough to know that the variance does not exceed that limit: or, equivalently, that the precision is greater than $\underline{\chi}^2_\alpha(n)/S^2$. However, in most cases a finite interval is wanted and the usual practice is to obtain it by removing $\frac{1}{2}\alpha$ from the upper and lower ends of the

† The upper ones only are tabulated for $100\alpha = 10$ and $1/10$ %.

χ^2-distribution; that is, an equal amount of probability from each tail. The result is

$$\pi(\underline{\chi}^2_{\frac{1}{2}\alpha}(n) < S^2/\theta < \overline{\chi}^2_{\frac{1}{2}\alpha}(n)\,|\,\mathbf{x}) = \beta,$$

or
$$\pi(S^2/\overline{\chi}^2_{\frac{1}{2}\alpha}(n) < \theta < S^2/\underline{\chi}^2_{\frac{1}{2}\alpha}(n)\,|\,\mathbf{x}) = \beta. \tag{10}$$

But this is arbitrary and although it provides a valid confidence interval there is no justification for preferring it to any other. In order to obtain confidence intervals satisfying the rule given in §5.2 that values inside the interval should have higher density than those outside, it is first necessary to decide which parameter is to be used since, as explained in §5.2, the rule is not invariant under a change, say from θ to θ^{-1}. Since $\ln\theta$ has been assumed uniformly distributed in the prior distribution, so that its density is constant and no value of $\ln\theta$ is more likely than any other, it seems natural to use $\ln\theta$ in the posterior distribution. Tables of $\underline{\chi}^2(n)$ and $\overline{\chi}^2(n)$ such that

$$\pi(S^2/\overline{\chi}^2(n) < \theta < S^2/\underline{\chi}^2(n)\,|\,\mathbf{x}) = \beta, \tag{11}$$

and values of $\ln\theta$ in the interval have higher density than those outside are given in the Appendix. (The dependence on α has been temporarily omitted.)

Example

Consider the example of §5.1. Suppose the ten readings cited there had been made on a material of conductivity known to be 16, with an instrument free from systematic error but of unknown variance. The sum of squares about $\mu = 16$, $S^2 = \sum_{i=1}^{n} (x_i - \mu)^2$, is equal to 16·70, and $n = 10$. The upper and lower 5 % points of the χ^2-distribution with 10 degrees of freedom are $\overline{\chi}^2_{0\cdot05}(10) = 18\cdot31$ and $\underline{\chi}^2_{0\cdot05}(10) = 3\cdot94$. Hence we can say that, with confidence coefficient 95 %,

$$\theta < 16\cdot70/3\cdot94 = 4\cdot24;$$

or the precision is most likely greater than 0·236; or the standard deviation is most likely below 2·06. The upper and lower $2\frac{1}{2}$ % points of the same distribution are 20·48 and 3·25 so that (10) becomes

$$\pi(16\cdot70/20\cdot48 < \theta < 16\cdot70/3\cdot25\,|\,\mathbf{x}) = 0\cdot95$$

or the variance most likely lies between 0·815 and 5·14. The shortest confidence interval for the log-variance is obtained from the tables just referred to in the Appendix: still with $\alpha = 0{\cdot}05$, $\bar{\chi}^2(10) = 21{\cdot}73$ and $\underline{\chi}^2(10) = 3{\cdot}52$ so that (11) says that the variance most likely lies between 0·769 and 4·74. There is not much difference between the intervals given by (10) and (11). Notice that $s^2 = S^2/n = 1{\cdot}67$, which, being a point estimate of θ, is naturally near the centre of these intervals. Notice also that had we been in the position contemplated at the beginning of this section where the instrument had been claimed to have lower variance than the old (with variance unity) then the claim would hardly have been substantiated since $\theta = 1$ is quite a probable value for the variance. We return to this point later in discussing significance tests.

5.4. Interval estimates for the normal mean and variance

Again the data are a random sample from a normal distribution but now both the mean and variance are unknown. The general ideas in §5.1 extend without difficulty to the case where θ is a finite set of real numbers $\boldsymbol{\theta} = (\theta_1, \theta_2, \ldots, \theta_s)$; here $s = 2$. The only change is that it will be necessary to consider the joint density of θ_1 and θ_2, both prior and posterior, instead of the univariate densities.

If a random variable, usually in this context denoted by t, has a density proportional to

$$(1+t^2/\nu)^{-\frac{1}{2}(\nu+1)} \tag{1}$$

for all t, and some $\nu > 0$, it is said to have *Student's t-distribution with ν degrees of freedom*, or simply a *t-distribution*.

Theorem 1. *Let $\mathbf{x} = (x_1, x_2, \ldots, x_n)$ be a random sample of size n from $N(\theta_1, \theta_2)$ and the prior distributions of θ_1 and $\ln \theta_2$ be independent and both uniform over $(-\infty, \infty)$. Then the posterior distribution of θ_1 is such that $n^{\frac{1}{2}}(\theta_1 - \bar{x})/s$ has a t-distribution with $\nu = n-1$ degrees of freedom, where*

$$s^2 = \sum_{i=1}^{n} (x_i - \bar{x})^2/\nu. \tag{2}$$

The joint prior density of θ_1 and θ_2 may, because they are independent, be obtained by taking the product of the separate densities of θ_1 and θ_2, which are constant and proportional to

θ_2^{-1} respectively. Hence the joint prior density is proportional to θ_2^{-1}.

The likelihood is

$$(2\pi\theta_2)^{-\frac{1}{2}n} \exp\left[-\sum_{i=1}^{n}(x_i-\theta_1)^2/2\theta_2\right], \tag{3}$$

so that the joint posterior density is

$$\pi(\theta_1, \theta_2 \mid \mathbf{x}) \propto \theta_2^{-\frac{1}{2}(n+2)} \exp\left[-\sum_{i=1}^{n}(x_i-\theta_1)^2/2\theta_2\right].$$

It is convenient to rewrite $\sum_{i=1}^{n}(x_i-\theta_1)^2$ in an alternative form. We have

$$\begin{aligned}\sum_{i=1}^{n}(x_i-\theta_1)^2 &= \sum_{i=1}^{n}(x_i-\bar{x}+\bar{x}-\theta_1)^2 \\ &= \sum_{i=1}^{n}(x_i-\bar{x})^2+n(\bar{x}-\theta_1)^2 = \nu s^2+n(\bar{x}-\theta_1)^2\end{aligned}$$

so that

$$\pi(\theta_1, \theta_2 \mid \mathbf{x}) \propto \theta_2^{-\frac{1}{2}(n+2)} \exp\left[-\{\nu s^2+n(\bar{x}-\theta_1)^2\}/2\theta_2\right]. \tag{4}$$

To obtain the posterior density of θ_1 it is necessary to integrate (4) with respect to θ_2 (equation 3.2.6). This is easily done using theorem 5.3.2 with $m = \frac{1}{2}(n+2)$ and $A = \frac{1}{2}\{\nu s^2+n(\bar{x}-\theta_1)^2\}$. The result is

$$\begin{aligned}\pi(\theta_1 \mid \mathbf{x}) &\propto \{\nu s^2+n(\bar{x}-\theta_1)^2\}^{-\frac{1}{2}n} \\ &\propto \{1+n(\bar{x}-\theta_1)^2/\nu s^2\}^{-\frac{1}{2}n}.\end{aligned} \tag{5}$$

This is the density for θ_1: to obtain that of $t = n^{\frac{1}{2}}(\bar{x}-\theta_1)/s$ we use theorem 3.5.1. The Jacobian of the transformation from θ_1 to t is a constant, and hence

$$\pi(t \mid \mathbf{x}) \propto \{1+t^2/\nu\}^{-\frac{1}{2}(\nu+1)},$$

which is (1), proving the theorem.

Theorem 2. *Under the same conditions as in theorem* 1 *the posterior distribution of* $\nu s^2/\theta_2$ *is* χ^2 *with* ν *degrees of freedom.*

To obtain the posterior density of θ_2 it is only necessary to integrate (4) again, this time with respect to θ_1: that is, we have to evaluate

$$\theta_2^{-\frac{1}{2}(n+2)} e^{-\nu s^2/2\theta_2} \int_{-\infty}^{\infty} \exp\{-n(\bar{x}-\theta_1)^2/2\theta_2\}\, d\theta_1.$$

The integral is the usual normal integral proportional to $\theta_2^{\frac{1}{2}}$. Hence

$$\pi(\theta_2 \mid \mathbf{x}) \propto e^{-\nu s^2/2\theta_2}\theta_2^{-\frac{1}{2}\nu-1}, \tag{6}$$

and a comparison with equation 5.3.2 immediately establishes the result.

A parameter which enters a likelihood but about which it is not desired to make inferences is called a *nuisance parameter*. For example, θ_2 in theorem 1 and θ_1 in theorem 2 are both nuisance parameters.

Example

The situation of this section, where both the parameters of a normal distribution are unknown, is of very common occurrence and its investigation by a statistician writing under the pseudonym of 'Student' in 1908 is a milestone in statistical theory. It is often reasonable to assume that the data come from some normal distribution, usually because of past experience with similar data, but that the mean and variance are unknown. The theory is then so simple that, regrettably, it is often assumed that the distributions are normal without any better reason for this assumption than the convenience of the results that flow from it. The example of a new measuring instrument considered in §5.3 is one where, granted the normality, the theory of the present section applies. As previously explained any instrument may have bias (systematic error) and random error and these would be represented by the mean minus the known true value of the conductivity, and the variance, and they would typically both be unknown. The analysis of the last section applied when the bias was known.

Student's distribution

Student's t-distribution is not obviously related to any of the distributions previously studied, except the Cauchy distribution (3.6.5) which it equals when $\nu = 1$. Its density is symmetrical about the origin, where it is a maximum, and tends to zero as $t \to \pm\infty$ like $t^{-(\nu+1)}$. The mean is therefore zero, provided $\nu > 1$, otherwise the integral diverges. The missing constant of proportionality in (1) is easily found by integration from $-\infty$ to

$+\infty$, or, because of the symmetry, by doubling the integral from 0 to $+\infty$. The substitution $t^2/\nu = x/(1-x)$ with

$$dt/dx = \tfrac{1}{2}\nu^{\frac{1}{2}}x^{-\frac{1}{2}}(1-x)^{-\frac{3}{2}}$$

gives

$$2\int_0^\infty (1+t^2/\nu)^{-\frac{1}{2}(\nu+1)}\,dt = \nu^{\frac{1}{2}}\int_0^1 x^{-\frac{1}{2}}(1-x)^{\frac{1}{2}\nu-1}\,dx$$

$$= \nu^{\frac{1}{2}}(-\tfrac{1}{2})!\,(\tfrac{1}{2}\nu-1)!/(\tfrac{1}{2}\nu-\tfrac{1}{2})!.$$

This last result follows from the Beta-integral

$$\int_0^1 x^m(1-x)^n\,dx = m!\,n!/(m+n+1)! \tag{7}$$

for $m, n > -1$. (This standard result should be learnt if not already known, as it is frequently needed in statistical calculations.) We saw (§2.5) that $(-\frac{1}{2})! = \sqrt{\pi}$, so that the t-distribution has density

$$\frac{[\frac{1}{2}(\nu-1)]!}{\sqrt{(\nu\pi)}\,(\frac{1}{2}\nu-1)!}\,\frac{1}{(1+t^2/\nu)^{\frac{1}{2}(\nu+1)}}. \tag{8}$$

A similar substitution enables the variance to be found: provided $\nu > 2$ the result is $\nu/(\nu-2)$, otherwise the integral diverges. More important than these results is the behaviour of the distribution as $\nu \to \infty$. Since, from a standard result,

$$\lim_{\nu\to\infty}(1+t^2/\nu)^{-\frac{1}{2}(\nu+1)} = e^{-\frac{1}{2}t^2}$$

it follows that the density tends to the standardized normal density. Thus, for large ν, the distribution is approximately normal. This is certainly not true for small ν since, as we have seen, with $\nu \leqslant 2$ the spread is so great that the variance does not exist. The distribution function is extensively tabulated. Lindley and Miller (1961) give $t_\alpha(\nu)$, the upper $100\alpha\,\%$ point of the distribution on ν degrees of freedom, for $100\alpha = 5, 2\frac{1}{2}, 1, \frac{1}{2}$ and other values. Because of the symmetry about the origin the lower $100\alpha\,\%$ point is $-t_\alpha(\nu)$, and, as with the standardized normal distribution, $p(|t| \geqslant t_{\frac{1}{2}\alpha}(\nu)) = \alpha$. It is worth observing that $t_\alpha(\nu)$ is greater than the corresponding $100\alpha\,\%$ point of the standardized normal distribution, which is to be expected since the variance of $t\,(=\nu/(\nu-2))$ exceeds the normal value of unity. As $\nu \to \infty$ the percentage points of the t-distribution approach quite rapidly those of the normal distribution.

Prior distribution

Since there are two unknown parameters (that is two parameters about which the degrees of belief have to be expressed) it is necessary to consider a joint distribution of them (§3.2). We have already discussed at length in the earlier sections of this chapter the prior distribution of either one when the other is known, $\pi(\theta_1 | \theta_2, H)$, and $\pi(\theta_2 | \theta_1, H)$. Thus, if $\theta_2 = \sigma^2$ we took $\pi(\theta_1 | \sigma^2, H)$ to be $N(\mu_0, \sigma_0^2)$. But we did not say, because we did not need to, whether μ_0 or σ_0^2 would change with σ^2. To do this we would have to answer the question: if someone came along and convinced you that the variance of each observation was not σ^2, but τ^2, say, would this change your beliefs about θ_1? In many situations it would not and therefore we can take the conditional distribution $\pi(\theta_1 | \theta_2, H)$ to be the same for all θ_2; that is, θ_1 and θ_2 are independent. Similar remarks apply, of course, to $\pi(\theta_2 | \theta_1, H)$ provided the precision, θ_2^{-1}, is not affected by the true value θ_1 being measured. We therefore suppose θ_1 and θ_2 to be independent, when their joint distribution can be defined by their separate distributions. In the case of imprecise prior knowledge we saw that it was a reasonable approximation to take θ_1 and $\ln \theta_2$ to be uniformly distributed over the whole real line. This explains the choice of prior distribution in the theorems. These choices are only reasonable in the case of vague knowledge about independent mean and variance: or, in the spirit of theorem 5.2.1, when the densities of θ_1 and $\ln \theta_2$ are appreciably constant over the effective range of the likelihood function and not too large elsewhere.

Likelihood

The rearrangement of the likelihood function to obtain (4) is illuminating, besides being necessary for the integration leading to (6). In the case where only the mean was unknown we saw (§5.1) that the statistician could discard the sample provided that he retained $\bar{x}$ and n: when only the variance was unknown (§5.3), he needed to retain $\Sigma(x_i - \mu)^2$ and n. Equation (4) shows that when both mean and variance are unknown and only the normality is assumed, $\bar{x}$, $s^2 = \sum_{i=1}^{n} (x_i - \bar{x})^2/(n-1)$ and n need be retained. (Notice that s^2 was defined differently in §5.3, where it denoted $\sum_{i=1}^{n} (x_i - \mu)^2/n$.) Again the statistician has achieved a substantial reduction of data (at least for all but very small n) by replacing n numbers by three. The statistics, $\bar{x}$ and s^2 are most easily calculated by first evaluating Σx_i and Σx_i^2 and then

$$\bar{x} = \Sigma x_i/n, \quad s^2 = [\Sigma x_i^2 - (\Sigma x_i)^2/n]/(n-1).$$

The latter result is easily verified (cf. theorem 2.4.1). $\bar{x}$ and s^2 are called the *sample mean* and *sample variance* respectively. Of course there is some ambiguity about what should be retained: Σx_i and Σx_i^2, together with n, would be enough. What is required is at least enough for the likelihood to be evaluated: $\bar{x}$ and s^2 are perhaps the most convenient pair of values.

Posterior distribution of the mean

Now consider theorem 1. The posterior distribution of θ_1 is given by equation (5). The most probable value is $\bar{x}$ and the density falls symmetrically as the difference between θ_1 and $\bar{x}$ increases, the rate of decrease depending mainly on s^2. The situation is perhaps best understood by passing to t and comparing the result with that of theorem 5.2.1 where the variance was known. The posterior distribution of θ_1 was $N(\bar{x}, \sigma^2/n)$, when θ_1 had constant prior density; and this may be expressed by saying:

(i) If the variance, σ^2, is known, $n^{\frac{1}{2}}(\theta_1 - \bar{x})/\sigma$ is $N(0, 1)$.

Compare this with the above theorem which says:

(ii) If the variance is unknown, $n^{\frac{1}{2}}(\theta_1 - \bar{x})/s$ has a t-distribution with $\nu = n-1$ degrees of freedom.

The parallelism between (i) and (ii) is obvious: σ is replaced by s and the normal by the t-distribution. Since σ^2 is unknown in the second situation it has to be replaced by a statistic and $s^2 = \Sigma(x_i - \bar{x})^2/(n-1)$ is a reasonable substitute. We saw in §5.3 that $\Sigma(x_i - \mu)^2/n \to \sigma^2$ as $n \to \infty$, but that statistic cannot be used here to replace σ^2 since μ is unknown, so $\Sigma(x_i-\bar{x})^2/n$ naturally suggests itself. In fact

$$\Sigma(x_i-\bar{x})^2 = \Sigma(x_i-\mu)^2-n(\mu-\bar{x})^2,$$

and since $\bar{x}\to\mu$ as $n\to\infty$, by the strong law of large numbers, $\Sigma(x_i-\bar{x})^2/(n-1)$ tends to σ^2 as $n\to\infty$. The use of $(n-1)$ instead of n is merely for later convenience (§6.1).

The knowledge contained in the posterior distribution of θ_1, when θ_2 is unknown, would be expected to be less than when θ_2 was known, since more knowledge is available prior to sampling in the latter case and s has to replace σ. This is reflected in the use of the t-distribution in place of the normal, for we have just seen that it has larger tails than the normal because the upper percentage points are higher (compare the use of the inverse of the variance to measure the precision: here the precision is $(\nu-2)/\nu < 1$). But as $n\to\infty$, so does ν, and the t-distribution approaches the normal. The lack of prior knowledge about the variance has little effect when the sample size is large.

The close relationship between (i) and (ii) enables confidence interval statements about θ_1 to be made in a similar manner to those in §5.2. For example, from (i) $(\bar{x}-1{\cdot}96\sigma/\sqrt{n}, \bar{x}+1{\cdot}96\sigma/\sqrt{n})$ is a 95% confidence interval for θ_1: the corresponding statement here is obtained from the t-distribution. We have with $\beta = 1-\alpha$,

$$p[-t_{\frac{1}{2}\alpha}(\nu) \leqslant t \leqslant t_{\frac{1}{2}\alpha}(\nu)] = \beta$$

or

$$p[-t_{\frac{1}{2}\alpha}(\nu) \leqslant n^{\frac{1}{2}}(\theta_1-\bar{x})/s \leqslant t_{\frac{1}{2}\alpha}(\nu)] = \beta,$$

giving

$$p[\bar{x}-t_{\frac{1}{2}\alpha}(\nu)s/\sqrt{n} \leqslant \theta_1 \leqslant \bar{x}+t_{\frac{1}{2}\alpha}(\nu)s/\sqrt{n}] = \beta. \qquad (9)$$

With $\beta = 0{\cdot}95$ this gives a 95% confidence interval for θ_1 which has the same structure as the case of known variance, with s for σ and $t_{0{\cdot}025}(\nu)$ for 1·96. For example, with $\nu = 10$, $t_{0{\cdot}025}(10) = 2{\cdot}23$; for $\nu = 20$, $t_{0{\cdot}025}(20) = 2{\cdot}09$. Thus if s is near σ, as would be expected, the confidence interval is longer when the variance is unknown than when it is known. Intervals like (9), which are symmetric about the origin are obviously the shortest ones and satisfy the rule of §5.2 because the t-density is unimodal and symmetric about zero.

Posterior distribution of the variance

The posterior distribution of θ_2 in theorem 2 is very similar to that in §5.3. We put the results together for comparison:

(iii) If the mean, μ, is known, $\Sigma(x_i-\mu)^2/\theta_2$ is χ^2 with n degrees of freedom (theorem 5.3.1 with $\nu_0 = 0$).

(iv) If the mean is unknown, $\Sigma(x_i-\bar{x})^2/\theta_2$ is χ^2 with $(n-1)$ degrees of freedom (theorem 2).

The effect of lack of knowledge of the mean is to replace μ by $\bar{x}$, a natural substitution, and to reduce the degrees of freedom of χ^2 by one. The mean and variance of χ^2 being ν and 2ν respectively, they are both reduced by lack of knowledge on the mean and the distribution is more concentrated. The effect on the distribution of θ_2, which is inversely proportional to χ^2, is just the opposite: the distribution is more dispersed. The mean and variance of θ_2 are proportional to $\nu/(\nu-2)$ and $2\nu^2/(\nu-2)^2\,(\nu-4)$ respectively (equations 5.3.5, 5.3.6), values which increase as ν decreases from n to $n-1$. This is the effect of the loss of information about μ. Confidence interval statements are made as in §5.3, with the degrees of freedom reduced by one and the sum of squares about the sample mean replacing the sum about the true mean.

Example

Consider again the numerical example of the conductivity measurements (§5.1). Suppose the ten readings are from a normal distribution of unknown mean and variance; that is, both the systematic and random errors are unknown. The values of $\bar{x}$ and s^2 are 16·57 and 1·490. Hence a confidence interval for θ_1 with coefficient 95 % is $16{\cdot}57 \pm 2{\cdot}26 \times (1{\cdot}490/10)^{\frac{1}{2}}$; that is $16{\cdot}57 \pm 0{\cdot}87$, which is larger than the interval $16{\cdot}57 \pm 0{\cdot}63$, obtained in §5.2 when $\sigma = 1$. A confidence interval for θ_2 with coefficient 95 % is $(13{\cdot}41/20{\cdot}31 < \theta < 13{\cdot}41/2{\cdot}95)$. This is obtained by inserting the values for $\overline{\chi}^2(9)$ and $\underline{\chi}^2(9)$ from the appendix, instead of $\overline{\chi}^2(10)$ and $\underline{\chi}^2(10)$ used in §5.3, and the value $\Sigma(x_i-\bar{x})^2 = 13{\cdot}41$. The result (0·660, 4·55) gives lower limits than when the mean is known (0·815, 5·14) because part of the variability of the data can now be ascribed to variation of the mean from the known value, $\mu = 16$, used in §5.3.

Joint posterior distribution

The two theorems concern the univariate distributions of mean and variance separately. This is what is required in most applications, the other parameter usually being a nuisance parameter that it is not necessary to consider, but occasionally the joint distribution is of some interest. It is given by equation (4), but, like most joint distributions, is most easily studied by considering the conditional distributions. From (4) and (6)

$$\pi(\theta_1 \mid \theta_2, \mathbf{x}) = \pi(\theta_1, \theta_2 \mid \mathbf{x})/\pi(\theta_2 \mid \mathbf{x})$$
$$\propto \theta_2^{-\frac{1}{2}} \exp[-n(\bar{x}-\theta_1)^2/2\theta_2], \tag{10}$$

that is, $N(\bar{x}, \theta_2/n)$, in agreement with the result of §5.1. But since the variance in (10) is θ_2/n this distribution depends on θ_2 and we see that, after sampling, θ_1 and θ_2 are not independent. The larger θ_2 is, the greater is the spread of the distribution of θ_1, and the less precise the knowledge of θ_1. This is sensible since each sample value, x_i, has variance θ_2 and has more scatter about μ the larger θ_2 is, and so provides less information about μ. If, for example, θ_2 doubles in value, it requires twice as many sample values to acquire the same precision (n/θ_2) about μ.

From (4) and (5), absorbing into the constant of proportionality terms which do not involve θ_2, we have

$$\pi(\theta_2 \mid \theta_1, \mathbf{x}) = \pi(\theta_1, \theta_2 \mid \mathbf{x})/\pi(\theta_1 \mid \mathbf{x})$$
$$\propto \theta_2^{-\frac{1}{2}(\nu+3)} \exp[-\{\nu s^2 + n(\bar{x}-\theta_1)^2\}/2\theta_2], \tag{11}$$

that is, $\{\nu s^2 + n(\bar{x}-\theta_1)^2\}/\theta_2$ is χ^2 with $\nu+1 = n$ degrees of freedom. Again this involves θ_1, as it must since θ_1 and θ_2 are not independent. The numerator of the χ^2 quantity is least when $\theta_1 = \bar{x}$ so that θ_2, which is this numerator divided by χ^2, has least mean and spread when $\theta_1 = \bar{x}$. If $\bar{x} = \theta_1$ then the sample mean has confirmed the value of the mean of the distribution, but otherwise $\bar{x}$ departs from θ_1 and this may be due to θ_2, hence the uncertainty of θ_2 increases. Although θ_1 and θ_2 are not independent their covariance is zero. This easily follows from (10), which shows that $\mathscr{E}(\theta_1 \mid \theta_2, \mathbf{x}) = \bar{x}$ for all θ_2, and equation

3.1.21. We shall see later (§7.1) that in large samples the joint distribution is normal so that the zero covariance will imply independence. The explanation is that in large samples the posterior density will only be appreciable in a small region around $\theta_1 = \bar{x}$, $\theta_2 = s^2$, as may be seen by considering the posterior means and variances of θ_1 and θ_2, and within this region the dependence of one on the other is not strong.

Tabulation of posterior distributions

It may be helpful to explain why, in the theorems of this section and the last, the results have been expressed not as posterior distributions of the parameter concerned but as posterior distributions of related quantities: thus, t instead of θ_1. The reason is that the final stage of an inference is an expression in numerical terms of a probability, and this means that the distribution function involved in the posterior distribution has to be tabulated and the required probability obtained from the tables; or inversely the value of the distribution obtained for a given probability. It is naturally convenient to have firstly, as few tables as possible, and secondly, tables with as few independent variables as possible. In inferences with normal samples it has so far been possible to use only three tables; those of the normal, t- and χ^2-distributions. The normal table involves only one variable, the probability, to obtain a percentage point. The t- and χ^2-distributions involve two, the probability and the degrees of freedom. This tabulation advantage of the normal distribution was explained in connexion with the binomial distribution (§2.5). The t- and χ^2-distributions have similar advantages. Consider, for example, θ_2 with a posterior density given by equation (6), dependent on ν and s^2. Together with the probability, this seems to require a table of triple entry to obtain a percentage point. But $\nu s^2/\theta_2$ has the χ^2-distribution, involving only a table of double entry, the probability and ν. The other variable, s^2, has been absorbed into $\nu s^2/\theta_2$. Similarly the mean and standard deviation of θ_1 can be absorbed into $n^{\frac{1}{2}}(\theta_1 - \bar{x})/s$. Of course it is still a matter of convenience whether, for example, the distribution of $\nu s^2/\theta_2 = \chi^2$ or χ^{-2} is tabulated. Historical reasons play a part but there is a good reason, as we shall see later (§6.1), why s^2, equation (2), should be defined by dividing by $(n-1)$ instead of n.

Notice that in the two cases of posterior distributions of means, the convenient quantity to consider is of the form: the difference between unknown mean and sample mean divided by the standard deviation of the sample mean, or an estimate of it if unknown. (Cf. (i) and (ii) above remembering that $\mathscr{D}^2(\bar{x}) = \sigma^2/n$ from theorem 3.3.3.) This property will be retained in other examples to be considered below. Thus confidence limits for the unknown mean will always be of the form: the sample mean plus or minus a multiple of the (estimated or known) standard deviation of that mean; the multiple depending on the normal, t-, or later (§6.3), Behrens's distribution.

5.5. Sufficiency

Denote by $\mathbf{x}$ any data and by $p(\mathbf{x}|\theta)$ the family of densities of the data depending on a single parameter θ. Then $p(\mathbf{x}|\theta)$, considered as a function of θ, is the likelihood of the data. Usually $\mathbf{x}$, as in the earlier sections of this chapter, will be a random sample from a distribution which depends on θ and $p(\mathbf{x}|\theta)$ will be obtained as in equation 5.1.1. Let $t(\mathbf{x})$ be any real-valued function of $\mathbf{x}$, usually called a *statistic* (cf. §2.4) and let $p(t(\mathbf{x})|\theta)$ denote the density of $t(\mathbf{x})$, which will typically also depend on θ. Then we may write

$$p(\mathbf{x}|\theta) = p(t(\mathbf{x})|\theta)\, p(\mathbf{x}|t(\mathbf{x}), \theta) \tag{1}$$

in terms of the conditional density of the data, given $t(\mathbf{x})$. Equation (1) is general and assumes only the existence of the necessary densities. But suppose the conditional density in (1) does not involve θ, that is

$$p(\mathbf{x}|\theta) = p(t(\mathbf{x})|\theta)\; p(\mathbf{x}|t(\mathbf{x})); \tag{2}$$

then $t(\mathbf{x})$ is said to be a *sufficient statistic* for the family of densities $p(\mathbf{x}|\theta)$, or simply, sufficient for θ. The reason for the importance of sufficient statistics is the following

Theorem 1. *(The sufficiency principle.) If $t(\mathbf{x})$ is sufficient for the family $p(\mathbf{x}|\theta)$; then, for any prior distribution, the posterior distributions given $\mathbf{x}$ and given $t(\mathbf{x})$ are the same.*

We have $\pi(\theta|\mathbf{x}) \propto p(\mathbf{x}|\theta)\,\pi(\theta)$, by Bayes's theorem,

$$= p(t(\mathbf{x})|\theta)\,p(\mathbf{x}|t(\mathbf{x}))\,\pi(\theta), \quad \text{by (2)},$$

$$\propto p(t(\mathbf{x})|\theta)\,\pi(\theta) \propto \pi(\theta|t(\mathbf{x})).$$

The essential point of the proof is that since $p(\mathbf{x}|t(\mathbf{x}))$ does not involve θ it may be absorbed into the constant of proportionality.

It follows that inferences made with $t(\mathbf{x})$ are the same as those made with $\mathbf{x}$. The following theorem is important in recognizing sufficient statistics.

***Theorem* 2.** *(Neyman's factorization theorem.) A necessary and sufficient condition for $t(\mathbf{x})$ to be sufficient for $p(\mathbf{x}|\theta)$ is that $p(\mathbf{x}|\theta)$ be of the form*

$$p(\mathbf{x}|\theta) = f(t(\mathbf{x}), \theta)\,g(\mathbf{x}) \tag{3}$$

for some functions f and g.

The condition is clearly necessary since (2) is of the form (3) with $f(t(\mathbf{x}), \theta) = p(t(\mathbf{x})|\theta)$ and $g(\mathbf{x}) = p(\mathbf{x}|t(\mathbf{x}))$. It only remains to prove that (3) implies (2). Integrate or sum both sides of (3) over all values of $\mathbf{x}$ such that $t(\mathbf{x}) = t$, say. Then the left-hand side will, from the basic property of a density, be the density of $t(\mathbf{x})$ for that value t, and

$$p(t|\theta) = f(t, \theta)\,G(t),$$

where $G(t)$ is obtained by integrating $g(\mathbf{x})$. This result holds for all t, so substituting this expression for f into (3) we have

$$p(\mathbf{x}|\theta) = p(t(\mathbf{x})|\theta)\,g(\mathbf{x})/G(t(\mathbf{x})).$$

But comparing this result with the general result (1) we see that $g(\mathbf{x})/G(t(\mathbf{x}))$ must be equal to the conditional probability of $\mathbf{x}$ given $t(\mathbf{x})$, and that it does not depend on θ, which is the defining property of sufficiency.

The definition of sufficiency extends without difficulty to any number of parameters and statistics. If $p(\mathbf{x}|\boldsymbol{\theta})$ depends on $\boldsymbol{\theta} = (\theta_1, \theta_2, \ldots, \theta_s)$, s real parameters, and

$$\mathbf{t}(\mathbf{x}) = (t_1(\mathbf{x}), t_2(\mathbf{x}), \ldots, t_r(\mathbf{x}))$$

is a set of r real functions such that

$$p(\mathbf{x}|\boldsymbol{\theta}) = p(\mathbf{t}(\mathbf{x})|\boldsymbol{\theta})\,p(\mathbf{x}|\mathbf{t}(\mathbf{x})), \tag{4}$$

then the statistics $t_1(\mathbf{x}), \ldots, t_r(\mathbf{x})$ are said to be *jointly sufficient statistics* for the family of densities $p(\mathbf{x}|\boldsymbol{\theta})$. The sufficiency principle obviously extends to this case: the joint posterior distributions of $\theta_1, \ldots, \theta_s$ are the same given $\mathbf{x}$ or given $\mathbf{t}(\mathbf{x})$. The factorization theorem similarly extends. In both cases it is only necessary to write $\boldsymbol{\theta}$ for θ and $\mathbf{t}$ for t in the proofs.

As explained above, we are often interested in the case where $\mathbf{x}$ is a random sample from some distribution and then the natural question to ask is whether, whatever be the size of sample, there exist jointly sufficient statistics, the number, r in the above notation, being the same for all sizes, n, of sample. The answer is provided by the following

Theorem 3. *Under fairly general conditions, if a family of distributions with densities $f(x_i|\boldsymbol{\theta})$ is to have a fixed number of jointly sufficient statistics $t_1(\mathbf{x}), \ldots, t_r(\mathbf{x})$, whatever be the size of the random sample $\mathbf{x} = (x_1, x_2, \ldots, x_n)$ from a distribution of the family, the densities must be of the form*

$$f(x_i|\boldsymbol{\theta}) = F(x_i)\, G(\boldsymbol{\theta}) \exp\left\{\sum_{j=1}^{r} u_j(x_i)\, \phi_j(\boldsymbol{\theta})\right\}, \tag{5}$$

where $F, u_1, u_2, \ldots, u_r$ are functions of x_i and $G, \phi_1, \phi_2, \ldots, \phi_r$ are functions of $\boldsymbol{\theta}$. Then

$$t_j(\mathbf{x}) = \sum_{i=1}^{n} u_j(x_i) \quad (j = 1, 2, \ldots, r). \tag{6}$$

If $f(x_i|\boldsymbol{\theta})$ is of the form (5) then it is easy to see that $t_1(\mathbf{x}), \ldots, t_r(\mathbf{x})$, defined by (6), are jointly sufficient because

$$\begin{aligned} p(\mathbf{x}|\boldsymbol{\theta}) &= \prod_{i=1}^{n} f(x_i|\boldsymbol{\theta}) \\ &= \left\{\prod_{i=1}^{n} F(x_i)\right\} G(\boldsymbol{\theta})^n \exp\left[\sum_{i=1}^{n}\sum_{j=1}^{r} u_j(x_i)\, \phi_j(\boldsymbol{\theta})\right] \end{aligned}$$

which satisfies the multiparameter form of equation (3), with $g(\mathbf{x}) = \prod_{i=1}^{n} F(x_i)$. It is more difficult to show that the distributions (5) are the only ones giving sufficient statistics for any size of sample, and we omit the proof. The result, in that direction, will not be needed in the remainder of the book.

Any distribution which can be written in the form (5) is said to belong to the *exponential family*.

Equation (2) is obtained from (1) by supposing the conditional probability does not depend on θ. Another specialization of (1) is to suppose the distribution of $t(\mathbf{x})$ does not depend on θ, that is

$$p(\mathbf{x}\,|\,\theta) = p(t(\mathbf{x}))\,p(\mathbf{x}\,|\,t(\mathbf{x}), \theta), \tag{7}$$

when $t(\mathbf{x})$ is said to be an *ancillary statistic* for the family of densities $p(\mathbf{x}\,|\,\theta)$, or simply, ancillary for θ. The definition extends to higher dimensions in the obvious way.

Example: binomial distribution

Though we have already met examples in earlier sections in connexion with a normal distribution, a computationally simpler case is a random sequence of n trials with a constant probability θ of success (§1.3). In other words, $x_i = 1$ (success) or 0 (failure) with

$$f(1\,|\,\theta) = \theta, \quad f(0\,|\,\theta) = 1-\theta, \tag{8}$$

and we have a random sample from the (discrete) density (8). To fix ideas suppose that the results of the n trials are, in order,

$$\mathbf{x} = (1\ 0\ 1\ 1\ 1\ 0\ 0\ 1\ 0\ 0\ 1\ 1).$$

Then the density for this random sample, or the likelihood, is

$$p(\mathbf{x}\,|\,\theta) = \theta^7(1-\theta)^5. \tag{9}$$

Now consider the statistic $t(\mathbf{x}) = \sum_{i=1}^{n} x_i$, the sum of the x_i's, in this case, 7, the number of successes. We know the density of $t(\mathbf{x})$ from the results of §2.1 to be binomial, $B(12, \theta)$, so that

$$p(t(\mathbf{x})\,|\,\theta) = \binom{12}{7}\,\theta^7(1-\theta)^5. \tag{10}$$

Hence it follows from the general result (1) that

$$p(\mathbf{x}\,|\,t(\mathbf{x}), \theta) = \binom{12}{7}^{-1}, \tag{11}$$

which does not depend on θ. Hence $t(\mathbf{x})$, the total number of successes, is sufficient for θ.

Equation (11) means that all random samples with 7 successes and 5 failures, there are $\binom{12}{7}$ of them, have equal probability, $\binom{12}{7}^{-1}$, whatever be the value of θ. To appreciate the importance of this lack of dependence on θ we pass to the sufficiency principle, theorem 1. Suppose for definiteness that θ has a prior distribution which is uniform in the interval [0, 1]; that is, $\pi(\theta) = 1$ for $0 \leqslant \theta \leqslant 1$, the only possible values of θ. Then, given the sample, the posterior density is proportional to $\theta^7(1-\theta)^5$. But, given not the whole sample but only the number of successes out of the 12 values, the posterior density is proportional to $\binom{12}{7}\,\theta^7(1-\theta)^5$, or $\theta^7(1-\theta)^5$ because $\binom{12}{7}$ may be absorbed into the missing constant of proportionality. Hence the two posterior distributions are the same and since they express the beliefs about θ it follows that we believe exactly the same whether we are given the complete description of the 12 values or just the total number of successes. In everyday language this total number is enough, or sufficient, to obtain the posterior beliefs. In other words, the actual order in which the successes and failures occurred is irrelevant, only their total number matters. We say that the total number of successes is a sufficient statistic for the random sequence of n trials with constant probability of success.

Recognition of a sufficient statistic

It was possible in this example to demonstrate sufficiency because we knew the distribution of $t(\mathbf{x}) = \sum_{i=1}^{n} x_i$. In other cases it might be difficult to find the distribution of $t(\mathbf{x})$. The integration or summation might be awkward; and, in any case, one would not like to have to find the distribution of a statistic before testing it for sufficiency. This is where Neyman's factorization theorem is so valuable: it gives a criterion for sufficiency without reference to the distribution of $t(\mathbf{x})$. In the example we see that the likelihood (9) can be written in the form (3) as the product of a function of $t(\mathbf{x}) = \sum_{i=1}^{n} x_i$ and θ and a function of $\mathbf{x}$; indeed, it is already so written with $g(\mathbf{x})$ in (3) trivially equally to 1. Consequently, $t(\mathbf{x})$ is sufficient and its distribution theory is irrelevant. It is also possible to define sufficiency by (3) and deduce (1) and hence theorem 1.

Finally note that the densities (8) are members of the exponential family. For we may write

$$f(x \mid \theta) = (1-\theta)\exp[x\ln\{\theta/(1-\theta)\}] \tag{12}$$

for the only possible values of x, 0 and 1. This is of the form (5) with $F(x) = 1$, $G(\theta) = (1-\theta)$, $u_1(x) = x$, $\phi_1(\theta) = \ln\{\theta/(1-\theta)\}$ and $r = 1$. Notice that $\phi_1(\theta)$ is the logarithm of the odds in favour of $x = 1$ against $x = 0$.

Examples: normal distribution

The idea of sufficiency has occurred in the earlier sections of this chapter and we now show that the statistics considered there satisfy the definition of sufficiency. In §5.1, where the family $N(\theta, \sigma^2)$ was considered, the likelihood was (equation 5.1.9)

$$p(\mathbf{x} \mid \theta) \propto \exp[-\tfrac{1}{2}(\bar{x}-\theta)^2(n/\sigma^2)],$$

where the constant of proportionality depended only on $\mathbf{x}$ (and σ, but this is fixed throughout the argument). So it is immediately of the form (3) with $t(\mathbf{x}) = \bar{x}$, and the sample mean is sufficient. The density is of the exponential form since

$$\begin{aligned} f(x \mid \theta) &= (2\pi\sigma^2)^{-\frac{1}{2}}\exp[-(x-\theta)^2/2\sigma^2] \\ &= [(2\pi\sigma^2)^{-\frac{1}{2}}\exp\{-\tfrac{1}{2}x^2/\sigma^2\}] \\ &\quad \times [\exp\{-\tfrac{1}{2}\theta^2/\sigma^2\}]\exp(x\theta/\sigma^2), \end{aligned} \tag{13}$$

which is of the form (5) with $u_1(x) = x$, $\phi_1(\theta) = \theta/\sigma^2$, $r = 1$ and F and G given by the functions in square brackets.

In §5.3 the family was $N(\mu, \theta)$ and the likelihood (equation 5.3.3) was proportional to $\exp[-S^2/2\theta]\theta^{-\frac{1}{2}n}$ so that

$$S^2 = \Sigma(x_i-\mu)^2$$

is sufficient. Again the family is of the exponential form since

$$f(x \mid \theta) = (2\pi\theta)^{-\frac{1}{2}}\exp[-(x-\mu)^2/2\theta], \tag{14}$$

which is already in the form (5) with $u_1(x) = (x-\mu)^2$, $\phi_1(\theta) = -1/2\theta$, $r = 1$, $G(\theta) = (2\pi\theta)^{-\frac{1}{2}}$ and $F(x) = 1$.

Finally, in §5.4, both parameters of the normal distribution were considered and the family was $N(\theta_1, \theta_2)$. Here, with two parameters, the likelihood (equation 5.4.3) can be written (cf. equation 5.4.4)

$$p(\mathbf{x} \mid \theta_1, \theta_2) = (2\pi\theta_2)^{-\frac{1}{2}n} \exp[-\{\nu s^2 + n(\bar{x}-\theta_1)^2\}/2\theta_2], \quad (15)$$

showing that s^2 and $\bar{x}$ are jointly sufficient for θ_1 and θ_2. Again the family is of the exponential form since

$$f(x \mid \theta_1, \theta_2) = [(2\pi\theta_2)^{-\frac{1}{2}} \exp\{-\tfrac{1}{2}\theta_1^2/\theta_2\}] \exp\{x\theta_1/\theta_2 - x^2/2\theta_2\}, \quad (16)$$

which is of the form (5) with $u_1(x) = x$, $u_2(x) = x^2$, $\phi_1(\boldsymbol{\theta}) = \theta_1/\theta_2$, $\phi_2(\boldsymbol{\theta}) = -1/2\theta_2$, $r = s = 2$, $G(\boldsymbol{\theta})$ equal to the expression in square brackets and $F(x) = 1$. Notice the power of the factorization theorem here: we have deduced the joint sufficiency of s^2 and $\bar{x}$ without knowing their joint distribution.

Minimal sufficient statistics

Consider next the problem of uniqueness of a sufficient statistic: can there exist more than one sufficient statistic? Take the case of a single statistic; the general case of jointly sufficient statistics follows similarly. The key to this is found by thinking of the factorization theorem as saying: given the value of a sufficient statistic, $t(\mathbf{x})$, but not the value of $\mathbf{x}$, the likelihood function can be written down, apart from a constant factor which does not involve θ. This is not true for a general statistic. Suppose that $s(\mathbf{x})$ is another function of the sample values and that $t(\mathbf{x})$ is a function of $s(\mathbf{x})$. Then $s(\mathbf{x})$ must also be sufficient since given $s(\mathbf{x})$ one can calculate $t(\mathbf{x})$ (this is what is meant by saying that $t(\mathbf{x})$ is a function of $s(\mathbf{x})$) and hence the likelihood, so that the latter can be found from $s(\mathbf{x})$. If the situation is reversed and $s(\mathbf{x})$ is a function of a sufficient statistic $t(\mathbf{x})$ then $s(\mathbf{x})$ is not necessarily sufficient. To show this consider the binomial example. With the sample as before let $s(\mathbf{x}) = 1$ or 0 according as Σx_i is even or odd. This is a function of $t(\mathbf{x}) = \Sigma x_i$, a sufficient statistic, and yet is clearly not sufficient, as it is not enough to know whether the number of successes was odd or even to be able to write down the likelihood function. On the other hand, if $s(\mathbf{x})$ is a one-to-one function of $t(\mathbf{x})$; that is, if $t(\mathbf{x})$ is also a function of $s(\mathbf{x})$—to one value of s corresponds one value of t and conversely—then $s(\mathbf{x})$ is certainly sufficient. So we can think of a sequence of functions $t_1(\mathbf{x})$, $t_2(\mathbf{x})$, ..., each one of which is a function of the previous member of the sequence: if $t_j(\mathbf{x})$ is sufficient, so is $t_i(\mathbf{x})$ for $i \leqslant j$. Typically there will be a last member, $t_k(\mathbf{x})$, which is sufficient; no $t_i(\mathbf{x})$, $i > k$, being sufficient. Now every time we move along this sequence, from $t_i(\mathbf{x})$ to $t_{i+1}(\mathbf{x})$, we gain a further reduction of data (cf. §5.1) since $t_{i+1}(\mathbf{x})$, being a function of $t_i(\mathbf{x})$, is a reduction of $t_i(\mathbf{x})$: it can be found from $t_i(\mathbf{x})$ but not necessarily conversely. So $t_k(\mathbf{x})$ is the best sufficient statistic amongst $t_1(\mathbf{x})$, ..., $t_k(\mathbf{x})$ because it achieves the maximum reduction of data. These considerations lead us to adopt the definition: a statistic $t(\mathbf{x})$ which is sufficient and no function of which, other than a one-to-one function, is sufficient is a *minimal sufficient statistic*. It is minimal sufficient statistics that are of interest to us: they are unique except for a one-to-one functional change—in the last example above $\bar{x}$, s^2 and Σx_i, Σx_i^2 are both minimal sufficient—and represent the maximal reduction of data that is possible. In this book, in accord with common practice, we shall omit the adjective minimal, since non-minimal sufficient statistics are not of interest. The whole sample $\mathbf{x}$ is usually a non-minimal sufficient statistic. It is beyond the level of this book to prove that minimal sufficient statistics exist.

Equivalence of sufficient statistics and sample

The above arguments all amount to saying that nothing is lost by replacing a sample by a sufficient statistic. We now show that this is true in a stronger sense: namely, given the value of a sufficient statistic it is possible to produce a sample identical in probability structure with the original sample. The method is to consider the distribution $p(\mathbf{x} \mid t(\mathbf{x}))$ which is completely known: it involves no unknown parameters. Consequently $\mathbf{x}$ can be produced using tables of random sampling numbers in the way discussed in §3.5. For example, in the case of a random sequence of trials with 7 successes out of 12, we saw (equation (11)) that all the $\binom{12}{7}$ sequences with that property were equally likely. So one of them can be chosen at random and the resulting sequence will, whatever be the value of θ, have the same distribution as the original sample, namely that given by equation (9). To take another example, if, in sampling from a normal distribution someone had calculated the median, he might feel that he had some advantage over someone who had lost the sample and had retained only $\bar{x}$ and s^2. But this is not so: if the second person wants the median he can produce a sample from $\bar{x}$ and s^2 using $p(\mathbf{x}|\bar{x}, s^2)$ and calculate the median for that sample. It will have the same probability distribution as the first person's median whatever be the values of θ_1 and θ_2. Of course, the person with the sufficient statistics has lost something if the true distribution is not of the family considered in defining sufficiency. For the definition of sufficiency is relative to a family of distributions. What is sufficient for the family $N(\theta, \sigma^2)$ (namely $\bar{x}$) is not sufficient for the wider family $N(\theta_1, \theta_2)$. Generally the wider the family the more complicated the sufficient statistics. In sampling from (8) the order is irrelevant: but had the probability of success in one trial been dependent on the result of the previous trial, as with a Markov chain (§§4.5, 4.6), then the order would be relevant.

Number of sufficient statistics

It is typically true that the number of statistics (r in the above notation) is not less than the number of parameters s, and it is commonly true that $r = s$, as in all our examples so far. An example with $r > s$ is provided by random samples from a normal distribution with known coefficient of variation, v; that is, from $N(\theta, v^2\theta^2)$. The density is

$$\begin{aligned} f(x|\theta) &= (2\pi v^2\theta^2)^{-\frac{1}{2}} \exp[-(x-\theta)^2/2v^2\theta^2] \\ &= (2\pi v^2\theta^2)^{-\frac{1}{2}} e^{-1/2v^2} \exp\left[-\frac{x^2}{2v^2}\frac{1}{\theta^2}+\frac{x}{v^2}\frac{1}{\theta}\right], \end{aligned} \tag{17}$$

which is of the form (5) but with $r = 2$ and $s = 1$: $\bar{x}$ and s^2 are jointly sufficient for the single parameter θ. Of course, if the distribution is rewritten in terms, not of $\boldsymbol{\theta}$, but of $\phi_1, \ldots, \phi_r$

(equation (5)), then necessarily the numbers of statistics and parameters are equal.

Exponential family

Two comments on theorem 3 are worth making. Suppose we are sampling from an exponential family with prior distribution $\pi(\theta)$ and consider the way in which the beliefs change with the size of sample. We suppose, purely for notational simplicity, that $r = s = 1$ so that

$$f(x_i|\theta) = F(x_i)\,G(\theta)\exp[u(x_i)\,\phi(\theta)].$$

Then
$$\pi(\theta|\mathbf{x}) \propto \pi(\theta)\,G(\theta)^n \exp[t(\mathbf{x})\,\phi(\theta)], \tag{18}$$

with $t(\mathbf{x}) = \sum_{i=1}^{n} u(x_i)$, after a random sample of size n. As n and $t(\mathbf{x})$ change, the family of densities of θ generated by (18) can be described by two parameters, n and $t(\mathbf{x})$. This is because the functional form of $\pi(\theta|\mathbf{x})$ is always known, apart from these values. Consequently the posterior distribution always belongs to a family dependent on two parameters, one of which, n, changes deterministically, and the other, $t(\mathbf{x})$, changes randomly as sampling proceeds. In sampling from an exponential family there is thus a natural family of distributions of θ to consider; namely that given by (18). Furthermore, it is convenient to choose $\pi(\theta)$ to be of the form $G(\theta)^a e^{b\phi(\theta)}$, for suitable a and b, because then the prior distributions fit in easily with the likelihood. This is part of the reason for the choice of prior distributions in theorems 5.1.1 and 5.3.1. For general r and s, the joint posterior density of the s parameters depends on $(r+1)$ parameters, n and the statistics $\sum_{i=1}^{n} u_j(x_i)$.

The distributions (18) are said to belong to the family which is *conjugate* to the corresponding member of the exponential family. The nomenclature is due to Raiffa and Schlaifer (1961). One advantage in using the conjugate family is that methods based on it essentially reduce to the methods appropriate to vague prior knowledge, like most of the methods developed in this book. The point has been mentioned in §5.2 and examples are available in the alternative proof of theorem 6.6.1 and in

§7.2. If we have samples from the exponential family (still considering $r = s = 1$) and decide the prior distribution is adequately represented by the conjugate distribution (18) with $n = a$, $t(\mathbf{x}) = b$, then the observations are adequately summarized by the sample size, n, and $\sum_{i=1}^{n} u(x_i)$, and the posterior distribution is of the conjugate family with parameters $a+n$ and $b+\sum_{i=1}^{n} u(x_i)$. This posterior distribution is equivalent to that which would have been obtained with a sample of size $a+n$ yielding sufficient statistic $b+\sum_{i=1}^{n} u(x_i)$ with prior knowledge $\pi(\theta)$ (in (18)). If $\pi(\theta)$ corresponds to vague prior knowledge, we may use the methods available for that case.

Regularity conditions

A second comment on theorem 3 concerns the first few words 'under fairly general conditions'. We have no wish to weary the reader with the precise conditions, but the main condition required for the result to obtain is that the range of the distribution does not depend† on θ. If the range is a function of θ then the situation can be quite different even in simple cases. Consider, for example, a uniform distribution in the interval $(0, \theta)$ with, as usual, θ as the unknown parameter. Here

$$f(x_i \mid \theta) = \theta^{-1}, \quad 0 \leqslant x_i \leqslant \theta;$$

and is otherwise zero, and

$$p(\mathbf{x} \mid \theta) = \theta^{-n}, \quad \text{if} \quad 0 \leqslant x_i \leqslant \theta \quad \text{for all } i;$$

and is otherwise zero. Consequently the likelihood is θ^{-n}, provided $\theta > \max_i x_i$, and is otherwise zero. Hence the likelihood depends only on $\max_i x_i$ which is sufficient, and yet the density is not of the exponential family. Difficulties of this sort usually arise where the range of possible values of the random variable depends on the unknown parameter. In the other cases studied in this chapter the range is the same for all parameter values.

† Readers familiar with the term may like to know that difficulties arise when the distributions are not *absolutely continuous* with respect to each other.

Ancillary statistics

Ancillary statistics do not have the same importance as sufficient statistics. Their principal use is to say that the distribution of $t(\mathbf{x})$ is irrelevant to inferences about θ, because the likelihood, $p(\mathbf{x} \mid \theta)$, is proportional to $p(\mathbf{x} \mid t(\mathbf{x}), \theta)$. In other words, $t(\mathbf{x})$ may be supposed constant. Important examples of their use are given in theorems 7.6.2 and 8.1.1. A simple example is provided by considering a scientist who wishes to take a random sample from a distribution with density $f(x \mid \theta)$. Suppose that the size of the random sample he is able to take depends upon factors such as time at his disposal, money available, etc.: factors which are in no way dependent on the value of θ, but which may determine a probability density, $p(n)$, for the sample size. Then the density of the sample is given by

$$p(\mathbf{x} \mid \theta) = p(n) \prod_{i=1}^{n} f(x_i \mid \theta).$$

It follows that $t(\mathbf{x}) = n$ is an ancillary statistic. Consequently the scientist can suppose n fixed at the value he actually obtained and treat the problem as one of a random sample of fixed size from $f(x \mid \theta)$. In particular, if $f(x \mid \theta)$ is normal, the methods of this chapter will be available.

Nuisance parameters

Finally, notice something that has *not* been defined. If there is more than one parameter; say two, θ_1 and θ_2, we have not defined the notion of statistics sufficient or ancillary for one of the parameters, θ_1, say, when θ_2 is a nuisance parameter. No satisfactory general definition is known. For example, it is not true that, in sampling from $N(\theta_1, \theta_2)$, $\bar{x}$ is sufficient for θ_1, for the posterior distribution of θ_1 involves s, so without it the distribution could not be found. However, something can be done. Suppose (cf. (1))

$$p(\mathbf{x} \mid \theta_1, \theta_2) = p(t(\mathbf{x}) \mid \theta_1, \theta_2) p(\mathbf{x} \mid t(\mathbf{x}), \theta_2), \qquad (19)$$

so that θ_1 is not present in the second probability on the right-hand side. Then, *given* θ_2, $t(\mathbf{x})$ is *sufficient* for θ_1: for, if θ_2 were known the only unknown parameter is θ_1 and (2) obtains. So we can speak of sufficiency for one parameter, given the other. In the normal example just cited $\bar{x}$ is sufficient for θ_1, given θ_2.

Similarly, if

$$p(\mathbf{x}|\theta_1, \theta_2) = p(t(\mathbf{x})|\theta_2)\, p(\mathbf{x}|t(\mathbf{x}), \theta_1, \theta_2), \tag{20}$$

then, given θ_2, $t(\mathbf{x})$ is *ancillary* for θ_1. Hence it is possible to speak of ancillary statistics for one parameter given the other. If (19) can be specialized still further to

$$p(\mathbf{x}|\theta_1, \theta_2) = p(t(\mathbf{x})|\theta_1)\, p(\mathbf{x}|t(\mathbf{x}), \theta_2), \tag{21}$$

then, given θ_2, $t(\mathbf{x})$ is sufficient for θ_1; and, given θ_1, $t(\mathbf{x})$ is ancillary for θ_2.

Both the original definitions made no mention of prior information, but if this is of a particular type then further consequences emerge. Suppose, with a likelihood given by (21), θ_1 and θ_2 have independent prior distributions. The posterior distribution is then

$$\pi(\theta_1, \theta_2|\mathbf{x}) \propto [p(t(\mathbf{x})|\theta_1)\, \pi(\theta_1)]\, [p(\mathbf{x}|t(\mathbf{x}), \theta_2)\, \pi(\theta_2)], \tag{22}$$

and θ_1 and θ_2 are still independent. Furthermore, if only the distribution of θ_1 is required $t(\mathbf{x})$ provides all the information: if only that of θ_2 is required, then $t(\mathbf{x})$ may be supposed constant. Or, for independent prior distributions, $t(\mathbf{x})$ is sufficient for θ_1 and ancillary for θ_2. But unlike the usual definitions of sufficient and ancillary statistics these depend on certain assumptions about the prior distribution.

5.6. Significance tests and the likelihood principle

Suppose we have some data, $\mathbf{x}$, whose distribution depends on several parameters, $\theta_1, \theta_2, \ldots, \theta_s$, and that some value of θ_1, denoted by $\bar{\theta}_1$, is of special interest to us. Let the posterior density of θ_1 be used to construct a confidence interval for θ_1 with coefficient β (on data $\mathbf{x}$). If this interval does *not* contain the value $\bar{\theta}_1$ then the data tend to suggest that the true value of θ_1 departs significantly from the value $\bar{\theta}_1$ (because $\bar{\theta}_1$ is not contained in the interval within which one has confidence that the true value lies) at a level measured by $\alpha = 1-\beta$. In these circumstances we will say the data are *significant at the α level*; α is the *level of significance* and the procedure will be called a *significance test*. The statement that $\theta_1 = \bar{\theta}_1$ is called the *null hypothesis*. Any statement that $\theta_1 = \theta_1^* \neq \bar{\theta}_1$ is an *alternative hypothesis*. The definitions extend to any subset of the parameters $\theta_1, \theta_2, \ldots, \theta_u$ with $u \leqslant s$ using the *joint* confidence set of $\theta_1, \ldots, \theta_u$, and values $\bar{\theta}_1, \ldots, \bar{\theta}_u$. An alternative interpretation of a confidence interval is an interval of those null hypotheses which are not significant on the given data at a prescribed significance level $\alpha = 1-\beta$.

The likelihood principle

If two sets of data, $\mathbf{x}$ and $\mathbf{y}$, have the following properties:

(i) their distributions depend on the same set of parameters;

(ii) the likelihoods of these parameters for the two sets are the same;

(iii) the prior densities of the parameters are the same for the two sets;

then any statement made about the parameters using $\mathbf{x}$ should be the same as those made using $\mathbf{y}$.

The principle is immediate from Bayes's theorem because the posterior distributions from the two sets will be equal.

Example of a significance test

Like the closely related concept of a confidence interval a significance test is introduced for practical rather than mathematical convenience. Consider the situation of §5.1, where the data are samples from $N(\theta, \sigma^2)$, so that there is only one parameter, and consider in particular the numerical illustration on the measurement of conductivity of an insulating material, with the uniform prior distribution ($\sigma_0 \to \infty$). There may well be a standard laid down to ensure that the conductivity is not too high: for example that the conductivity does not exceed 17, otherwise the insulation properties would be unsatisfactory. This value of 17 is the value $\bar{\theta}$ referred to in the definition: the null hypothesis. The main interest in the results of the experiment lies in the answer to the question: 'Is the conductivity below 17?' If it is, the manufacturer or the inspector (or the consumer) probably does not mind whether it is 15 or 16. Consequently the main feature of the posterior distribution that matters is its position relative to the value 17. Let us first consider the reasonableness of 17 as a possible value of the conductivity. The 95 % confidence interval for the mean is $16{\cdot}57 \pm 1{\cdot}96 \times (1/\sqrt{10})$; that is (15·95, 17·19) so that we can say that the true conductivity, granted that the measuring instrument is free from bias, most likely lies in this interval. But this interval includes the value 17 so it is possible that this material does not meet the specification. In the language above, the result is *not* significant at the 5 % level. Had an 80 % confidence interval been used the values would have been (16·17, 16·97), since $\Phi(-1{\cdot}28) = 0{\cdot}10$ (cf. equation 2.5.14). This does not include the value 17 so that the result *is* significant at the 20 % level. This is not normally regarded as a low enough level of significance. The values 5, 1 and 0·1 % are the levels commonly used, because they provide substantial amounts of evidence that the true value is not $\bar{\theta}_1$, when the result is significant. Occasionally the *exact significance level* is quoted: by this is meant the critical level such that levels smaller than it will not give significance, and values above it will. In the numerical example we require to find α such that

$$16{\cdot}57 + \lambda_\alpha(1/\sqrt{10}) = 17, \tag{1}$$

where $2\Phi(-\lambda_\alpha) = \alpha$: or, in the language of §5.3, λ_α is the upper $\frac{1}{2}\alpha$ point of the standardized normal distribution. From (1) the value of λ_α is 1·36 and the tables show that

$$\Phi(-1{\cdot}36) = 0{\cdot}087,$$

so that $\alpha = 0{\cdot}174$. The result is just significant at the level, 17·4 %. This agrees with what we have just found: the result is significant at 20 % but not at 5 %.

Just as there is some arbitrariness about a confidence interval, so there must be arbitrariness about a significance test (cf. §5.2). Thus, in this example, we could have used a one-sided confidence interval from $-\infty$ to some value and defined significance with respect to it. Indeed this might be considered more appropriate since all we want to do is to be reasonably sure that the true value does not exceed 17. At the 5 % level this interval extends to $16{\cdot}57+1{\cdot}64\,(1/\sqrt{10}) = 17{\cdot}09$ so we think that the conductivity is most likely less than 17·09 which would not normally be enough evidence to dismiss the doubts that the conductivity might exceed 17. If this test is used the exact significance level is the posterior probability that θ exceeds 17. Since θ is $N(16{\cdot}57, 0{\cdot}1)$ this value is

$$1-\Phi\left(\frac{17{\cdot}00-16{\cdot}57}{1/\sqrt{10}}\right) = 1-\Phi(1{\cdot}36) = 0{\cdot}087$$

(compare the calculation above) so that the exact significance level is 8·7 %, half what it was with the symmetric shortest confidence interval. In this example this is probably the best way of answering the question posed above: by quoting the relevant posterior probability that the material is unsatisfactory and avoiding significance level language completely. There is discernible in the literature a tendency for significance tests to be replaced by the more informative confidence intervals, and this seems a move in the right direction.

It must be emphasized that the significance level is, even more than a confidence interval, an incomplete expression of posterior beliefs. It is used because the complete posterior distribution is too complicated. A good example of its use will be found in §8.3 where many parameters are involved. In the simple cases so far discussed there seems no good reason for using significance tests.

Prior knowledge

The type of significance test developed here is only appropriate when the prior knowledge of θ is vague (in the sense discussed in §5.2). In particular, the prior distribution in the neighbourhood of the null value $\overline{\theta}$ must be reasonably smooth for the tests to be sensible. In practical terms this means that there is no reason for thinking that $\theta = \overline{\theta}$ any more than any other value near $\overline{\theta}$. This is true in many applications, but there are situations where the prior probability that $\theta = \overline{\theta}$ is appreciably higher than that for values near $\overline{\theta}$. For example, if θ is the parameter in genetics that measures the linkage between two genes, the value $\theta = \bar{\theta} = 0$ corresponds to no linkage, meaning that the genes are on different chromosomes; whereas θ near $\bar{\theta}$ means that there is some linkage, so that the genes are on the same chromosome, but relatively well separated on it. In this case the prior distribution may be of the mixed discrete and continuous type with a discrete concentration of probability on $\theta = \bar{\theta}$ and a continuous distribution elsewhere. The relevant quantity to consider here is the posterior probability that $\theta = \bar{\theta}$. Significance tests of this type have been considered by Jeffreys (1961) but are not much used. They will not be mentioned further in this book, but their introduction provides a further opportunity to remind the reader that the methods developed in this book are primarily for the situation of vague prior knowledge, and have to be adapted to take account of other possibilities (see, for example, §7.2).

Decision theory

It is also well to notice what a significance test is *not*. It is not a recipe for action in the sense that if a result is significant then one can act with reasonable confidence (depending on the level) as if θ was not $\bar{\theta}$. For example, the manufacturer of insulating material may not wish to send to his customers any material with $\theta \geqslant 17$. Since he does not know θ exactly he may have to decide whether or not to send it on the evidence of the data. He therefore has to act in one of two ways: to send it to the customer or to retain it for further processing, without exact knowledge of θ. We often say he has to take one of two *decisions*. The problem of action or decision is not necessarily answered by retaining it iff the test is significant. The correct way to decide is discussed in *decision theory*: which is a mathematical theory of how to make decisions in the face of uncertainty about the true value of parameters. The theory will not be discussed in detail in this book but since it is so closely related to inference a brief description is not out of place.

The first element in decision theory is a set of parameters $\boldsymbol{\theta}$ which describes in some way the material of interest and about which the decisions have to be made. In the conductivity example there is a single parameter, namely the conductivity, and it is required to decide whether or not it exceeds 17 in value. The second element is a set of decisions d which contains all the possible decisions that might be taken. In the example there are two decisions, to despatch or retain the material. Notice that the set of decisions is supposed to contain all the decisions that could be taken and not merely some of them: or to put it differently, we have, in the theory, to choose *amongst* a number of decisions. Thus it would not be a properly defined decision problem in which the only decision was whether to go to the cinema, because if the decision were not made (that is, one did not go to the cinema) one would have to decide whether to stay at home and read, or go to the public-house, or indulge in other activities. All the possible decisions, or actions, must be included in the set. Our example is probably inadequate in this respect since it does not specify just what is going to happen if the material is to be rejected. Is it, for instance, to be scrapped completely or to be reprocessed?

The two elements, the decisions d and the parameters $\boldsymbol{\theta}$, are related to one another in the sense that the best decision to take will depend on the value of the parameter. In the example it will clearly be better to reject if $\theta \geqslant 17$ and otherwise accept. We now discuss the form the relationship must take. If $\boldsymbol{\theta}$ were known then clearly all that is needed is a knowledge of the best decision for that value of $\boldsymbol{\theta}$. But decision theory deals with the case where $\boldsymbol{\theta}$ is unknown: it can then be shown that a relationship of such a form, namely knowing the best d for each $\boldsymbol{\theta}$, between d and $\boldsymbol{\theta}$ is inadequate. This can be seen in the example: one could only expect to be able to make a sensible decision if one knew how serious it was to retain material whose conductivity was below 17, and to despatch material to the customer whose conductivity was in excess of 17. If the latter was not serious the manufacturer might argue that he could 'get away with it' and send out material of inadequate insulation. If the seriousness were changed, for example by the introduction of a law making it an offence to sell material of conductivity in excess of 17, he might well alter his decision.

A relationship between d and $\boldsymbol{\theta}$ stronger than knowledge of the best d for each $\boldsymbol{\theta}$ is required. It has been shown (see §1.6) that the relationship can always be described in the following way. For each pair of values, d and $\boldsymbol{\theta}$, there is a real number $U(d, \boldsymbol{\theta})$ called the *utility* of making decision d when the true value is $\boldsymbol{\theta}$. $U(d, \boldsymbol{\theta})$ is called a *utility function*. It describes, in a way to be elaborated on below, the desirability or usefulness of the decision d in the particular set of circumstances described by $\boldsymbol{\theta}$. In particular, the value of d which maximizes $U(d, \boldsymbol{\theta})$ for a fixed $\boldsymbol{\theta}$ is the best decision to take if the situation is known to have that value of $\boldsymbol{\theta}$. Generally if $U(d, \boldsymbol{\theta}) > U(d', \boldsymbol{\theta})$ then, if $\boldsymbol{\theta}$ obtains, d is preferred to d'. In the example, where $\boldsymbol{\theta} = \theta$ is the conductivity, d is the decision to despatch and d' to retain, a utility function of the following form might be appropriate, at least in the neighbourhood of $\bar{\theta} = 17$:

$$U(d, \theta) = a + c(\bar{\theta} - \theta),$$
$$U(d', \theta) = a + c'(\theta - \bar{\theta}),$$

where a, c and c' are positive constants. Then, for $\theta > \bar{\theta}$, $U(d', \theta) > U(d, \theta)$ so that it is best to retain; and for $\theta < \bar{\theta}$ the opposite inequality obtains and it is best to despatch. Notice that if θ differs from $\bar{\theta}$ by an amount h, say, in either direction, then the two utilities differ by the same amount, namely $(c + c')h$. Consequently, the mistake of retention when $\theta = \bar{\theta} - h$ is just as serious (in the sense of losing utility) as is the mistake of despatch when $\theta = \bar{\theta} + h$. This would probably not be reasonable if there was a law against marketing insulating material of an unsatisfactory nature: the mistake of despatching poor material might then involve more serious consequences than that of retaining good workmanship.

The scale upon which utility is to be measured has to be carefully considered. In particular, consider what is meant by saying that one course of action has twice the utility of another. Specifically, in a given set of circumstances, defined by $\boldsymbol{\theta}$, let d_1 and d_2 be two courses of action with utilities 1 and 2 respectively, and let d_0 be a course of action with utility zero. Suppose that between d_0 and d_2 the course of action is chosen by the toss of a fair coin. If it falls heads then d_0 is taken, if it falls tails then d_2 is taken. The expected utility (in the usual sense of expectation) is half the utility of d_0 plus half the utility of d_2, or $\frac{1}{2}.0 + \frac{1}{2}.2 = 1$.

Hence this random choice between d_0 and d_2 has the same expected utility as d_1, namely 1, and the random choice is considered as of the same usefulness as d_1. In other words, one would be indifferent between d_1 and the random choice between d_0 and d_2. Utility is measured on a scale in which the certainty of 1 is as good as a fifty-fifty chance of 2 or 0. There would be nothing inconsistent in saying that the utility of obtaining £1 was 1 but that of obtaining £2 was $1\frac{1}{2}$, for an evens chance of £2 (yielding utility, $\frac{3}{4}$) may not be considered as desirable as the certainty of £1. Utility is not necessarily the same as money.

The discussion of the utility scale shows how the utility can be used to arrive at a decision in a sensible manner when the true value of $\boldsymbol{\theta}$ is not completely known. The important quantity is the expected utility. If there is some uncertainty about $\boldsymbol{\theta}$ it follows, from our earlier considerations, that there exists a probability distribution for $\boldsymbol{\theta}$, and hence we need to consider the expected utility, where the expectation is with respect to this distribution. Let $\pi(\boldsymbol{\theta})$ denote our knowledge of $\boldsymbol{\theta}$ expressed in the usual way as a probability density. The expected utility of a decision d is then, by definition,

$$\int U(d, \boldsymbol{\theta})\, \pi(\boldsymbol{\theta})\, d\boldsymbol{\theta} = \overline{U}(d), \tag{2}$$

say. The best decision to take is that which maximizes $\overline{U}(d)$. In the example of a utility function cited above the expected utility of despatch is

$$\overline{U}(d) = a + \int c(\overline{\theta} - \theta)\pi(\theta)\, d\theta = a + c(\overline{\theta} - \theta_1)$$

and of retention is $a + c'(\theta_1 - \overline{\theta})$, where θ_1 is the expected value of θ, or the mean of the density $\pi(\theta)$. $\overline{U}(d) > \overline{U}(d')$ iff $\theta_1 < \overline{\theta}$. Hence the material is despatched iff the expected value of θ is below the critical value. This is sensible because, as we remarked, mistakes on either side of the critical value are equally serious with this utility function.

It remains to discuss the role of observations in decision theory. Suppose that a set of observations, $\mathbf{x}$, is obtained and that, as a result, the knowledge of $\boldsymbol{\theta}$ is changed by Bayes's theorem to the posterior distribution given $\mathbf{x}$, $\pi(\boldsymbol{\theta} \mid \mathbf{x})$. The decision must now be made according to the new expected utility

$$\int U(d, \boldsymbol{\theta})\, \pi(\boldsymbol{\theta}\,|\,\mathbf{x})\, d\boldsymbol{\theta} = \bar{U}(d, \mathbf{x}), \tag{3}$$

say; that decision being taken which maximizes this quantity. On intuitive grounds one expects the decision made on the basis of (3) to be better than that made using (2), for otherwise the observations have not been of any value. This feeling is expressed mathematically by saying that the expected value of the maximum of (3) exceeds the maximum of (2). The expected value is

$$\int \max_{d} \bar{U}(d, \mathbf{x})\, \pi(\mathbf{x})\, d\mathbf{x}, \tag{4}$$

where $\pi(\mathbf{x})$ is the distribution of $\mathbf{x}$ anticipated before the observations were made: that is, in the light of the prior knowledge concerning $\boldsymbol{\theta}$. (Compare the discussion on beliefs about the sample in §5.1.) Precisely

$$\pi(\mathbf{x}) = \int p(\mathbf{x}\,|\,\boldsymbol{\theta})\, \pi(\boldsymbol{\theta})\, d\boldsymbol{\theta},$$

where $p(\mathbf{x}\,|\,\boldsymbol{\theta})$ is the density of the observation given $\boldsymbol{\theta}$ in the usual way. This mathematical result is easily established.

Inference and decision theory

The reader is referred to the list of suggestions for further reading if he wishes to pursue decision theory beyond this brief account. It should be clear that the inference problem discussed in the present book is basic to any decision problem, because the latter can only be solved when the knowledge of $\boldsymbol{\theta}$ is adequately specified. The inference problem is just the problem of this specification: the problem of obtaining the posterior distribution. Notice that the posterior distribution of $\boldsymbol{\theta}$ that results from an inference study is exactly what is required to solve *any* decision problem involving $\boldsymbol{\theta}$. The role of inference is to provide a statement of the knowledge of $\boldsymbol{\theta}$ that is adequate for making decisions which depend on $\boldsymbol{\theta}$. This it does by providing the posterior distribution either in its entirety or in some convenient summary form, such as a confidence statement.

The person making the inference need not have any particular decision problem in mind. The scientist in his laboratory does not consider the decisions that may subsequently have to be made concerning his discoveries. His task is to describe accurately what is known about the parameters in question.

[The remainder of this section is not necessary for an understanding of the rest of the book. It is included so that the person who learns his statistics from this book can understand the language used by many other scientists and statisticians.]

Non-Bayesian significance tests

The concept of a significance test is usually introduced in a different way and defined differently. We shall not attempt a general exposition but mainly consider the significance test for a normal mean when the variance is known, the case just discussed numerically. This will serve to bring out the major points of difference. Suppose, then, that we have a random sample of size n from $N(\theta, \sigma^2)$ and we wish to test the null hypothesis that $\theta = \overline{\theta}$. The usual argument runs as follows. Since $\bar{x}$ is sufficient it is natural to consider it alone, without the rest of the data. $\bar{x}$ is known to have a distribution which is $N(\mu, \sigma^2/n)$ if μ is the true, but unknown, value of θ (cf. §§3.3, 3.5). Suppose the true value is the null hypothesis value, $\overline{\theta}$, then $\bar{x}$ is $N(\overline{\theta}, \sigma^2/n)$ and so we can say that, with probability 0·95, $\bar{x}$ will lie within about $2\sigma/\sqrt{n}$ of $\overline{\theta}$. (Notice that this probability is a frequency probability, and is not normally thought of as a degree of belief.) Hence we should be somewhat surprised, if μ were $\overline{\theta}$, to have $\bar{x}$ lie outside this interval. If it does lie outside then one of two things must have happened: either (i) an event of small probability (0·05) has taken place, or (ii) μ is not $\overline{\theta}$. If events of small probability are disregarded then the only possibility is that μ is not $\overline{\theta}$. The data would seem to signify, whenever $\bar{x}$ differs from $\overline{\theta}$ by more than $2\sigma/\sqrt{n}$, that μ is not $\overline{\theta}$, the strength of the significance depending on the (small) probability that has been ignored. Hence, if $|\bar{x}-\overline{\theta}| > 2\sigma/\sqrt{n}$, it is said that the result is significant at the 5 % level. Notice that this interval is exactly the same as that obtained using Bayes's theorem and confidence intervals to construct the significance test: for the confidence interval for θ is $\bar{x} \pm 2\sigma/\sqrt{n}$. This identification comes from the parallelism between (*a*) and (*b*) in §5.1.

The general procedure, of which this is an example, is to take a statistic $t(\mathbf{x})$, which is not necessarily, or even usually, sufficient; to find the probability distribution of $t(\mathbf{x})$ when $\theta_1 = \bar{\theta}_1$ and to choose a set of values of $t(\mathbf{x})$ which, when $\theta_1 = \bar{\theta}_1$, are more probable under some plausible alternative hypotheses than under the null hypothesis. If the observed value of $t(\mathbf{x})$ belongs to this set then, either $\theta_1 = \bar{\theta}_1$ and an improbable event has occurred, or $\theta_1 \neq \bar{\theta}_1$. The result is said to be significant at a level given by the probability of this set. There exists a great deal of literature on the choice of $t(\mathbf{x})$ and the set. Confidence limits can be obtained from significance tests using the alternative definition of the limits given above.

There are many criticisms that can be levelled against this approach but we mention only two. First, the probability quoted in the example of a significance test is a frequency probability derived from random sampling from a normal distribution: if one was to keep taking samples from the distribution the histogram of values of $\bar{x}$ obtained would tend to $N(\theta, \sigma^2/n)$. But the interpretation of the probability is in terms of degree of belief, because the 5 %, or 1 %, is a measure of how much belief is attached to the null hypothesis. It is used as if 5 % significance meant the posterior probability that θ is near $\bar{\theta}$ is 0·05. This is not so: the distortion of the meaning is quite wrong in general. It may, as with $N(\theta, \sigma^2)$, be justifiable, but this is not always so.

Significance tests and the likelihood principle

The second criticism is much more substantial. The use of significance tests based on the sampling distribution of a statistic $t(\mathbf{x})$ is in direct violation of the likelihood principle. The principle is self-evident once it has been admitted that degrees of belief obey the axioms of probability; yet its practical consequences are far-reaching. Since it is not our purpose to provide an account of significance tests based on the sampling distribution of a statistic we shall use an example to show how the usual significance test violates the likelihood principle. Our example may seem a little curious but this is because it is necessary to take a case where the choice of $t(\mathbf{x})$ is unambiguous so that there is no doubt about the significance test and therefore about the significance level. We shall quote the *exact* level defined above.

Suppose a random sample of one is taken from $B(4, \theta)$: or equivalently we have a random sequence of four trials with constant probability, θ, of success. By the sufficiency argument only x, the number of successes is relevant, so that x may take one of 5 values: 0, 1, 2, 3, 4. Suppose scientist 1 can observe the value of x. Suppose scientist 2 can merely observe whether the number of successes is 1, or not: in other words, he will observe A if $x = 1$ and $\bar{A}$ if x is 0, 2, 3 or 4. Now let them observe the same random value and suppose scientist 2 observes A and therefore scientist 1 observes $x = 1$, so that they both have the same information. (Of course, had scientist 2 observed $\bar{A}$, they would not have been similarly placed, but that is irrelevant.) Then the likelihood for both scientists is $4\theta(1-\theta)^3$ and hence, by the likelihood principle their inferences should be the same. Nevertheless, had the above type of significance test based on a sampling distribution been used with $\theta = \frac{1}{2}$ as the null hypothesis and $\frac{1}{4} \leqslant \theta < \frac{1}{2}$ as the alternatives, scientist 1 would have declared the result significant at 500/16 % and scientist 2 at 400/16 %. The explanation of the difference is that scientist 1 would have included the points 0 and 1 in the set of values, whereas scientist 2 would only have included A. When $\theta = \frac{1}{2}$, scientist 1's set has probability $1/16+4/16 = 5/16$, whereas scientist 2's set has probability 4/16. Larger differences could be obtained with numerically more complicated examples.

Similar objections can be raised to confidence intervals derived from such significance tests and that is why they are not used in this book. Nevertheless, in the situations commonly studied in statistics the results given here agree with those obtained by other methods. The reason is the existence of parallel properties of $t(\mathbf{x})$ and θ similar to those mentioned in §5.1 ((*a*) and (*b*)). Only in §6.3 will there be a case where the results are different.

Suggestions for further reading

As already stated in the text the treatment of statistics adopted in this book is unorthodox in that the approach is unusual, though the results are broadly the same as provided by the usual methods. Every reader ought to look at one of the orthodox books. He will find much overlap with our treatment, though a few topics (like unbiased, minimum variance estimation) which loom large in those books receive little attention here, and others (like posterior distributions) are scarcely considered by the orthodox school. The two approaches complement each other and together provide a substantially better understanding of statistics than either approach on its own. Among the orthodox books at about the level of the present one mention can be made of Alexander (1961), Birnbaum (1962), Brunk (1960), Hogg and Craig (1959), Hoel (1960), Tucker (1962) and Fraser (1958).

Statistics is essentially a branch of applied mathematics in the proper† meaning of the word 'applied' and no satisfactory understanding of the subject can be obtained without some acquaintance with the applications. These have largely been omitted from the present work, for reasons explained in the preface, so that every reader ought to look at at least one book written, not for the mathematician, but for the user of statistics. The number of these is legion and cover applications to almost every branch of human knowledge. We mention only one which seems extremely good, that edited by Davies (1957).

Little statistical work is possible without a set of tables of the more common distributions. A small set, adequate for most of the methods described in this book is that by Lindley and Miller (1961). Two larger collections are those by Fisher and Yates (1963) and Pearson and Hartley (1958). An index to statistical tables has recently been issued by Greenwood and Hartley (1962).

The field of decision theory is covered in two books, Schlaifer (1959) and Raiffa and Schlaifer (1961), along Bayesian lines,

† As distinct from its meaning in the subject Applied Mathematics as taught in British Universities, which is confined to applications to physics.

broadly similar to those in the present book. Other treatments are given by Chernoff and Moses (1959) at an elementary level, by Weiss (1961) at an intermediate level, and by Blackwell and Girshick (1954) at an advanced level.

The most important name in modern statistics is that of Fisher and his works can always be read with interest and profit. The early book on methods, Fisher (1958) could replace Davies's mentioned above. His later views on the foundations of statistics are contained in Fisher (1959).

One great work that everyone ought to read at some time is Jeffreys (1961). This book covers the whole field from the foundations to the applications and contains trenchant criticisms of the orthodox school. The present book owes much to the ideas contained therein.

Exercises

1. The following are ten independent experimental determinations of a physical constant. Find the posterior distribution of the constant and give 95 % confidence limits for its value:

11·5, 11·7, 11·7, 11·9, 12·0, 12·1, 12·2, 12·2, 12·4, 12·6.

(Camb. N.S.)

2. The Young's modulus of a fibre determined from extension tests is $1{\cdot}2 \times 10^8$ c.g.s. units, with negligible error. To determine the possible effects of fibre anistropy on elastic properties, the effective Young's modulus is estimated from bending tests; in this experiment there is appreciable random error. Eight tests gave the following values:

1·5, 1·2, 1·0, 1·4, 1·3, 1·3, 1·2, 1·7

$\times 10^8$ c.g.s. units.

Is there reason to think that the extension and bending tests give different average results? (Camb. N.S.)

3. A new method is suggested for determining the melting-point of metals. Seven determinations are carried out with manganese, of which the melting-point is known to be 1260 °C, and the following values obtained:

1267, 1262, 1267, 1263, 1258, 1263, 1268 °C.

Discuss whether these results provide evidence for supposing that the new method is biased. (Camb. N.S.)

4. The value of θ is known to be positive and all positive values of θ are thought to be equally probable. A random sample of size 16 from $N(\theta, 1)$ has mean 0·40. Obtain a 95 % confidence interval for θ.

5. A manufacturer is interested in a new method of manufacture of his product. He decides that it will cost £10,000 to change over to the new method and that if it increases the yield by θ he will gain θ times £8000. (Thus it will not be worth using it unless $\theta > 1{\cdot}25$.) His knowledge of θ is confined to the results of 16 independent determinations of θ that gave a sample mean of 2·00 and an estimate of variance of 2·56. Find the probability distribution of his expected profit if he changes over to the new method.

6. For fixed s, the random variable r has expectation s and variance σ_1^2, the same for all s. The random variable s has expectation μ and variance σ_2^2. Find the unconditional expectation and variance of r.

For each of several rock samples from a small region the strength of the remanant magnetism is determined in the laboratory. The laboratory error in determining the magnetism of a sample is known to be free from bias and have variance $\sigma_1^2 = 0{\cdot}36$. There may, however, be variations between samples of unknown variance σ_2^2. For the determinations for 10 samples given below carry out a significance test to see if there is evidence for suspecting this second source of variation, and if so give limits between which σ_2^2 most probably lies.

11·80, 9·23, 11·73, 11·78, 10·21, 9·65, 10·62, 9·76, 8·81, 9·66.

(The sum of the values is 103·25, the sum of the squares about the mean is 11·07825; the units are conventional.) (Camb. N.S.)

7. A random sample of size n is to be taken from $N(\theta, \sigma^2)$, σ^2 known. The prior distribution of θ is $N(\mu_0, \sigma_0^2)$. How large must n be to reduce the posterior variance of θ to σ_0^2/k? $(k > 1)$.

8. The following two random samples were taken from distributions with different means but common variance. Find the posterior distribution of this variance:

9·1, 9·3, 9·9, 8·8, 9·4;

10·6, 10·4, 9·3, 10·1, 10·1, 10·0, 10·7, 9·6.

(Camb. N.S.)

9. The following are the lengths in inches of six tubes taken as a random sample from the output of a mass production process:

11·93, 11·95, 12·01, 12·02, 12·03, 12·06.

Estimate the mean and standard deviation of the probability distribution of the length of a tube. The tubes are used in batches of ten, being fixed, end to end, to form a single large tube; it being a requirement that the total length of each large tube shall not exceed 10 ft. 1 in. Find a point estimate of the frequency with which a batch will have to be sent back to be made up afresh. (Camb. N.S.)

10. A parameter θ has a prior distribution which is $N(\mu_0, \sigma_0^2)$, and an observation x is $N(\theta, \sigma^2)$ with σ^2 known.

An experimenter only retains x if

$$|x - \mu_0| < 2\sigma_0:$$

otherwise he takes an independent observation y which is also $N(\theta, \sigma^2)$ and ignores the value of x. Given the value of y and the knowledge that $|x-\mu_0| \geqslant 2\sigma_0$ (but not what the value of x was), obtain the posterior distribution of θ. Sketch the form of this distribution when $y = \mu_0$.

(Wales Maths.)

11. A random sample of size m from $N(\theta, \sigma^2)$, with σ^2 known and θ having a uniform prior distribution, yields a sample mean $\bar{x}$. Show that the distribution of the mean $\bar{y}$ from a second independent random sample of size n from the same distribution, given the value of $\bar{x}$; that is, $p(\bar{y}\,|\,\bar{x}, \sigma^2)$; is

$$N(\bar{x}, \sigma^2(m^{-1}+n^{-1})).$$

A scientist using an apparatus of known standard deviation 0·12 takes nine independent measurements of the same quantity and obtains a mean of 17·653. Obtain limits between which a tenth measurement will lie with 99 % probability.

12. It is required to test the hypothesis that $\mu = \mu_0$, using a random sample of 30 observations. To save time in calculation, the variance is estimated from the first 10 observations only, although the mean is estimated from all 30. Find the appropriate test. (Lond. B.Sc.)

13. Two scientists have respectively beliefs about θ which are

$$N(\mu_i, \sigma_i^2) \quad (i = 1, 2).$$

On discussing the reasons for their beliefs, they decide that the separate pieces of information which led them to their beliefs are independent. What should be their common beliefs about θ if their knowledge is pooled?

14. $x_1, x_2, \ldots, x_n$ form a random sample from a rectangular distribution over the interval $(\alpha-\beta, \alpha+\beta)$. Discuss the joint posterior distribution of α and β under reasonable assumptions about their prior distribution. What are the sufficient statistics for α and β? (Camb. Dip.)

15. The length of life t of a piece of equipment before it fails for the first time is

$$e^{-\theta t}\theta^r t^{r-1}/(r-1)!,$$

where r is known but θ is not. The prior distribution of θ is uniform in $(0, \infty)$. n pieces are independently tested and, after a time T, m of them have failed at times $t_1, t_2, \ldots, t_m$ and the remaining $(n-m)$ are still working. Find the posterior distribution of θ in terms of the incomplete Γ-function

$$I(x, r) = \int_0^x t^r e^{-t} dt \Big/ \int_0^\infty t^r e^{-t} dt.$$

Obtain 95 % confidence limits for θ in the case $r = 1$.

16. The Pareto distribution has density function

$$\theta L^\theta / x^{1+\theta}$$

for $x > L$ where θ is a positive parameter. A random sample of size n is available from a Pareto distribution with $L = 1$. Show there exists a sufficient statistic, and if the prior distribution of $\ln\theta$ is uniform, show that the posterior distribution of θ is $\Gamma(n, n\ln z)$, where z is the geometric mean of the observations. Hence describe how confidence limits for θ can be obtained.

17. Find sufficient statistics for random samples from the multivariate normal distribution (§ 3.5).

18. Show that

$$f(x|\theta) = \frac{\theta^2}{\theta+1}(x+1)e^{-x\theta} \quad (x > 0)$$

is, for any $\theta > 0$, a density function. A random sample of size n is taken from this density giving values $x_1, x_2, \ldots, x_n$. Irrespective of your prior knowledge of θ, what quantities would you calculate from the sample in order to obtain the posterior distribution of θ?

19. The set of observations $X = (x_1, x_2, \ldots, x_n)$ has a probability density which is known except for the values of two parameters θ_1, θ_2. Prove that, if $t_1 = t_1(X)$ is sufficient for θ_1 when θ_2 is known and $t_2 = t_2(X)$ is sufficient for θ_2 when θ_1 is known, then $T = (t_1, t_2)$ is sufficient for $\theta = (\theta_1, \theta_2)$.
(Manch. Dip.)

20. Prove that if $x_1, x_2, \ldots, x_n$ is a random sample of size n from $N(\mu, \sigma^2)$, then $x_i - \bar{x}$ is independent of $\bar{x}$ and hence $\Sigma(x_i-\bar{x})^2$ of $\bar{x}$.

Use this result to prove an extension of the result of § 5.3 that $\Sigma(x_i-\mu)^2/\sigma^2$ is χ^2 with n degrees of freedom; namely that $\Sigma(x_i-\bar{x})^2/\sigma^2$ is χ^2 with $(n-1)$ degrees of freedom.

21. The breakdowns of a machine occur in such a way that the probability of a breakdown in the interval $(t, t+\delta t)$ is $\lambda(t)\delta t + o(\delta t)$ independently of stoppages prior to t. The machine is observed in the interval $(0, T)$ and the time of breakdowns recorded. If $\lambda(t) = \alpha e^{\beta t}$ obtain a test of the hypothesis that $\beta = 0$ (i.e. that the breakdown rate does not change with time).
(Camb. Dip.)

22. In a large town, the size n of families, is distributed with probability density function

$$p(n) = (1-\rho)\rho^n \quad (0 < \rho < 1),$$

and in each family the chance of a male child is p independent of other children. Given a series of k families, m of which are known to contain $r_1, \ldots, r_m$ boys (with $r_i > 0, i = 1, \ldots, m$) but whose sizes are unknown, whilst the rest contain no boys, obtain the posterior distribution of ρ.
(Lond. Dip.)

23. Bacterial organisms are known to be present in a large volume of water but it is not known whether there are many or few. Accordingly it is decided to examine samples each of 10 cm^3 of water for the presence (or absence) of the organism, to go on taking samples until k samples have been found containing the organism and then to stop. Find the distribution of the number of samples, n, necessary to achieve k positive results, and the mean and standard deviation of this distribution. Explain how to set up confidence limits for p, the proportion of infected samples, so that one may be 100α % certain that p will lie within these limits.

(Lond. Dip.)

24. A Geiger counter has a probability $\theta\,\delta u + O(\delta u^2)$ of clicking during the time interval u, $u + \delta u$, independent of u and all previous clicks.

If the average interval between consecutive members of a sequence of 11 clicks is 2·2 sec, show that θ may be asserted to lie in the range 0·25 $(\text{sec})^{-1}$ to 0·71 $(\text{sec})^{-1}$ with a 90 % level of confidence. (Camb. N.S.)

25. Certain radioactive substances emit particles at random at an average rate λ. n intervals between such emissions are observed for each of two such substances, the mean intervals being $\bar{x}$ and $\bar{y}$ respectively. Explain how you could test whether the rates of emission are the same.

(Leic. Gen.)

26. Observations are made as follows on a Poisson process in which events occur at unknown rate θ per unit time. In the time period $(0, u]$ it is observed that m events occur at instants $x_1 < x_2 < \ldots < x_m$, whereas in the time interval $(u, 2u]$ it is observed only that n events occur, the instants of occurrence not being recorded. What is the (minimal) sufficient statistic for θ and what is its sampling distribution? (Lond. M.Sc.)

6

INFERENCES FOR SEVERAL NORMAL DISTRIBUTIONS

The previous chapter was concerned with the problems of inference that arise when a single sample is taken from a normal distribution: in this chapter similar problems are considered when two or more samples are taken from possibly different normal distributions. We shall continue to use uniform prior distributions for the mean and the logarithm of the variance, thinking of them as approximations to distributions representing little prior knowledge of the parameters, as in theorem 5.2.1.

6.1. Comparison of two means

Theorem 1. *Let* $\mathbf{x}_1 = (x_{11}, x_{12}, \ldots, x_{1n_1})$ *be a random sample of size* n_1 *from* $N(\theta_1, \sigma_1^2)$ *and* $\mathbf{x}_2 = (x_{21}, x_{22}, \ldots, x_{2n_2})$ *be an independent random sample of size* n_2 *from* $N(\theta_2, \sigma_2^2)$, *where* σ_1 *and* σ_2 *are known. Then if the prior distributions of* θ_1 *and* θ_2 *are independent and both uniform over* $(-\infty, \infty)$, *the posterior distribution of* $\delta = \theta_1 - \theta_2$ *is* $N(\bar{x}_1 - \bar{x}_2, \sigma_1^2/n_1 + \sigma_2^2/n_2)$, *where* $\bar{x}_1$ *and* $\bar{x}_2$ *are the respective means of the two samples.*

The joint prior distribution of θ_1 and θ_2 is everywhere constant and the likelihood of the two samples is (equation 5.1.9) proportional to

$$\exp\left[-\frac{n_1(\bar{x}_1-\theta_1)^2}{2\sigma_1^2}-\frac{n_2(\bar{x}_2-\theta_2)^2}{2\sigma_2^2}\right], \tag{1}$$

so that (1) is also proportional to the joint posterior distribution of θ_1 and θ_2 given $\mathbf{x}_1$ and $\mathbf{x}_2$. This is a product of two factors, one involving θ_1 only, one involving θ_2 only, and hence (§3.1) θ_1 and θ_2 are independent. Furthermore, they are clearly $N(\bar{x}_1, \sigma_1^2/n_1)$ and $N(\bar{x}_2, \sigma_2^2/n_2)$ respectively. (This could be deduced directly from the corollary to theorem 5.1.1 with $\sigma_0 \to \infty$.) Since, if θ_2 is normal so is $-\theta_2$, it follows from theorem 3.5.5 that $\delta = \theta_1 - \theta_2$ is also normal with mean equal to the difference of the means and variance equal to the sum of the variances.

Theorem 2. *Let* $\mathbf{x}_1$ *be a random sample of size* n_1 *from* $N(\theta_1, \phi)$ *and* $\mathbf{x}_2$ *be an independent random sample of size* n_2 *from* $N(\theta_2, \phi)$. *Then if the prior distributions of* θ_1, θ_2 *and* $\ln \phi$ *are independent and uniform over* $(-\infty, \infty)$, *the posterior distribution of* $\nu s^2/\phi$ *is* χ^2 *with* ν *degrees of freedom, where*

$$\nu_i s_i^2 = S_i^2 = \sum_{j=1}^{n_i} (x_{ij} - \bar{x}_i)^2, \quad \nu_i = n_i - 1 \quad (i = 1, 2) \tag{2}$$

and

$$\nu s^2 = S^2 = S_1^2 + S_2^2, \quad \nu = \nu_1 + \nu_2. \tag{3}$$

[Note that the variances of the two normal distributions from which samples are taken are supposedly known to be equal.]

The joint prior density of θ_1, θ_2 and ϕ is proportional to ϕ^{-1} and the likelihoods of the two samples are given by equation 5.4.3 (with the change of a few suffixes to distinguish the two samples). The likelihoods may be rearranged in the form (5.4.4) and multiplying these expressions together we see that the joint posterior distribution of θ_1, θ_2 and ϕ is

$$\pi(\theta_1, \theta_2, \phi \mid \mathbf{x}_1, \mathbf{x}_2) \propto \phi^{-\frac{1}{2}(n_1+n_2+2)} \exp[-\{n_1(\bar{x}_1 - \theta_1)^2 + n_2(\bar{x}_2 - \theta_2)^2 + S_1^2 + S_2^2\}/2\phi]. \tag{4}$$

To obtain the posterior distribution of ϕ it is only necessary to integrate (4) with respect to θ_1 and θ_2. This is easily done since the two integrals are the usual normal ones (compare the passage to equation 5.4.6) and the result is

$$\pi(\phi \mid \mathbf{x}_1, \mathbf{x}_2) \propto e^{-\nu s^2/2\phi} \phi^{-\frac{1}{2}\nu - 1}. \tag{5}$$

A comparison with equation 5.3.2 establishes the result.

Theorem 3. *Under the same conditions as in theorem 2 the posterior distribution of* $\delta = \theta_1 - \theta_2$ *is such that*

$$t = \{(\bar{x}_1 - \bar{x}_2) - \delta\}/s\{1/n_1 + 1/n_2\}^{\frac{1}{2}} \tag{6}$$

has Student's t-distribution with ν *degrees of freedom.*

The situations in theorem 3 and in theorem 1 (with $\sigma_1 = \sigma_2$) are the same except that ϕ, unknown in theorem 3, is known in theorem 1 to be equal to the common value of σ_1 and σ_2. Hence theorem 1, in the notation of this theorem, says that

$$\pi(\delta \mid \phi, \mathbf{x}_1, \mathbf{x}_2) \quad \text{is} \quad N(\bar{x}_1 - \bar{x}_2, \quad \phi\{1/n_1 + 1/n_2\}).$$

Furthermore, $\pi(\phi \mid \mathbf{x}_1, \mathbf{x}_2)$ is known from theorem 2, so combining these results we have

$$\begin{aligned}
\pi(\delta, \phi \mid \mathbf{x}_1, \mathbf{x}_2) & \\
&= \pi(\delta \mid \phi, \mathbf{x}_1, \mathbf{x}_2)\, \pi(\phi \mid \mathbf{x}_1, \mathbf{x}_2) \\
&\propto \phi^{-\frac{1}{2}(\nu+3)} \exp\left[-\left\{\left(\frac{1}{n_1}+\frac{1}{n_2}\right)^{-1} [\delta - (\bar{x}_1 - \bar{x}_2)]^2 + \nu s^2\right\} \Big/ 2\phi\right] \\
&= \phi^{-\frac{1}{2}(\nu+3)} \exp\left[-(t^2/\nu + 1)\, \nu s^2 / 2\phi\right], \qquad (7)
\end{aligned}$$

on substituting the expression (6) for t. The integration with respect to ϕ is easily carried out using theorem 5.3.2 with the result that

$$\pi(\delta \mid \mathbf{x}_1, \mathbf{x}_2) \propto \{1 + t^2/\nu\}^{-\frac{1}{2}(\nu+1)}. \qquad (8)$$

The Jacobian of the transformation from δ to t is a constant so that a comparison with equation 5.4.1 establishes the result.

A few definitions will be useful. S_1^2 is called the *sum of squares*, ν_1 the *degrees of freedom* and $s_1^2 = S_1^2/\nu_1$ the *mean square*, for the first sample; with similar definitions for the second sample. S^2 is called the *within sum of squares* and ν the *within degrees of freedom*. $s^2 = S^2/\nu$ is the *mean square* for ϕ. These terms will be used again later (§6.5).

Comparative experiments

Comparative experiments, in which several samples are compared, are much more common than single sample experiments in which the comparison is with some standard. For example, a scientist wishing to investigate the qualities of a new variety of wheat would not merely sow some fields with it and obtain yields, the liability to disease and other factors, since he might obtain a good yield because it was a good year and freedom from disease because it was generally a disease-free year. He would sow neighbouring fields, or plots, some with the new variety and some with one or more varieties which had been used for several years and whose behaviour was well known.

The new variety would be judged by comparison with the older ones: he is using what scientists call a *control*. This is a comparative experiment with two or more samples. The experiments used to illustrate the single sample techniques in chapter 5 were all absolute experiments (to measure the conductivity of the material (§5.1)) or experiments to compare new material with a standard (to compare the precision of a new instrument with the standard (§5.3)). In some branches of science standards are not easy to obtain and it is necessary to use controls. Even where standards are available a comparative experiment may be preferable because the error is less: a point to be discussed in detail under paired comparisons below.

Case of known variances

Theorem 1 is the simplest case of a two-sample experiment and is a direct extension of the corresponding result for the single-sample experiment (theorem 5.2.1). The variances of the two normal distributions are supposed known and the two samples are independent. It is clear from the likelihood (1) that the two sample means are jointly sufficient for the population means. The joint posterior distribution of θ_1 and θ_2 is normal with means $\bar{x}_1$ and $\bar{x}_2$ and variances σ_1^2/n_1 and σ_2^2/n_2, and, since they are independent, correlation zero. It is usually, however, just the difference $\delta = \theta_1 - \theta_2$ that is of interest, and theorem 3.5.5 shows that it is also normally distributed. Consequently confidence intervals for it may be constructed in exactly the same way as with the single value θ_1. Thus with $\tau^2 = \sigma_1^2/n_1 + \sigma_2^2/n_2$, the posterior variance of δ, a 95 % confidence interval for δ is $(\bar{x}_1 - \bar{x}_2) \pm 1{\cdot}96\tau$. It is quite usual for the value $\delta = \bar{\delta} = 0$ to be of especial interest because it corresponds to the two means being equal. For example, if the first sample is taken from the new material and the second from the control, the null hypothesis $\delta = 0$ would say that the mean of the new material was the same as that of the control, whereas if it were not zero then there would be some difference between them. The result will be significant at the 5 % level if $|\bar{x}_1 - \bar{x}_2|/\tau$ exceeds $1{\cdot}96$. If it is only of interest to know whether the new material is an improvement over the control, in the sense of having higher mean, then only values of $\delta > 0$ are of interest and a one-sided confidence interval would be used leading to a different significance test. Thus, we would be fairly certain that $\delta > (\bar{x}_1 - \bar{x}_2) - 2{\cdot}33\tau$, using a 99 % confidence interval $[\Phi(2{\cdot}33) = 0{\cdot}99]$ and the result would be judged significant at the 1 % level if this interval does not include $\delta = 0$, that is, if $(\bar{x}_1 - \bar{x}_2)/\tau > 2{\cdot}33$. But a significance test is rarely adequate in this situation because if the result is significant (that is, you feel fairly sure that the new material is better than the control) then you naturally want to know how much better. Even if it is not significant the additional knowledge of the confidence interval is valuable because it provides a warning of what values of δ, apart from zero, are reasonable. If the confidence interval is too wide you may feel obliged to take additional measurements before being content that the new material is not different from the control. We again issue the warning that a significance test is not a method of making decisions: it only expresses one's degree of belief about the null hypothesis. If the scientist has to decide whether or not to grow the new variety he should use the methods of decision theory, as explained in §5.6. Similarly, if he has prior reason to believe that δ is 0, or is very near zero, this prior information should be incorporated into the analysis. For example, if θ_1 and θ_2 have a joint prior normal distribution which is such that $\delta = \theta_1 - \theta_2$ has mean zero and small variance, theorem 6.6.3 may be used to obtain a posterior normal distribution for δ that incorporates this prior knowledge.

Case of unknown variance

Theorem 3 is an extension of theorem 1 to the case where the variances are unknown and is similar to the extension of the single sample case with variance known (theorem 5.2.1) to the case of unknown variance (theorem 5.4.1). In both extensions the normal distribution is replaced by the t-distribution but otherwise the methods are the same. Theorem 2 is also similar to the single sample variance result (theorem 5.4.2) leading again to a χ^2-distribution. But before discussing the simplicity and elegance of these extensions it must be emphasized that in both theorems it is assumed that the two normal distributions have the same variance, even though the common value is unknown. This severe assumption is often likely to be satisfied in practical applications. For example, the same measuring instrument may be used in both samples, giving the same precision; or the variability in the two samples may be due to common causes, as with wheat, where it is unlikely that the two varieties would differ in their variability over a field.† The corresponding result where the variances are not assumed equal will be given in §6.3, and will show that there is not much difference between the two situations (variances equal or unequal) as far as the difference of means is concerned. The elegance and simplicity of the results when the variances are equal, the beautiful extensions to several samples (§6.4) and the analysis of variance (§6.5), make the assumption a most convenient one.

We discuss theorem 3 first. In the case of known, equal variances, $\sigma_1^2 = \sigma_2^2 = \sigma^2$, theorem 1 says that

$$\{(\bar{x}_1 - \bar{x}_2) - \delta\}/\sigma\{1/n_1 + 1/n_2\}^{\frac{1}{2}} \tag{9}$$

has a posterior distribution $N(0, 1)$. The quantity t, equation (6), is the same as (9) except that s replaces σ. So we have a parallelism between the two situations closely similar to that existing in the single sample case, 5.4(i) and (ii). Confidence intervals and significance tests for $\delta = 0$ may be constructed in the same way as for known σ, with the substitution of s for σ and Student's distribution for the normal. These are the same replacements as were needed in passing from 5.4(i) to (ii) and need not be discussed again. What do merit attention, however, are the values of s and ν, the degrees of freedom for t. There are two sources for the posterior knowledge of the variance; namely the two samples. In §5.4 we saw that the sum of squares of the sample values about their mean divided by the number of observations, n, was a reasonable statistic to replace the variance. In fact we divided by $(n-1)$ for a reason which will appear in a moment. Similarly, in the two sample case $S_1^2/\nu_1 = s_1^2$ and $S_2^2/\nu_2 = s_2^2$ are both reasonable estimates of the variance and we

† With wheat, it is usually the coefficient of variation that stays constant; but if the means do not differ too much the variances will not either.

naturally combine them. A suggestion would be to take their average $\frac{1}{2}(s_1^2+s_2^2)$ but the analysis of the theorem shows that the weighted average $(\nu_1 s_1^2+\nu_2 s_2^2)/(\nu_1+\nu_2) = s^2$ is the more convenient quantity to use because, from equation (4), s^2, $\bar{x}_1$ and $\bar{x}_2$ are jointly sufficient. So what we do (in terms of the definitions above) is to take the sums of squares for the two samples and add them together obtaining the within sum of squares; to take the degrees of freedom for the two samples and add them together obtaining the within degrees of freedom; and divide the former by the latter to obtain the mean square to replace the variance. This simple procedure of addition of sums of squares and degrees of freedom generalizes to more complicated situations. (Notice that the sums of squares are always sums of squares about the sample means; the last four words are understood in speaking of sums of squares. Sums of squares about the origin are called *uncorrected* (§6.5).)

We can now explain the term 'degrees of freedom'. The sum

$$S_1^2 = \sum_{i=1}^{n_1} (x_{1i}-\bar{x}_1)^2$$

is the sum of the squares of n_1 terms and therefore appears at first glance to have n_1 parts which can vary. But the terms are constrained (to use a mechanical term) to add to zero, since $\sum_{i=1}^{n_1} (x_{1i}-\bar{x}) = 0$, so that once values are assigned to (n_1-1) of them, the last is then fixed. To continue the analogy with mechanics, it is rather like a mechanical system of n_1 parts with only (n_1-1) of them free to vary because of a single constraint, and we say, as we would of the mechanical system, that it has (n_1-1) degrees of freedom. Similarly, S_2^2 has (n_2-1) degrees of freedom and S^2, the sum of n_1+n_2 terms with two constraints has $n_1+n_2-2 = (n_1-1)+(n_2-1)$ degrees of freedom. Again this idea extends to more complicated situations. The extension will also explain the use of the adjective 'within'.

It is also possible to see (at the cost of a little algebra which is omitted) why the degrees of freedom were used to divide the sum of squares, and not n. Suppose $s^{*2} = \Sigma(x_i-\bar{x})^2/n$ had been used in §5.4 instead of s^2 and t^* had been defined as t with s^* for s.

Then the posterior distribution of t^* would have been proportional to

$$(1+t^{*2}/n)^{-\frac{1}{2}n}. \quad (10)$$

If, in the two-sample case, $s^{*2} = (S_1^2+S_2^2)/(n_1+n_2)$ had been used instead of s^2 then the posterior distribution of t^*, defined as t with s^* for s, would have been proportional to

$$\{1+t^{*2}/(n_1+n_2)\}^{-\frac{1}{2}(n_1+n_2-1)}. \quad (11)$$

Now (10) and (11) are not of the same form, with n_1+n_2 replacing n, and new tables would be needed for the situation of this section: whereas (8) is the same as expression 5.4.1, and the same tables suffice.

An alternative proof of theorem 3 starts from equation (4). In this we can change variables to $\delta = \theta_1 - \theta_2$ and $\theta_1+\theta_2$, say, and integrate with respect to $\theta_1+\theta_2$ to obtain (7), the joint density of δ and ϕ. The argument using conditional probabilities avoids this integration, or rather utilizes the fact that it has already been done in the proof of theorem 1, in effect.

Theorem 2 is a natural extension of theorem 5.4.2. The degrees of freedom still determine which χ^2-distribution is appropriate and the quantity which, when divided by the variance, has this distribution is still the sum of squares, νs^2, here the total sum of squares. Inferences about the variance can therefore be made using the χ^2-distribution in exactly the same way as for a single sample.

Paired comparisons

Whenever a theorem is used one should make sure that the conditions of the theorem are reasonably well satisfied. This is particularly true of theorems 1 and 3, and we now illustrate a possible misuse of them. As we have already explained a common use of the result is in experiments that compare a control and a new treatment. One way of designing such experiments can be illustrated on a method used to examine the effect of a paint in preventing corrosion. Pieces of metal had one half treated with the paint and the other left in the usual state appropriate to the use that the metal was to be put. The pieces were placed in a wide range of positions, differing in exposure to weather, etc., and at the end of a suitable period of time measurements were made of the corrosion, x_{1i}, on the untreated part and also, x_{2i}, on the treated part of the ith piece. It is not unreasonable to suppose that the x_{1i} form a random sample from $N(\theta_1, \phi)$ where θ_1 is the average corrosion and ϕ is a measure of the variability under different conditions of exposure, etc. Similarly, the x_{2i} form a random sample from $N(\theta_2, \phi)$, assuming the same variability for painted and unpainted metal. We wish to investigate $\theta_1 - \theta_2$, the average reduction in corrosion due to painting. At first glance it would appear that theorem 3 could be used, but this is not so because the two random samples $\{x_{1i}\}$ and $\{x_{2i}\}$ are not independent. Since x_{1i} and x_{2i} refer to the corrosion on two parts of a piece of metal subject to identical conditions except for the painting, they are likely to be much closer together than, say, x_{1i} and x_{2j} $(j \neq i)$ which were in different conditions of exposure. Consequently the conditions of the theorem are not satisfied and the likelihood would be different because of the correlations between x_{1i} and x_{2i}.

Inferences about $\theta_1 - \theta_2$ may be made in the following way. The random variables $z_i = x_{1i} - x_{2i}$ have expectations $\theta_1 - \theta_2$ and variances ϕ_1, say, and might perhaps be assumed normally distributed. (If x_{1i} and x_{2i} have a bivariate normal distribution this will follow from theorem 3.5.5.) If so they would form a random sample from a normal distribution, since the separate pieces of metal are probably independent, and we would wish to make inferences about the mean of this distribution. This is a single sample problem and theorem 5.4.1 enables the usual confidence limits, or significance test for the mean being zero, to be found with the aid of the t-distribution. Notice that this analysis does not assume that x_{1i} and x_{2i} have the same variances.

Although this method is valid, it is not obvious that it is a complete inference about $\theta_1 - \theta_2$ from the whole of the data. The argument amounts to considering only the differences between the two values on the same plate and takes no account of the separate values. Although it is intuitively obvious that no information about $\theta_1 - \theta_2$ is lost by this procedure the argument does need further justification.

This is provided by remarking that we can write, with $w_i = x_{1i}+x_{2i}$, omitting reference to the parameters, and the rest of the notation obvious,

$$p(\mathbf{x}_1, \mathbf{x}_2) = p(\mathbf{z}, \mathbf{w}) = p(\mathbf{z})p(\mathbf{w}|\mathbf{z}).$$

Whilst $p(\mathbf{z})$ depends on δ, $p(\mathbf{w}|\mathbf{z})$ typically will not, and, whatever parameters the latter does depend on, they will usually not occur in $p(\mathbf{z})$ and will, prior to the observations, be independent of those in $p(\mathbf{z})$. Consequently as far as inferences about δ are concerned we may confine attention to $p(\mathbf{z})$, absorbing $p(\mathbf{w}|\mathbf{z})$ into the constant of proportionality. (Compare the discussion on ancillary statistics in §5.5.) If x_{1i} and x_{2i}, and hence z_i and w_i, have a bivariate normal distribution, w_i, for fixed z_i, will depend on α, β, σ^2 (equations 3.2.13 and 3.2.14) whilst z_i will depend on μ_1 and σ_1^2 (in the notation of §3.2). μ_1 is here δ, and provided $(\alpha, \beta, \sigma^2)$ and (μ_1, σ_1^2) are independent the inference using $\mathbf{z}$ only will not lose any information.

Two practical points of interest emerge from the experiment. First, the quantity $\theta_1-\theta_2$, whose posterior distribution, given the observed differences, has been found, is the average reduction over the conditions of the experiment, and may not be the most sensible thing to consider. For example, suppose the effect of the paint is to reduce the corrosion by a fixed percentage irrespective of the amount of the corrosion. Then $\theta_2 = \lambda\theta_1$ and the difference is $\theta_1(1-\lambda)$. But θ_1 may be different if the experiment is done again and is, in any case, irrelevant in assessing the effect of the paint. The quantity of interest is λ and this may be studied by taking logarithms of the readings: the differences of these will have an approximate mean value of $\ln\lambda$ and inferences can be made about that, although it must be remembered that the logarithms might not have a normal distribution. It is important, in applications, to make sure that one is applying the theoretical results to the quantity of interest.

Secondly, one might ask why this experimental design was used at all. Why not take a random sample of painted pieces and a second, independent, random sample of unpainted pieces? The reason is that the precision of the determination of the effect, whether measured by the inverse of the variance of the posterior distribution or by some other means, such as considering the width of the final confidence interval, depends on the true variance of the normal distribution (or distributions) from which the samples have been taken. The x_{1i} and the x_{2i} will have a variance which includes the variation from place to place whereas the differences $x_{1i}-x_{2i}$ will have a variance, which only involves variations due to the instrument measuring the corrosion and to any minor differences in the two parts of the same piece of metal. Consequently the effect of taking the differences is to reduce, probably very considerably, the variance, and hence increase the precision of the determination of the effect of painting. As the readings are taken in pairs the method is termed '*paired comparisons*'. We shall later (§§6.4, 6.5) see how to extend this idea and to analyse the variability in some experiments into different parts in order to remove the larger parts and permit a more sensitive analysis.

6.2. Comparison of two variances

If a random variable, usually in this context denoted by F, has a density proportional to

$$F^{\frac{1}{2}\nu_1-1}/(\nu_2+\nu_1 F)^{\frac{1}{2}(\nu_1+\nu_2)} \tag{1}$$

for $F \geqslant 0$, and zero otherwise, where ν_1 and ν_2 are positive, it is said to have *Fisher's*† *F-distribution with* ν_1 *and* ν_2 *degrees of freedom*, or simply an *F-distribution*. We shall often speak of an $F(\nu_1, \nu_2)$ distribution. Notice that the order of reference to the degrees of freedom is relevant: $F(\nu_1, \nu_2)$ is not the same as $F(\nu_2, \nu_1)$.

Theorem 1. *Let* $\mathbf{x}_i = (x_{i1}, x_{i2}, \ldots, x_{in_i})$ *be a random sample of size* n_i *from* $N(\theta_i, \phi_i)$ $(i = 1, 2)$, *with* $\mathbf{x}_1$ *and* $\mathbf{x}_2$ *independent; let the prior distributions of* θ_1, θ_2, $\ln\phi_1$ *and* $\ln\phi_2$ *be independent and each uniform over* $(-\infty, \infty)$. *Then the posterior distribution of* $(s_1^2/s_2^2)/(\phi_1/\phi_2)$ *is* $F(\nu_1, \nu_2)$, *where* s_1^2, s_2^2, ν_1 *and* ν_2 *are as in theorem* 6.1.2.

Because of the independence, both of the samples and of the prior distributions, the posterior distributions of ϕ_1 and ϕ_2 are independent‡ and their separate distributions are given by

† Some writers refer to it as Snedecor's F-distribution.

‡ Compare the argument with which the proof of theorem 6.1.1 began.

theorem 5.4.2, that is $\nu_i s_i^2/\phi_i$ is χ^2 with ν_i degrees of freedom ($i = 1, 2$). Hence (equation 5.3.2)

$$\pi(\phi_1, \phi_2 | \mathbf{x}_1, \mathbf{x}_2) \propto \phi_1^{-\frac{1}{2}\nu_1-1}\phi_2^{-\frac{1}{2}\nu_2-1}\exp\{-\nu_1 s_1^2/2\phi_1 - \nu_2 s_2^2/2\phi_2\}. \quad (2)$$

Let $F = (s_1^2/s_2^2)/(\phi_1/\phi_2)$; then the joint density of F and ϕ_2 is (by theorem 3.5.2, with Jacobian equal to $s_1^2\phi_2/s_2^2\phi_1^2$) proportional to

$$F^{\frac{1}{2}\nu_1-1}\phi_2^{-\frac{1}{2}(\nu_1+\nu_2)-1}\exp\{-(\nu_2+\nu_1 F)s_2^2/2\phi_2\}. \quad (3)$$

The integration with respect to ϕ_2 is easily carried out using theorem 5.3.2 and the result is

$$\pi(F | \mathbf{x}_1, \mathbf{x}_2) \propto F^{\frac{1}{2}\nu_1-1}/(\nu_2+\nu_1 F)^{\frac{1}{2}(\nu_1+\nu_2)}$$

as required.

***Corollary**. Let the conditions be as in the theorem except that the means θ_1, θ_2 are known, equal to μ_1 and μ_2 respectively. Then the posterior distribution of $(\bar{s}_1^2/\bar{s}_2^2)/(\phi_1/\phi_2)$ is $F(n_1, n_2)$, where*

$$n_i \bar{s}_i^2 = \sum_{j=1}^{n_i} (x_{ij} - \mu_i)^2 \quad (i = 1, 2). \quad (4)$$

The posterior distribution of ϕ_1 is now such that $n_i\bar{s}_i^2/\phi_i$ is χ^2 with n_i degrees of freedom (theorem 5.3.1 with $\nu_0 = 0$), and the result follows exactly as in the proof of the theorem.

The F-distribution

With the F-distribution we meet the last of the trio (χ^2, t and F) that has played so important a part in modern statistics. It is the most important of the three, and, indeed, the other two are special cases of it as we shall see below. The missing constant of proportionality in (1) is easily found by integration from 0 to ∞, which is carried out by substituting $x = \nu_1 F/(\nu_2 + \nu_1 F)$, with $dF/dx = \nu_2/[\nu_1(1-x)^2]$. Then

$$\int_0^\infty F^{\frac{1}{2}\nu_1-1}(\nu_2+\nu_1 F)^{-\frac{1}{2}(\nu_1+\nu_2)}dF$$

$$= \int_0^1 x^{\frac{1}{2}\nu_1-1}(1-x)^{\frac{1}{2}\nu_2-1}dx/\nu_1^{\frac{1}{2}\nu_1}\nu_2^{\frac{1}{2}\nu_2}$$

$$= \nu_1^{-\frac{1}{2}\nu_1}\nu_2^{-\frac{1}{2}\nu_2}(\tfrac{1}{2}\nu_1-1)!\,(\tfrac{1}{2}\nu_2-1)!/[\tfrac{1}{2}(\nu_1+\nu_2)-1]!.$$

This last result follows from the B-integral (equation 5.4.7). Hence the $F(\nu_1, \nu_2)$ distribution has the density

$$\frac{[\frac{1}{2}(\nu_1+\nu_2)-1]!}{(\frac{1}{2}\nu_1-1)!\,(\frac{1}{2}\nu_2-1)!}\,\nu_1^{\frac{1}{2}\nu_1}\nu_2^{\frac{1}{2}\nu_2}\,\frac{F^{\frac{1}{2}\nu_1-1}}{(\nu_2+\nu_1F)^{\frac{1}{2}(\nu_1+\nu_2)}}. \tag{5}$$

For $\nu_1 > 2$ the density (5) is zero at $F = 0$, increases to a maximum at $F = \nu_2(\nu_1-2)/\nu_1(\nu_2+2)$ and then decreases to zero. For large ν_1 and ν_2 the maximum is at about $F = 1$. The mean and variance are easily found from (5) using the same substitution as produced the missing constant in (1). The results are $\nu_2/(\nu_2-2)$ for the mean, provided $\nu_2 > 2$, and

$$2\nu_2^2(\nu_1+\nu_2-2)/\nu_1(\nu_2-4)\,(\nu_2-2)^2$$

for the variance, provided $\nu_2 > 4$. For large degrees of freedom these values are approximately 1 and $2(\nu_1+\nu_2)/\nu_1\nu_2$. If $\nu_1 \leqslant 2$ the density has the maximum at $F = 0$ and decreases to zero as $F \to \infty$. In the case $\nu_1 = 1$ the substitution $F = t^2$ in (5) gives the density of Student's t-distribution with ν_2 degrees of freedom (equation 5.4.8). Since the t-distribution is symmetric it follows that tables of the F-distribution with $\nu_1 = 1$ are equivalent to tables of the t-distribution. The reason for this connexion between the two distributions will appear later (§6.4).

It follows from the proof of the theorem that

$$\nu_1F/\nu_2 = (\nu_1s_1^2/\phi_1)/(\nu_2s_2^2/\phi_2)$$

is the ratio of two independent quantities which are χ^2 with ν_1 and ν_2 degrees of freedom respectively: hence the nomenclature for ν_1 and ν_2. Now consider what happens as $\nu_2 \to \infty$. The size of the second sample increases and, as explained in §5.3, the knowledge of ϕ_2 becomes more and more precise so that we effectively know the true value of ϕ_2, namely the limit of s_2^2. Hence as $\nu_2 \to \infty$ we approach the case where ϕ_2 is known, and then only the first sample is relevant. So if $\phi_2 = \sigma_2^2$, s_2^2 tends to σ_2^2 and we see that ν_1F tends to $\nu_1s_1^2/\phi_1$ which we know to be χ^2 with ν_1 degrees of freedom. Hence $\nu_1F(\nu_1, \infty)$ is χ^2 with ν_1 degrees of freedom. Thus the statement above that t and χ^2 are both special cases of F is substantiated.

The upper percentage points of the F-distribution have been extensively tabulated: see, for example, Lindley and Miller (1961). They give values $\bar{F}_\alpha(\nu_1, \nu_2)$ such that if F is $F(\nu_1, \nu_2)$ then

$$p(F > \bar{F}_\alpha(\nu_1, \nu_2)) = \alpha \qquad (6)$$

for $100\alpha = 5, 2\frac{1}{2}, 1$ and $0{\cdot}1$, and a wide range of degrees of freedom. Notice that it requires a whole page to tabulate the values for a single value of α. This is a table of triple entry: three values α, ν_1 and ν_2 are needed to determine a single value of F. With the normal distribution a table of single entry (α) sufficed: with t and χ^2 a table of double entry (α, ν) was needed. With more complex problems more extensive tables are needed and the inference problems cannot be considered solved until they have been provided. Approximate devices to avoid these difficulties will be discussed later (§7.1). Fortunately, with the F-distribution it is not necessary to tabulate the lower percentage points as well. The reason is that $F^{-1} = (s_2^2/s_1^2)/(\phi_2/\phi_1)$ has, by interchanging the first and second samples in the theorem, also an F-distribution with the degrees of freedom similarly interchanged, namely ν_2 and ν_1. Consequently, if F is $F(\nu_1, \nu_2)$

$$p(F < x) = p(F^{-1} > x^{-1})$$

and this equals α, from (6), when $x^{-1} = \bar{F}_\alpha(\nu_2, \nu_1)$. Thus

$$\underline{F}_\alpha(\nu_1, \nu_2) = \{\bar{F}_\alpha(\nu_2, \nu_1)\}^{-1}, \qquad (7)$$

where $p(F < \underline{F}_\alpha(\nu_1, \nu_2)) = \alpha$. We therefore have the rule: to find the lower point, interchange the degrees of freedom and take the reciprocal of the upper point for those values. Of course this result may be established purely algebraically, without appeal to the theorem, directly from the density, equation (5).

Comparison of two variances

The theorem refers to independent samples from normal distributions and the prior distributions represent ignorance about all four parameters. It is not assumed, as in the previous section, that the two variances are equal. Such a situation might arise when comparing two measuring instruments for precision. They give rise to the two samples, one for each instrument, and one would be particularly interested in knowing if ϕ_1

were equal to ϕ_2, so that a possible null hypothesis of interest is $\phi_1 = \phi_2$. This hypothesis could be expressed by saying $\phi_1 - \phi_2 = 0$ but it is more convenient to use $\phi_1/\phi_2 = 1$ because the posterior distribution of the ratio is easier to handle (for tabulation purposes) than that of the difference. It is also consistent with the use of the logarithms of the variances as the parameters having uniform prior distributions to consider the posterior distribution of the difference of their logarithms; that is, the logarithms of their ratio. Since ϕ_1 and ϕ_2 are known, by theorem 5.4.2, to have independent posterior distributions related to χ^2, the only point of substance in the proof is to find the distribution of the ratio of independent χ^2's. The result is the F-distribution if the degrees of freedom divide the χ^2 variables. Precisely, $\nu_i s_i^2/\phi_i = \chi_i^2$ and $F = (\chi_1^2/\nu_1)/(\chi_2^2/\nu_2)$. (Compare the relation to χ^2 discussed above.) Notice that s_i^2 is an estimate of ϕ_i (§5.3) so that F is the ratio of the ratio of estimates of variance to the ratio of the population variances. For this reason it is sometimes called a *variance-ratio*.

It is now possible to make confidence interval statements about the ratio of ϕ_1 to ϕ_2. For example, from (6), with

$$F = (s_1^2/s_2^2)/(\phi_1/\phi_2)$$

we have

$$\pi(\phi_1/\phi_2 < (s_1^2/s_2^2)/\bar{F}_\alpha(\nu_1, \nu_2) \mid \mathbf{x}_1, \mathbf{x}_2) = \alpha. \tag{8}$$

Similarly,

$$\pi(\phi_1/\phi_2 > (s_1^2/s_2^2)/\underline{F}_\alpha(\nu_1, \nu_2) \mid \mathbf{x}_1, \mathbf{x}_2) = \alpha, \tag{9}$$

and hence, from (7),

$$\pi(\phi_1/\phi_2 > \bar{F}_\alpha(\nu_2, \nu_1)\,(s_1^2/s_2^2) \mid \mathbf{x}_1, \mathbf{x}_2) = \alpha. \tag{10}$$

From (8) and (10) a confidence limit of finite extent for ϕ_1/ϕ_2 with coefficient $\beta = 1-\alpha$ is given by

$$(s_1^2/s_2^2)/\bar{F}_{\frac{1}{2}\alpha}(\nu_1, \nu_2) < \phi_1/\phi_2 < \bar{F}_{\frac{1}{2}\alpha}(\nu_2, \nu_1)\,(s_1^2/s_2^2).$$

This interval will not be the shortest in the sense of §5.2 but tables for the shortest interval have not been calculated.

The main use for the theorem is to provide a significance test for the null hypothesis that $\phi_1 = \phi_2$. Sometimes the alternative hypotheses of principal interest are that $\phi_1 > \phi_2$: for example, if the first sample is taken using a control instrument, and the second using a new instrument, one may only be interested in the latter if it is more accurate. The confidence limit required would be that ϕ_1/ϕ_2 exceeds some value, so would be obtained from (8). The result would be significant if the confidence interval did not contain the value 1. That is, if the interval in (8) did contain this value. Hence it is significant at the level α if

$$s_1^2/s_2^2 > \bar{F}_\alpha(\nu_1, \nu_2). \tag{11}$$

The final inequality is simply that the ratio of the sample variances exceeds the upper 100α % point of the F-distribution with ν_1 and ν_2 degrees of freedom: the first degree of freedom refers to the numerator of the sample ratio. This one-sided significance test will appear again later (§6.4) in an important problem.

The corollary is not often useful since it is unusual for the means to be known but not the variances. There is little need to comment on it except to remark that we again have the F-distribution, and the only difference is that the degrees of freedom are equal to the sample sizes and not one less than them. Otherwise confidence intervals and significance tests are as before.

6.3. General comparison of two means

If t_1 and t_2 are independent random variables each distributed in Student's t-distribution, with ν_1 and ν_2 degrees of freedom respectively, and if $\tilde{\omega}$ is a constant, representing an angle between 0 and 90 degrees, the random variable

$$d = t_1 \cos\tilde{\omega} - t_2 \sin\tilde{\omega} \tag{1}$$

will be said to have *Behrens's distribution* with ν_1 and ν_2 degrees of freedom and angle $\tilde{\omega}$.

Theorem 1. *Let the conditions be the same as those of theorem 6.2.1; then the posterior distribution of $\delta = \theta_1 - \theta_2$ is such that*

$$d = \{\delta - (\bar{x}_1 - \bar{x}_2)\}/(s_1^2/n_1 + s_2^2/n_2)^{\frac{1}{2}} \tag{2}$$

has Behrens's distribution with ν_1 and ν_2 degrees of freedom, and angle $\tilde{\omega}$ given by

$$\tan\tilde{\omega} = (s_2/\sqrt{n_2})/(s_1/\sqrt{n_1}). \tag{3}$$

Again, because of the independence both of the samples and of the prior distributions, the posterior distributions of θ_1 and θ_2 are independent and their separate distributions are given by theorem 5.4.1. That is, $n_i^{\frac{1}{2}}(\theta_i - \bar{x}_i)/s_i = t_i$, say, has Student's t-distribution with ν_i degrees of freedom ($i = 1, 2$). We have

$$\theta_i - \bar{x}_i = t_i s_i/\sqrt{n_i} \quad (i = 1, 2)$$

so that $$(\theta_1 - \theta_2) - (\bar{x}_1 - \bar{x}_2) = t_1(s_1/\sqrt{n_1}) - t_2(s_2/\sqrt{n_2}),$$

or, dividing by $(s_1^2/n_1 + s_2^2/n_2)^{\frac{1}{2}}$ and using (2) and (3),

$$d = t_1 \cos\tilde{\omega} - t_2 \sin\tilde{\omega},$$

which, on comparison with (1), proves the theorem.

Behrens's distribution

We shall not attempt any detailed study of Behrens's distribution. It will suffice to remark that it is symmetrical about $d = 0$, that its moments follow easily from those of Student's distribution and that, as ν_1 and ν_2 both tend to infinity it tends to normality. This final result follows because the t-distribution tends to normality (§5.4) and the difference of normal variables is also normal (theorem 3.5.5). Some percentage points of the distribution are to be found in Fisher and Yates (1963). These tables give values $d_\alpha(\nu_1, \nu_2, \tilde{\omega})$ such that

$$\pi(|d| > d_{\frac{1}{2}\alpha}(\nu_1, \nu_2, \tilde{\omega}) \mid \mathbf{x}_1, \mathbf{x}_2) = \alpha \tag{4}$$

and because of the symmetry

$$\pi(d > d_\alpha(\nu_1, \nu_2, \tilde{\omega}) \mid \mathbf{x}_1, \mathbf{x}_2) = \alpha. \tag{5}$$

Comparison of two means

Theorem 1 differs from theorem 6.1.3 only that in the latter the two variances are assumed equal with a uniform prior distribution for the logarithm of their common value, whereas here the two variances are supposed independent with each logarithm uniformly distributed. In both situations we seek the posterior distribution of the differences between the means, $\theta_1 - \theta_2 = \delta$. We recall that in using the earlier theorem the two estimates of variance, s_1^2 and s_2^2, from the two samples were pooled to provide a single estimate s^2 given by $\nu s^2 = \nu_1 s_1^2 + \nu_2 s_2^2$ with $\nu = \nu_1 + \nu_2$. The quantity t was then obtained by taking the quantity used when the common variance, σ^2, was known (equation 6.1.9) and replacing σ in it by the estimate s. In the present situation the two variances must be separately estimated, by s_1^2 and s_2^2, and the quantity used when the variances are known

$$\{\delta - (\bar{x}_1 - \bar{x}_2)\}/(\sigma_1^2/n_1 + \sigma_2^2/n_2)^{\frac{1}{2}}$$

(from theorem 6.1.1 where it is shown to be $N(0, 1)$) continues to be used, but with s_1^2 and s_2^2 replacing σ_1^2 and σ_2^2: this gives (2). Unfortunately its distribution is complicated, and furthermore depends on three parameters, so that a table of quadruple entry is needed. The change of the percentage points with $\tilde{\omega}$, and with ν_1 and ν_2 when these are both large, is small so that rather coarse grouping of the arguments is possible in tabulation. Notice that the statistic d is still (see §5.4) of the form: the difference between δ (the unknown value) and the difference of sample means, divided by the estimated standard deviation of this difference. For $\mathscr{D}^2(\bar{x}_1 - \bar{x}_2) = \sigma_1^2/n_1 + \sigma_2^2/n_2$, since the samples, and hence $\bar{x}_1$ and $\bar{x}_2$, are independent. Confidence limits for δ are therefore provided in the usual way by taking the sample difference plus or minus a multiple, depending on Behrens's distribution, of the estimated standard deviation of this difference. Thus, from (4), writing d_α for $d_\alpha(\nu_1, \nu_2, \tilde{\omega})$,

$$\pi(\bar{x}_1 - \bar{x}_2 - d_{\frac{1}{2}\alpha}(s_1^2/n_1 + s_2^2/n_2)^{\frac{1}{2}} \leqslant \delta \leqslant \bar{x}_1 - \bar{x}_2 + d_{\frac{1}{2}\alpha}(s_1^2/n_1 + s_2^2/n_2)^{\frac{1}{2}} \,|\, \mathbf{x}_1, \mathbf{x}_2) = \beta, \quad (6)$$

where, as usual, $\alpha = 1 - \beta$. A significance test of $\delta = 0$ at level α is obtained by declaring the result to be significant if this interval does not contain the origin.

Relationship to Student's problem

It is a moot point which of theorems 6.1.3 and the present one should be used in practical problems. Fortunately the differences are not very great; that is $t_\alpha(\nu_1 + \nu_2)$ is not usually very different from $d_\alpha(\nu_1, \nu_2, \tilde{\omega})$, and the differences between pooling the variances or using them separately are usually small, at least if the sample sizes are nearly equal. Another argument that is sometimes used is the following. In an obvious abbreviated notation the joint posterior distribution of δ and ϕ_1/ϕ_2 may be written

$$\pi(\delta, \phi_1/\phi_2 | \mathbf{x}) = \pi(\delta | \phi_1/\phi_2, \mathbf{x})\, \pi(\phi_1/\phi_2 | \mathbf{x}). \tag{7}$$

Now $\pi(\phi_1/\phi_2 | \mathbf{x})$ is known (theorem 6.2.1). Suppose that this has its maximum value near $\phi_1/\phi_2 = 1$ and that it decreases sharply from the maximum: in other words, a confidence interval for ϕ_1/ϕ_2 is a small interval containing 1. Then an integration of (7) with respect to ϕ_1/ϕ_2, in order to obtain the marginal distribution of δ, will only have an appreciable integrand near $\phi_1/\phi_2 = 1$ and will therefore be approximately $\pi(\delta | 1, \mathbf{x})$ which gives Student's distribution (theorem 6.1.3). Consequently, one can first investigate the ratio of variances and, if this suggests that they are about equal, use Student's method based on the assumption that they are. The adequacy of the approximation has not been investigated. Often it is held enough that the test for $\phi_1/\phi_2 = 1$ be not significant for Student's method to be used: that is, one merely looks to see if the confidence interval contains 1, not how small it is. There seems little point in doing this here since Behrens's distribution is tabulated, but in more complicated situations the procedure of using a preliminary significance test has much to recommend it (see §6.5). Undoubtedly the simplicity and elegant extensions (§§6.4, 8.3) of the equal variances result make it more attractive in practical situations and it is used today far more often than Behrens's result. There is, however, one other reason for this which we now discuss.

[The remainder of this section uses the ideas developed at the end of §5.6 and, like those remarks, may be omitted.]

Non-Bayesian significance test

Behrens's result is the first example in this book of a situation where the confidence limits and significance test derived from a Bayesian approach differ from those obtained using a significance test based on the sampling distribution of a statistic. It was also the first important example to arise historically, and has been the subject of much discussion. The statistic used in the alternative approach is still d, which we now write $d(\mathbf{x})$, where $\mathbf{x} = (\mathbf{x}_1, \mathbf{x}_2)$, to indicate that it is as a function of $\mathbf{x}$, and not δ, that it interests us. Unfortunately the distribution of $d(\mathbf{x})$ under the null hypothesis, that is, when $\delta = 0$, depends on σ_1^2/σ_2^2, the ratio of the two population variances. It is not therefore possible to find a set of values of d such that the probability is α that $d(\mathbf{x})$ belongs to this set, when $\delta = 0$, irrespective of the unknown value of σ_1^2/σ_2^2. It is necessary to be able to do this in order that the level, α, quoted be correct whatever the value of the ratio. Instead the procedure is to declare the result significant if $|d(\mathbf{x})| \geqslant g(s_1^2/s_2^2)$ where g is some function chosen so that the probability is constant when $\delta = 0$. It is not known if such a function exists, but an adequate approximation has been given by Welch and is tabulated in Pearson and Hartley (1958). The resulting test is different from that of this section. To make matters more complicated, and more unsatisfactory for the user of statistical results, Fisher has derived Behrens's result by an argument different from ours, but one which does not find ready acceptance by any substantial group of statisticians. In view of the remarks in §5.6 about significance tests based on the sampling distribution of a statistic, it seems clear that Behrens's solution to the problem is correct, granted either the appropriateness of the prior distributions, or the acceptance of Fisher's argument.

6.4. Comparison of several means

In theorem 6.1.3 a significance test, using Student's t-distribution, was developed to test the null hypothesis that two means of normal distributions were equal when their unknown variances were equal. In the present section the significance test is generalized to several means. The proof of the theorem is somewhat involved and the reader is advised not to try to understand all the details at first, only returning to study them when he has seen how the result is used.

Theorem 1. *Let* $\mathbf{x}_i = (x_{i1}, x_{i2}, \ldots, x_{in})$ $(i = 1, 2, \ldots, r)$ *be* r *independent random samples each of size* n *from* $N(\theta_i, \phi)$, *and* $\theta_1, \theta_2, \ldots, \theta_r$ *and* $\ln\phi$ *have uniform and independent prior distribu-*

tions over $(-\infty, \infty)$. *Then a significance test at level* α *of the null hypothesis that all the* θ_i *are equal is provided by declaring the data significant if*

$$F = \frac{n \sum_i (x_{i.} - x_{..})^2/\nu_1}{\sum_{i,j} (x_{ij} - x_{i.})^2/\nu_2} \tag{1}$$

exceeds $\bar{F}_\alpha(\nu_1, \nu_2)$, *the upper* $100\alpha\,\%$ *point of the F-distribution with* ν_1 *and* ν_2 *degrees of freedom (equation 6.2.6).*

The notation in (1) *is*

$$x_{i.} = \sum_{j=1}^{n} x_{ij}/n, \quad x_{..} = \sum_{i=1}^{r} x_{i.}/r = \sum_{i,j} x_{ij}/rn, \tag{2}$$

and

$$\nu_1 = r-1, \quad \nu_2 = r(n-1). \tag{3}$$

The joint posterior distribution of all the parameters is clearly

$$\pi(\theta_1, \theta_2, \ldots, \theta_r, \phi \,|\, \mathbf{x}) \\ \propto \phi^{-\frac{1}{2}(nr+2)} \exp\left[-\left\{\sum_i n(x_{i.} - \theta_i)^2 + \sum_i S_i^2\right\}/2\phi\right], \tag{4}$$

a direct generalization of equation 6.1.4, where

$$S_i^2 = \sum_{j=1}^{n} (x_{ij} - x_{i.})^2.$$

$x_{i.}$ is a new notation for $\bar{x}_i$ and $\mathbf{x}$ denotes $(\mathbf{x}_1, \mathbf{x}_2, \ldots, \mathbf{x}_r)$. It is convenient to write S^2 for $\sum_i S_i^2$. Now change from the random variables $\theta_1, \theta_2, \ldots, \theta_r$ to $\theta, \lambda_1, \lambda_2, \ldots, \lambda_r$, where $\theta_i = \theta + \lambda_i$ and $\Sigma\lambda_i = 0$; leaving ϕ unaffected. The point of the change is that we are interested in differences between the θ_i which can be expressed in terms of the λ_i: the null hypothesis is that

$$\lambda_1 = \lambda_2 = \ldots = \lambda_r = 0. \tag{5}$$

The Jacobian of the transformation is constant (theorem 3.5.2) so that, provided $\Sigma\lambda_i = 0$,

$$\pi(\theta, \lambda_1, \lambda_2, \ldots, \lambda_r, \phi \,|\, \mathbf{x}) \\ \propto \phi^{-\frac{1}{2}(nr+2)} \exp\left[-\left\{\sum_i n(x_{i.} - \theta - \lambda_i)^2 + S^2\right\}/2\phi\right]. \tag{6}$$

The summation over i may be written

$$\begin{aligned} & nr\theta^2 - 2\theta n\Sigma(x_{i.} - \lambda_i) + n\Sigma(x_{i.} - \lambda_i)^2 \\ &= nr\theta^2 - 2\theta nrx_{..} + n\Sigma(x_{i.} - \lambda_i)^2, \quad \text{since } \Sigma\lambda_i = 0, \\ &= nr(\theta - x_{..})^2 + n\Sigma(x_{i.} - \lambda_i)^2 - nrx^2_{..} \\ &= nr(\theta - x_{..})^2 + n\Sigma(x_{i.} - x_{..} - \lambda_i)^2, \text{ using } \Sigma\lambda_i = 0 \text{ again.} \end{aligned} \tag{7}$$

On substituting this expression into (6) the integration with respect to θ is easily carried out using the normal integral, with the result that

$$\pi(\lambda_1, \lambda_2, \ldots, \lambda_r, \phi \mid \mathbf{x}) \propto \phi^{-\frac{1}{2}(nr+1)} \exp\left[-\{n \sum_i (x_{i.} - x_{..} - \lambda_i)^2 + S^2\}/2\phi\right]. \tag{8}$$

Integrating with respect to ϕ, using theorem 5.3.2, we obtain

$$\pi(\lambda_1, \lambda_2, \ldots, \lambda_r \mid \mathbf{x}) \propto \{n \sum_i (x_{i.} - x_{..} - \lambda_i)^2 + S^2\}^{-\frac{1}{2}(nr-1)}, \tag{9}$$

provided $\Sigma\lambda_i = 0$.

Hence the joint density of the λ_i has spherical symmetry about the value

$$(x_{1.} - x_{..}, x_{2.} - x_{..}, \ldots, x_{r.} - x_{..})$$

in the $(r-1)$-dimensional space with $\Sigma\lambda_i = 0$. Furthermore, the density decreases as the distance of $\boldsymbol{\lambda} = (\lambda_1, \lambda_2, \ldots, \lambda_r)$ from the centre of symmetry increases. Consequently, a confidence set for the parameters λ_i would, according to the rule suggested at the end of § 5.2, consist of a sphere with centre at the centre of symmetry and a radius, Λ_0, so chosen that the probability, according to (9), of $\boldsymbol{\lambda}$ lying in the sphere was β, where β is the confidence coefficient. But this probability is exactly the probability that $\Lambda \leqslant \Lambda_0$, where $\Lambda^2 = \Sigma(x_{i.} - x_{..} - \lambda_i)^2$. Consequently, in order to obtain the relationship between β and Λ_0 it is only necessary to find the distribution of Λ^2 and to choose Λ_0 so that $\pi(\Lambda^2 \leqslant \Lambda_0^2 \mid \mathbf{x}) = \beta$.

In order to find this probability it is necessary to integrate (9) over all values such that $\Lambda^2 \leqslant \Lambda_0^2$. This will be the distribution function of Λ^2. Instead let us find the density of Λ^2; that is, let us find the probability that $\Lambda_0^2 \leqslant \Lambda^2 \leqslant \Lambda_0^2 + \delta$, where δ is small; the result will be, to order δ, the required density times δ (cf.

§2.2). Now Λ^2 satisfies these inequalities provided the values λ_i lie in between two spheres with centres at the centre of symmetry and squared radii Λ_0^2 and $\Lambda_0^2+\delta$, and in this region, called an annulus, the joint density of the λ_i is sensibly constant at a value proportional to

$$\{n\Lambda_0^2+S^2\}^{-\frac{1}{2}(nr-1)} \tag{10}$$

(from (9)). Hence the required probability, namely the integral of (9) over the annulus, will be equal to the constant value of the integrand, (10), times the volume of the annulus. The volume of a sphere of radius Λ_0 in $(r-1)$ dimensions (because $\Sigma\lambda_i = 0$) is proportional to $\Lambda_0^{r-1} = (\Lambda_0^2)^{\frac{1}{2}(r-1)}$, so that the volume of the annulus is proportional to

$$(\Lambda_0^2+\delta)^{\frac{1}{2}(r-1)}-(\Lambda_0^2)^{\frac{1}{2}(r-1)} \propto \delta(\Lambda_0^2)^{\frac{1}{2}(r-3)},$$

to order δ. Hence we have

$$\pi(\Lambda^2\,|\,\mathbf{x}) \propto (\Lambda^2)^{\frac{1}{2}(r-3)}/\{n\Lambda^2+S^2\}^{\frac{1}{2}(nr-1)}. \tag{11}$$

The substitution

$$\Phi = \frac{n\Lambda^2/(r-1)}{S^2/r(n-1)} = \frac{n\Lambda^2/\nu_1}{S^2/\nu_2}$$

gives

$$\pi(\Phi\,|\,\mathbf{x}) \propto \Phi^{\frac{1}{2}\nu_1-1}/(\nu_2+\nu_1\Phi)^{\frac{1}{2}(\nu_1+\nu_2)},$$

so that the posterior distribution of Φ is $F(\nu_1, \nu_2)$ (equation 6.2.1).

A confidence set for the λ's with confidence coefficient $\beta = 1-\alpha$ is therefore provided by the set of λ's for which $\Phi \leqslant \bar{F}_\alpha(\nu_1, \nu_2)$. Hence the null hypothesis that all the λ's are zero will be judged significant if the point (5) does not belong to this set. When (5) obtains $\Phi = F$, in the notation of (1), so that the result is significant if $F > \bar{F}_\alpha(\nu_1, \nu_2)$, as required.

Theorem 2. *If the situation is as described in theorem 1 except that ϕ is known, equal to σ^2: then the equivalent significance test is provided by declaring the data significant if*

$$\chi^2 = n\sum_i (x_{i.}-x_{..})^2/\sigma^2 \tag{12}$$

exceeds $\bar{\chi}^2_\alpha(\nu_1)$, the upper 100α % *point of the χ^2-distribution with ν_1 degrees of freedom* (§5.3).

The proof is similar to that of theorem 1 as far as equation (8) which will now read

$$\pi(\lambda_1, \lambda_2, \ldots, \lambda_r \mid \mathbf{x}) \propto \exp\left[-n \sum_i (x_{i.} - x_{..} - \lambda_i)^2/2\sigma^2\right]$$

with $\Sigma\lambda_i = 0$. The argument continues on similar lines since this joint density also has spherical symmetry. The only difference is that the sensibly constant value over the annulus is $\exp[-n\Lambda_0^2/2\sigma^2]$ instead of (10). Hence instead of (11), we have

$$\pi(\Lambda^2 \mid \mathbf{x}) \propto (\Lambda^2)^{\frac{1}{2}(r-3)} \exp[-n\Lambda^2/2\sigma^2].$$

The substitution $\Phi = n\Lambda^2/\sigma^2$ gives

$$\pi(\Phi \mid \mathbf{x}) \propto \Phi^{\frac{1}{2}(r-3)} e^{-\frac{1}{2}\Phi},$$

so that the posterior distribution of Φ is $\chi^2(\nu_1)$ (equation 5.3.1). The same argument as used in theorem 1 produces the significance test.

Relationship with the t-test

In this section we discuss the theorem and its proof and develop some corollaries from it: the practical significance of the result will be discussed in the next section. The first point to notice is that, apart from the restriction to equal sizes of sample, a restriction that we shall remove below, the theorem is a generalization of the significance test of theorem 6.1.3 from two to r samples. To see this we first notice that the samples, as in the earlier result, are independent and normal with a common, but unknown, variance. The significance test in the two-sample case was performed by calculating a statistic t and declaring the result significant if $|t|$ exceeded $t_{\frac{1}{2}\alpha}(\nu)$, the upper $\frac{1}{2}100\alpha\,\%$ point of the t-distribution with ν degrees of freedom; that is, if $t^2 \geqslant t^2_{\frac{1}{2}\alpha}(\nu)$. The statistic t, in the notation of 6.1.3, was equal to $(\bar{x}_1 - \bar{x}_2)/(2s^2/n)^{\frac{1}{2}}$ and ν was $2(n-1)$, remembering that $n_1 = n_2 = n$, say. If we change to the notation of this section, $\bar{x}_i$ is $x_{i.}$ and

$$s^2 = \sum_i \sum_j (x_{ij} - x_{i.})^2/2(n-1)$$

so that, taking the square of t,

$$t^2 = \frac{n(x_{1.} - x_{2.})^2}{\sum_{i,j} (x_{ij} - x_{i.})^2/(n-1)}.$$

But it is easy to verify that $(x_{1.}-x_{2.})^2 = 2\sum_i (x_{i.}-x_{..})^2$ so that

$$t^2 = \frac{n\sum_i (x_{i.}-x_{..})^2/\nu_1}{\sum_{i,j}(x_{ij}-x_{i.})^2/\nu_2}, \tag{13}$$

with $\nu_1 = 1$ and $\nu_2 = 2(n-1) = \nu$, and hence $t^2 = F$ in the notation of (1). Consequently the statistic used here is the same as that in the earlier result. But we saw in §6.2 that if t has a t-distribution with ν degrees of freedom, t^2 has an $F(1, \nu)$ distribution, which is just the distribution used here since $\nu_1 = 1$ and $\nu_2 = \nu$. Hence for $r = 2$, the two significance tests are the same. The change from $\frac{1}{2}\alpha$ in $t^2_{\frac{1}{2}\alpha}(\nu)$ to α in $\bar{F}_\alpha(1, \nu)$ arises because in the squaring, the $\frac{1}{2}\alpha$ in each tail of the t-distribution gives α in the single tail of the $t^2 = F$ distribution.

Unequal sizes of samples

Theorem 1 generalizes without difficulty to the case of unequal sample sizes. We state it formally as

Corollary 1. *If the situation is as described in the theorem except that the ith sample has size n_i, then*

$$F = \frac{\sum_i n_i(x_{i.}-x_{..})^2/\nu_1}{\sum_{i,j}(x_{ij}-x_{i.})^2/\nu_2} > \bar{F}_\alpha(\nu_1, \nu_2) \tag{14}$$

provides a significance test of the null hypothesis that all the θ_i are equal; where

$$x_{i.} = \sum_{j=1}^{n_i} x_{ij}/n_i, \quad x_{..} = \sum_{i,j} x_{ij}\Big/\sum_i n_i \tag{15}$$

and

$$\nu_1 = r-1, \quad \nu_2 = \sum_i (n_i-1). \tag{16}$$

The proof is exactly the same as that of the theorem with a few minor algebraic complexities which tend to obscure the salient points of the argument, this being the reason for first proving it in the equal sample size case.

Distribution of a single mean

The practical reasons for the significance test being more important than the confidence set must await discussion in the next section but it is already clear that even the joint distribution of the λ's (equation (9)) is rather a complicated expression of degrees of belief, and it is natural to consider instead the function Λ^2, or equivalently Φ, since the joint density depends only on this function. The change to Φ is made so that the tables of the F-distribution are readily available. Although the joint distribution is not suitable for practical work it is possible to obtain the posterior distribution of any θ_i.

Corollary 2. *Under the conditions of the theorem the posterior distribution, $\pi(\theta_i|\mathbf{x})$, of θ_i is such that $n^{\frac{1}{2}}(\theta_i - x_{i.})/s$ has a t-distribution with ν_2 degrees of freedom, where*

$$s^2 = \sum_{i,j}(x_{ij} - x_{i.})^2/\nu_2, \tag{17}$$

the denominator of (1).

Start from equation (4) and integrate with respect to all the θ's except θ_i. Each integral is a normal integral and the result is obviously

$$\pi(\theta_i, \phi|\mathbf{x}) \propto \phi^{-\frac{1}{2}[nr+2-(r-1)]}\exp[-\{n(x_{i.} - \theta_i)^2 + S^2\}/2\phi]$$

$$= \phi^{-\frac{1}{2}(\nu_2+3)}\exp[-\{n(x_{i.} - \theta_i)^2 + \nu_2 s^2\}/2\phi]; \tag{18}$$

the result follows from this just as theorem 5.4.1 followed from equation 5.4.4 by integration with respect to ϕ. The present result is clearly a generalization of that theorem.

Distribution of the variance

The analysis has so far concentrated on the distribution of the λ's. Sometimes the posterior distribution of ϕ is of interest.

Corollary 3. *Under the conditions of the theorem the posterior distribution of ϕ, $\pi(\phi|\mathbf{x})$, is such that S^2/ϕ is χ^2 with ν_2 degrees of freedom.*

This follows immediately from (4) on integrating with respect to all the θ's. The result is

$$\pi(\phi|\mathbf{x}) \propto \phi^{-\frac{1}{2}[n(r-1)+2]}e^{-S^2/2\phi}, \tag{19}$$

and a comparison with equation 5.3.2 establishes the result.

Inferences about the unknown variance can now be made in

the usual way (§5.3) with the χ^2-distribution. For general sample sizes the result persists with ν_2 now defined by (16), S^2 being unaltered.

Distribution of linear functions

A useful extension of corollary 2 is possible using this last result.

Corollary 4. *Under the conditions of the theorem, if $c_1, c_2, \ldots, c_r$ are any constants, not all zero, the posterior distribution of $\Sigma c_i \theta_i$ is such that $n^{\frac{1}{2}} \sum_i c_i(\theta_i - x_{i.})/s(\Sigma c_i^2)^{\frac{1}{2}}$ has Student's t-distribution with ν_2 degrees of freedom.*

(Corollary 2 is the special case where $c_i = 1$ and all the other c's vanish.) The proof is very similar to that of theorem 6.1.3. If ϕ is known the θ_i are independent and normal with means $x_{i.}$ and variances ϕ/n, by the corollary to theorem 5.1.1 with $\sigma_0 \to \infty$. Hence the posterior distribution of $\Sigma c_i \theta_i$ is normal with mean $\Sigma c_i x_{i.}$ and variance $\Sigma c_i^2 \phi/n$ by theorem 3.5.5. Hence we know $\pi(\Sigma c_i \theta_i | \phi, \mathbf{x})$ and from corollary 3 we know $\pi(\phi | \mathbf{x})$, so that $\pi(\Sigma c_i \theta_i, \phi | \mathbf{x})$ follows on multiplication and $\pi(\Sigma c_i \theta_i | \mathbf{x})$ on integration with respect to ϕ, exactly as in the earlier result.

In particular, the posterior distribution of $\theta = \Sigma \theta_i / r$ can be found. With unequal sample sizes the t-statistic is

$$\Sigma c_i(\theta_i - x_{i.})/s(\Sigma c_i^2/n_i)^{\frac{1}{2}}.$$

Significance test for a subset of the means

From this corollary it is possible to pass to a more general result.

Corollary 5. *Under the conditions of the theorem, a significance test at level α of the null hypothesis that some of the θ_i are equal, say $\theta_1 = \theta_2 = \ldots = \theta_t$, is provided by declaring the data significant if*

$$F_i = \frac{n \sum_{i=1}^{t} (x_{i.} - x_{(t).})^2/(t-1)}{\sum_{i,j} (x_{ij} - x_{i.})^2/\nu_2} \tag{20}$$

exceeds $\bar{F}_\alpha(t-1, \nu_2)$; where $x_{(t).} = \sum_{i=1}^{t} x_{i.}/t$.

The joint posterior distribution of $\theta_1, \theta_2, \ldots, \theta_t$ and ϕ is provided by integrating (4) with respect to $\theta_{t+1}, \ldots, \theta_r$. The result is clearly

$$\pi(\theta_1, \theta_2, \ldots, \theta_t, \phi \mid \mathbf{x}) \propto \phi^{-\frac{1}{2}[(nr+2)-(r-t)]} \exp\left[-\left\{\sum_{i=1}^{t} n(x_{i.}-\theta_i)^2 + S^2\right\}\Big/ 2\phi\right], \quad (21)$$

and the result follows from (21) in the same way that the theorem followed from (4). Notice that the numerator of (20) contains only those samples whose means are in question, whereas the denominator contains contributions from all samples, as before. It is the same with the degrees of freedom: for the numerator they are reduced from $(r-1)$ to $(t-1)$ but they stay the same, at $r(n-1)$, in the denominator.

It is worth noting that it follows from the form of the likelihood, which is proportional to the expression on the right-hand side of (4) multiplied by ϕ, and the factorization theorem 5.5.2 that $(x_{1.}, x_{2.}, \ldots, x_{r.}, S^2)$ are jointly sufficient for $(\theta_1, \theta_2, \ldots, \theta_r, \phi)$. All the posterior distributions quoted are in terms of these statistics.

Theorem 2 covers the situation where the variance is known. It is possible to see that the test statistic proposed, equation (12), is the natural analogue of (1), used when σ^2 is unknown, because, since σ^2 does not have to be estimated, the denominator of (1) can be replaced by σ^2 (see §6.5). This theorem has corollaries analogous to those for theorem 1.

Derivation of the density

There is one part of the proof that involves a type of argument that has not been used before and is therefore worth some comment. This is the derivation of the density of Λ^2. On many occasions we have had to integrate a joint density in order to obtain the density of a single random variable; but in these cases the joint density has been expressed already in terms of the single random variable required and others. But here we are required to obtain $\pi(\Lambda^2 \mid \mathbf{x})$ by integration of $\pi(\lambda_1, \lambda_2, \ldots, \lambda_r \mid \mathbf{x})$ where the latter is not already expressed as a density of Λ^2 and other random variables. One method would be to rewrite

$$\pi(\lambda_1, \lambda_2, \ldots, \lambda_r \mid \mathbf{x})$$

as a joint density of Λ^2 and $(r-2)$ other variables, using the Jacobian of the transformation to do so (theorem 3.5.2), and then integrate out the $(r-2)$ unwanted variables. This type of argument has already been used in the proof to remove θ. The geometrical argument, which is commonly useful, avoids the necessity of transforming and calculating the Jacobian. Notice that we have worked in terms of Λ^2, not Λ, since the sign of Λ is irrelevant. The posterior density of Λ is proportional to (11) times Λ, since $d\Lambda^2 \propto \Lambda d\Lambda$ (cf. equation 3.5.1).

6.5. Analysis of variance: between and within samples

In this section the more practical aspects of theorem 6.4.1 and its corollaries are discussed. There are r independent, random samples, each of size n, from normal distributions with unknown means and an unknown common variance. The prior distributions are those which represent considerable ignorance as to the values of the unknown parameters. It is important to remember this assumption about the prior distribution: there are practical circumstances where it is not appropriate. This point will be mentioned again, particularly in connexion with theorem 6.6.3.

The expression $\sum_{i,j}(x_{ij}-x_{..})^2$ is called the *total sum of squares*, $\sum_{i,j}(x_{ij}-x_{i.})^2$ is called the *within sum of squares* (cf. §6.1) and $n\sum_i(x_{i.}-x_{..})^2$ is called the *between sum of squares*. We have the identity

$$\sum_{i,j}(x_{ij}-x_{..})^2 = \sum_{i,j}(x_{ij}-x_{i.})^2+n\sum_i(x_{i.}-x_{..})^2, \tag{1}$$

which follows immediately on writing the left-hand side as $\Sigma[(x_{ij}-x_{i.})+(x_{i.}-x_{..})]^2$ and expanding the square. In words, (1) says that the total sum of squares is the sum of the between and within sums of squares. The total sum of squares has associated with it $rn-1$ degrees of freedom, called the *total degrees of freedom*; the within sum of squares has $r(n-1) = \nu_2$, the *within degrees of freedom* (corollary 3 to theorem 6.4.1); and, to preserve the addition, the between sum of squares has $(r-1) = \nu_1$, the *between degrees of freedom*, in agreement with the fact that it is the sum of r squares with one constraint, $\Sigma(x_{i.}-x_{..}) = 0$. The ratio of any sum of squares to its degrees of freedom is called a *mean square*. It is then possible to prepare table 6.5.1. Such a table is called an *analysis of variance table* and one is said to have carried out an *analysis of variance*. The first two columns are additive, but the third column is not. The final column gives the value of F, defined in the last section, the ratio of the two mean squares, needed to perform the significance test.

TABLE 6.5.1

	Sum of squares	Degrees of freedom	Mean square	F
Between	$n\sum_{i}(x_{i.}-x_{..})^2$	$\nu_1 = (r-1)$	M_1	M_1/M_2
Within	$\sum_{i,j}(x_{ij}-x_{i.})^2$	$\nu_2 = r(n-1)$	$s^2 = M_2 = S^2/\nu_2$	—
Total	$\sum_{i,j}(x_{ij}-x_{..})^2$	$(nr-1)$	—	—

Practical example: basic analysis of variance

The discussion will be carried out using a numerical example. In the preparation of a chemical, four modifications of the standard method were considered and the following experiment was carried out to see whether the purity would be improved by use of any of the modifications. Batches of the chemical were prepared from similar material using each of the five methods (the standard one and the four modifications), six samples were taken from each batch and tested for purity. The results are given in table 6.5.2 in suitable units. In this case the measures were of the impurities and all were between 1·3 and 1·4: 1·3 was therefore subtracted from each reading and the results were then multiplied by 1000. Past experience was available to show that within a batch produced by the standard method (number 1 in the table) the distribution was approximately normal, and the same might therefore be supposed for the new batches. The changes in method probably did not affect the variability, which was due more to the raw material than the method. The samples were taken independently (to ensure this the five methods were used in a random order). Hence all the conditions on the probability of the observations assumed in theorem 6.4.1 seem reasonably satisfied. For the moment assume the prior distribution used there; we return to this point later. The calculations then proceed as follows:

(1) Calculate the totals for each method (table 6.2) and the grand total, 1047. We note that $r = 5$, $n = 6$ in the notation of §6.4.

TABLE 6.5.2

Method ...	1	2	3	4	5
	43	33	10	44	37
	41	2	24	29	21
	54	31	40	31	35
	57	23	37	44	30
	48	41	24	45	28
	63	27	30	28	47
Totals	306	157	165	221	198

(2) Calculate the uncorrected (§6.1) sum of squares of all the readings; $\sum_{i,j} x_{ij}^2 = 41{,}493$.

(3) Calculate the uncorrected sum of squares of the totals for each method, divided by the number of samples contributing to each total, here $n = 6$; $\sum_i (\sum_j x_{ij})^2/n = 38{,}926$ (to the nearest integer).

(4) Calculate the *correction factor* defined as the square of the grand total divided by the total number of readings, here $rn = 30$; $(\sum_{i,j} x_{ij})^2/rn = 36{,}540$ (to the nearest integer).

(5) Calculate the total sum of squares, (2) − (4); 4953.

(6) Calculate the between sum of squares, (3) − (4); 2386.

(7) Construct the analysis of variance table (table 6.3). (Notice that the within sum of squares is calculated by subtraction.) The upper percentage points of the F-distribution with 4 and 25 degrees of freedom are, from the tables, 2·76 (at 5 %), 4·18 (at 1 %) and 6·49 (at 0·1 %). The test of the null hypothesis that none of the modified methods has resulted in a change in the purity is therefore significant at 1 % (since $F = 5{\cdot}81 > 4{\cdot}18$) but not at 0·1 % ($F < 6{\cdot}49$). The result can perhaps better be

expressed by saying that one's degree of belief that the differences between the methods are not all zero is higher than 0·99 but is not as great as 0·999. One is quite strongly convinced that there are some differences between the five methods.

TABLE 6.5.3

	Sum of squares	Degrees of freedom	Mean square	F
Between	2386	4	596·5	5·81
Within	2567	25	102·7	—
Total	4953	29	—	—

Tests for individual differences

Having established grounds for thinking there exist differences, one must consider what differences there are, and the most important of these to look for is obviously that between the standard method and the new ones: but which of the new ones? Corollary 5 of §6.4 enables us to see what evidence there is for differences between the four new methods. The new F statistic, F_4, is calculated by the same methods as were used in forming table 6.5.3. The within sum of squares is unaltered (in the denominator of (6.4.20)) and the between sum of squares is calculated as before, but excluding method 1. In place of (3) above we have $\sum_{i>1} (\sum_j x_{ij})^2/n = 23{,}320$, and for (4), $(\sum_{i>1,j} x_{ij})^2/(r-1)n = 22{,}878$. Thus the between sum of squares, the difference of these, is 442 with 3 degrees of freedom, giving a mean square (the numerator of (6.4.20)) of 147·3, which is only a little larger than the within mean square and the F_4 value (which would have to reach 2·99 with 3 and 25 degrees of freedom for significance at 5 %) is certainly not significant. The posterior probability that there are no differences between the new methods is quite high. It looks, therefore, as if the difference suggested by the first test must be due to a difference between the standard method and the new ones, which themselves, so far as we can see, are equivalent. Corollary 4 of §6.4 enables this point to be investigated. Consider $\theta_1 - \frac{1}{4}(\theta_2 + \theta_3 + \theta_4 + \theta_5) = \bar{\theta}$ say: the difference between the standard method and the average performance of the new ones.

Since $\Sigma c_i^2 = 1{\cdot}25$, $x_{1.} - \frac{1}{4}(x_{2.} + x_{3.} + x_{4.} + x_{5.}) = 20{\cdot}125$, $\nu_2 = 25$ and $s^2 = 102{\cdot}7$ (within degrees of freedom and mean square), the confidence limits for $\bar{\theta}$ are

$$20{\cdot}125 \pm t_{\frac{1}{2}\alpha}(25) \times (102{\cdot}7 \times 1{\cdot}25/6)^{\frac{1}{2}}$$

with confidence coefficient $1-\alpha$. At 5 % the t-value is 2·06 and the limits are (10·6, 29·7). At 0·1 % with a t-value of 3·72 the limits are (2·9, 37·3), which exclude the origin, so that the difference is significant at 0·1 %.

The results of the experiment may be summarized as follows:

(i) There is no evidence of any differences between the four new methods.

(ii) The average effect of the new methods is to reduce the impurity as compared with the standard method. The effect is estimated to lie most likely (with 95 % confidence) between a reduction of 0·0106 and 0·0297, and almost certainly (99·9 % confidence) to lie between 0·0029 and 0·0373: the most probable value is 0·0201.

A better way to express this would be to say that this reduction $\bar{\theta}$ was such that

$$\{\bar{\theta} - 0{\cdot}0201\}/0{\cdot}00463$$

had a posterior distribution which was Student's t on 25 degrees of freedom.

Form of the F-statistic

The computational method is merely a convenient way of arranging the arithmetic in order to obtain the required F-ratio. But since the sufficient statistics (the $x_{i.}$ and S^2) are found in doing this, it is suitable for any inference that one wishes to make. Let us first look at this F-statistic. In the language of the definitions given above it is the ratio of the between mean square to the within mean square. An extension of the discussion in §6.1 shows that the within mean square, being the ratio of the sum of sums of squares for each sample, ΣS_i^2, and the sum of the separate sample degrees of freedom, is a natural estimate to replace the unknown ϕ; and this, whatever be the values of the θ_i. A similar discussion shows that *if the null hypothesis is true*, so that the $x_{i.}$ are independent and normal with the *same mean* and variance ϕ/n, then $\Sigma(x_{i.} - x_{..})^2$, divided by its degrees of freedom, $(r-1)$, is an estimate of ϕ/n. Hence when the null hypothesis is true the between mean square, $n\Sigma(x_{i.}-x_{..})^2/(r-1)$, is also an estimate of ϕ, and F is the ratio of two such estimates, and is therefore about 1. When the null hypothesis is not true the between mean square will be larger since the $x_{i.}$ will be more scattered because the θ_i are. Consequently F will be typically greater than 1. This shows that the test, which says that the null hypothesis is probably not true when F is large, is sensible on intuitive grounds.

Notice that the sums of squares are all calculated by subtracting a squared sum from an uncorrected sum of squares (cf. §5.4). Thus the between sum of squares is

$$n\Sigma(x_{i.}-x_{..})^2 = n\Sigma x_{i.}^2 - nrx_{..}^2$$
$$= \sum_i (\sum_j x_{ij})^2/n - (\sum_{i,j} x_{ij})^2/rn, \qquad (2)$$

and the total sum of squares is

$$\Sigma(x_{ij}-x_{..})^2 = \Sigma x_{ij}^2 - (\Sigma x_{ij})^2/rn. \qquad (3)$$

It is useful to notice that whenever a quantity in the form of a sum is squared, it is always, after squaring, divided by the number of x_{ij} that were added up to give the quantity. Thus, in (2), $(\sum_j x_{ij})^2$ is divided by n because $\sum_j x_{ij}$ is the sum of n terms. The within sum of squares could be found by calculating each S_i^2 and adding, but it is computationally easier to use (1).

The calculations proceed in essentially the same manner when the sample sizes are unequal. In (2) we have to calculate $\sum_i \{(\sum_j x_{ij})^2/n_i\}$, in agreement with the sentence after equation (3): but otherwise there is no change in the method. Notice, too, that the calculations are much simplified by changing origin and scale to reduce the magnitudes and avoid the troubles of negative values. A desk calculating machine is essential for all but the smallest analyses.

Extended analysis of variance table

The additivity of the sums of squares can be extended still further. We saw in the last section that the F-test was a generalization of the t-test, and a t-test could be put in an F form by considering $t^2 = F$. Consider the use just made of the t-statistic in order to assess the difference between the standard method and the others. The square of the t-statistic is (from corollary 4 to theorem 6.4.1)

$$n\{\Sigma c_i(\theta_i - x_{i.})\}^2/s^2(\Sigma c_i^2)$$

(with $c_1 = 1$, $c_i = -\frac{1}{4}$, $i > 1$) and a test of the null hypothesis that $\Sigma c_i\theta_i = \theta_1 - \frac{1}{4}(\theta_2+\theta_3+\theta_4+\theta_5) = \bar{\theta} = 0$ will be significant if

$$n\{\Sigma c_i x_{i.}\}^2/s^2(\Sigma c_i^2) > F_\alpha(1, \nu_2),$$

or

$$\frac{\{4\sum_j x_{1j} - \sum_j (x_{2j}+x_{3j}+x_{4j}+x_{5j})\}^2/120}{\Sigma(x_{ij}-x_{i.})^2/\nu_2} > F_\alpha(1, \nu_2). \quad (4)$$

The test criterion is therefore still of the form of one mean square (of a single term) divided by the within mean square. A little algebra also shows that the numerator of the left-hand side of (4) (which is also a sum of squares, since the degree of freedom is 1) is the difference between the between sum of squares in table 6.5.3 and the between sum of squares calculated in order to test the null hypothesis $\theta_2 = \theta_3 = \theta_4 = \theta_5$. The analysis of variance (table 6.5.3) may therefore be extended. The sum of squares between the standard and the rest, 1944, is the numerator of the left-hand side of (4), and is therefore easily calculated. Consequently we have an easier way of calculating the sum of squares between the rest; namely by subtraction from the between sum of squares (table 6.5.3). Table 6.5.4 provides the two F values which are needed for the separate tests. The first one has 1 and 25 degrees of freedom for which the upper 0·1 % value is 13·88 so that it is significant at this level, agreeing with our previous computations of the confidence interval using the t-distribution.

TABLE 6.5.4

		Sums of squares	Degrees of freedom	Mean squares	F
Between	Standard and the rest	1944	1	1944·0	18·93
	The rest	442	3	147·3	1·43
Within		2567	25	102·7	—
Total		4953	29	—	—

This breakdown of the total sum of squares into additive parts is capable of important generalizations, some of which will be touched on in later sections (§§8.5, 8.6), but it is mainly useful as a computational device. The real problem is to express concisely the salient features of the joint posterior distribution of the λ's (equation 6.4.9), and ideally this should lead at least to confidence interval statements. These are usually obtained by doing preparatory significance tests to see what differences, if any, are important and estimating these. The numerical example is typical in this respect: only the difference between the standard method and the rest was significant and the posterior density of that alone, instead of the joint density, needed to be quoted.

The variance estimation

It is sometimes useful to consider the variance ϕ. In virtue of corollary 3 to theorem 6.4.1 inferences about it can be made using the χ^2-distribution in the usual way (§5.3). The within sum of squares divided by ϕ is χ^2 with ν_2 degrees of freedom, so that

$$\underline{\chi}^2(25) < 2567\phi^{-1} < \overline{\chi}^2(25).$$

At the 5 % level, the upper and lower points are respectively 13·51 and 41·66, from the Appendix, so that the 95 % confidence limits for ϕ are (61·6, 190·0): the mean square being 102·7. The limits for the standard deviation, in the original units, are therefore (0·0079, 0·0138). Notice that there are usually a large number of degrees of freedom available for error so that ϕ is usually tolerably accurately determined.

If ϕ is known, equal to σ^2, then theorem 6.4.2 is applicable. There is then no need to use the within sum of squares, which only serves to estimate ϕ, since it is already known. Instead the ratio of the between *sum* of squares (not the between *mean* square) to σ^2 provides the necessary test statistic, which is referred to the χ^2-table with $(r-1)$ degrees of freedom. One way of looking at this is to say that when ϕ is known, equal to σ^2, we have effectively an infinity of degrees of freedom for the within mean square and we saw (§6.2) that in these circumstances $\nu_1 F(\nu_1, \infty)$ was the same as a χ^2-distribution. The other devices used in this section extend to this case with the infinite degrees of freedom replacing ν_2.

If σ^2 is known it is possible to provide a check on the data by using corollary 3 to theorem 6.4.1 to provide a test of the null hypothesis that $\phi = \sigma^2$ in the usual way.

Prior distributions

In any application of the analysis of variance it is important to remember the assumptions about the probability of the observations: independence, normality, and constancy of variance. But it is also as well to remember the assumptions about the prior knowledge, particularly concerning the means. These are supposed independent and uniformly distributed, where the latter distribution is an approximation for a distribution of the type discussed in §5.2. Whilst any θ may have this distribution it is not always reasonable to suppose them independent. For example, in the chemical illustration the impurity measure may well fluctuate substantially depending on the raw material. This fluctuation was minimized here by making up the five batches from similar material. Although one might be very vague about the average value of the θ's, or about any one of them, because of one's vagueness about the raw material, one might well feel that any two of the θ's would be close together, precisely because the batches had been made from similar material. In the next section we shall show how some allowance can be made for this in the case where ϕ is known, equal to σ^2, say.

6.6. Combination of observations

Our topic is again that of several independent samples from normal distributions, but we no longer consider differences of means. In the first two theorems the means are supposed equal and the samples are combined to provide an inference about the common value. In the third theorem another sort of combination is considered.

Theorem 1. *Let* $\mathbf{x}_i = (x_{i1}, x_{i2}, \ldots, x_{in_i})$ $(i = 1, 2, \ldots, r)$ *be* r *independent random samples of sizes* n_i *from* $N(\theta, \sigma_i^2)$, *where the* σ_i *are known but the common mean,* θ, *is unknown. Then if* θ *has* *a uniform prior distribution over* $(-\infty, \infty)$, *the posterior distribution of* θ *is* $N(x_{..}, (\Sigma w_i)^{-1})$ *where*

$$x_{..} = \Sigma w_i x_{i.}/\Sigma w_i \quad \textit{and} \quad w_i^{-1} = \sigma_i^2/n_i. \tag{1}$$

w_i *is called the* weight *of the ith sample, and* $x_{..}$ *the* weighted mean (§5.1).

The likelihood of θ from the ith sample is given, apart from the unnecessary constant of proportionality, by equation 5.1.9; and hence the likelihood from all r samples is the product of these. Thus, if $\mathbf{x}$ denotes the set of all x_{ij}, and $x_{i.} = \sum_j x_{ij}/n_i$,

$$\begin{aligned} p(\mathbf{x}\,|\,\theta) &\propto \exp[-\tfrac{1}{2}\Sigma w_i(x_{i.}-\theta)^2] \\ &\propto \exp[-\tfrac{1}{2}\Sigma w_i\theta^2 + \Sigma w_i x_{i.}\theta] \\ &\propto \exp[-\tfrac{1}{2}(\Sigma w_i)(\theta - \Sigma w_i x_{i.}/\Sigma w_i)^2]. \end{aligned} \tag{2}$$

Since $\pi(\theta)$ is constant, $\pi(\theta\,|\,\mathbf{x})$ is also proportional to (2), which proves the theorem. (An alternative proof is given below.)

Theorem 2. *Let the conditions be as in theorem 1 except that* $\sigma_i^2 = \phi\tau_i^2$, *where the* τ*'s are known but* ϕ *is not. Then if* $\ln\phi$ *has a uniform prior distribution over* $(-\infty, \infty)$ *independent of* θ, *the posterior distribution of* θ *is such that*

$$(\Sigma \tilde{w}_i)^{\frac{1}{2}}(\theta - \tilde{x}_{..})/\tilde{s} \tag{3}$$

has Student's t-distribution with $\nu = (\Sigma n_i - 1)$ *degrees of freedom, where*

$$\tilde{x}_{..} = \Sigma \tilde{w}_i x_{i.}/\Sigma \tilde{w}_i, \quad \tilde{w}_i^{-1} = \tau_i^2/n_i \tag{4}$$

and

$$\tilde{s}^2 = [\sum_{i,j}(x_{ij}-x_{i.})^2/\tau_i^2 + \sum_i \tilde{w}_i(x_{i.}-\tilde{x}_{..})^2]/\nu = [\sum_{i,j}(x_{ij}-\tilde{x}_{..})^2/\tau_i^2]/\nu. \tag{5}$$

The likelihood now involves ϕ which cannot be absorbed into the constant of proportionality as in the proof of theorem 1. From the normal density for each x_{ij} the likelihood is easily seen to be

$$p(\mathbf{x}\,|\,\theta, \phi) \propto \phi^{-\frac{1}{2}(\nu+1)} \exp[-\sum_{i,j}(x_{ij}-\theta)^2/2\tau_i^2\phi], \tag{6}$$

and hence $\pi(\theta, \phi\,|\,\mathbf{x})$ is proportional to the same expression multiplied by ϕ^{-1}. To simplify this, collect together like terms in the exponential in θ and θ^2 in order to have a single term involving θ (compare the argument leading to equation 5.4.4). The expression in square brackets is $-1/2\phi$ times

$$\theta^2 \sum_i n_i/\tau_i^2 - 2\theta \sum_{i,j} x_{ij}/\tau_i^2 + \sum_{i,j} x_{ij}^2/\tau_i^2$$

$$= \theta^2(\sum_i \tilde{w}_i) - 2\theta \sum_i x_{i.}\tilde{w}_i + \sum_{i,j}(x_{ij}-x_{i.})^2/\tau_i^2 + \sum_i n_i x_{i.}^2/\tau_i^2$$

$$= (\sum_i \tilde{w}_i)(\theta - \tilde{x}_{..})^2 + \sum_{i,j}(x_{ij}-x_{i.})^2/\tau_i^2 + \sum_i \tilde{w}_i x_{i.}^2 - (\sum_i \tilde{w}_i)\tilde{x}_{..}^2.$$

The terms not involving θ^2 clearly reduce to $\nu\tilde{s}^2$. Hence

$$\pi(\theta, \phi \mid \mathbf{x}) \propto \phi^{-\frac{1}{2}(\nu+3)} \exp[-\{(\sum_i \tilde{w}_i)(\theta - \tilde{x}_{..})^2 + \nu\tilde{s}^2\}/2\phi]. \quad (7)$$

The result follows from (7) in the same way that theorem 5.4.1 followed from equation 5.4.4. The alternative expression for $\tilde{s}^2$ in (5) follows easily (cf. equation 6.5.1).

In the following theorem we use vector and matrix notation. $\mathbf{x}$ denotes a column vector, and $\mathbf{x}'$ its transpose, a row vector, with elements $(x_1, x_2, \ldots, x_n)$. Capital letters denote square matrices (cf. §§3.3, 3.5).

Theorem 3. *If the random variables* $\mathbf{x}$ *have a multivariate normal distribution with expectations* $\mathscr{E}(\mathbf{x}) = \mathbf{A\theta}$, *where* $\mathbf{A}$ *is known but* $\boldsymbol{\theta} = (\theta_1, \theta_2, \ldots, \theta_p)$ *is not, and known dispersion matrix* $\mathbf{C}$*; and if the prior distribution of* $\boldsymbol{\theta}$ *is multivariate normal with expectations* $\mathscr{E}(\boldsymbol{\theta}) = \boldsymbol{\mu}_0$ *and dispersion matrix* $\mathbf{C}_0$*; then the posterior distribution of* $\boldsymbol{\theta}$ *is also multivariate normal with expectations*

$$\boldsymbol{\mu}_1 = \{\mathbf{C}_0^{-1} + \mathbf{A}'\mathbf{C}^{-1}\mathbf{A}\}^{-1}\{\mathbf{C}_0^{-1}\boldsymbol{\mu}_0 + \mathbf{A}'\mathbf{C}^{-1}\mathbf{x}\} \quad (8)$$

and dispersion matrix

$$\{\mathbf{C}_0^{-1} + \mathbf{A}'\mathbf{C}^{-1}\mathbf{A}\}^{-1}. \quad (9)$$

From the expression for the normal multivariate density (equation 3.5.17), written in the vector notation, we have, since the dispersion matrix is known,

$$p(\mathbf{x} \mid \boldsymbol{\theta}) \propto \exp\{-\tfrac{1}{2}(\mathbf{x} - \mathbf{A\theta})' \mathbf{C}^{-1}(\mathbf{x} - \mathbf{A\theta})\} \quad (10)$$

and

$$\pi(\boldsymbol{\theta}) \propto \exp\{-\tfrac{1}{2}(\boldsymbol{\theta} - \boldsymbol{\mu}_0)' \mathbf{C}_0^{-1}(\boldsymbol{\theta} - \boldsymbol{\mu}_0)\}. \quad (11)$$

Hence $\pi(\boldsymbol{\theta} \mid \mathbf{x})$ is proportional to the product of these two exponentials. The resulting single exponential we manipulate in the same way as in the last theorem, discarding into the omitted constant any multiplicative terms not involving θ. Thus,

$$\pi(\boldsymbol{\theta}|\mathbf{x}) \propto \exp\{-\tfrac{1}{2}[\boldsymbol{\theta}'(\mathbf{C}_0^{-1}+\mathbf{A}'\mathbf{C}^{-1}\mathbf{A})\boldsymbol{\theta}-2\boldsymbol{\theta}'(\mathbf{C}_0^{-1}\boldsymbol{\mu}_0+\mathbf{A}'\mathbf{C}^{-1}\mathbf{x})]\}$$
$$\propto \exp\{-\tfrac{1}{2}[(\boldsymbol{\theta}-\boldsymbol{\mu}_1)'(\mathbf{C}_0^{-1}+\mathbf{A}'\mathbf{C}^{-1}\mathbf{A})(\boldsymbol{\theta}-\boldsymbol{\mu}_1)]\}.$$

A comparison of this last result with the normal density, (11), establishes the result.

Combinations of inferences

In calling this section 'combination of observations' we are extending the original meaning of this phrase, which referred to the results of the type discussed in §3.3, and made no reference to degrees of belief. Our excuse for the extension is that the earlier results were (and indeed are) used in an inference sense which is justified by the above results. In amplification of this point consider a simple typical result obtained in §3.3. If x_1 and x_2 are two independent, unbiased determinations of an unknown μ, then $\frac{1}{2}(x_1+x_2)$ is another unbiased determination. Furthermore, $\frac{1}{2}(x_1+x_2)$ is typically a more precise determination than either x_1 or x_2 separately. Precisely, if

$$\mathscr{D}^2(x_i) = \sigma_i^2, \quad \mathscr{D}^2[\tfrac{1}{2}(x_1+x_2)] = \tfrac{1}{4}(\sigma_1^2+\sigma_2^2).$$

However, there are two things that remain to be discussed. What inferences can be made about μ given x_1 and x_2; and is $\frac{1}{2}(x_1+x_2)$ the best determination? If the additional assumption of normality of x_1 and x_2 is made we shall see how to answer both these questions. The inference part is simple since we have already seen how a statement about a normal random variable can easily be turned into a statement of belief about its mean. (Compare the statements (*a*) and (*b*) in §5.1.) Thus, in the example, μ, given $\frac{1}{2}(x_1+x_2)$, is normally distributed about $\frac{1}{2}(x_1+x_2)$ with variance $\frac{1}{4}(\sigma_1^2+\sigma_2^2)$ provided the prior knowledge of μ is vague. In answer to the second question we now show that the ordinary average, $\frac{1}{2}(x_1+x_2)$, is not necessarily the best determination of μ.

Theorem 1 provides an answer to the question: if several independent determinations are made of the same unknown with different precisions, then how should they be combined into a single determination, and what precision has this?

First consider any sample. In the language of §3.3 this forms a set of n_i determinations of an unknown (now denoted by θ) with equal precisions σ_i^{-2}. (In §5.1 the precision was defined as the inverse of the variance.) The results of §5.1 show that, because of the normality, the mean is sufficient and forms a determination of precision n_i/σ_i^2, the w_i of the theorem. So much is familiar to us. The new point is to discover how the separate determinations from the different samples should be combined. The answer is to use the weighted mean (§5.1) with the inverses of the variances as weights: that is, the precisions of each determination. This is intuitively sensible since the larger the variance the smaller the precision or weight.

Relationship with Bayes's theorem

The result is very similar to the corollary to theorem 5.1.1, where a weighted mean was used to combine prior knowledge with the determination. In fact, the present theorem can be deduced from the corollary in the following way. Consider the samples taken one at a time, and apply the corollary each time a new sample is taken. Initially, in the notation of the corollary, we let $\sigma_0^2 \to \infty$; the posterior distribution after the first sample being $N(x_{1.}, \sigma_1^2/n_1)$. Now proceed to the second sample. The prior distribution now appropriate; that is, the distribution expressing one's present beliefs about θ, is the posterior distribution derived from the first sample, $N(x_{1.}, \sigma_1^2/n_1)$. Use of the corollary with $\mu_0 = x_{1.}, \sigma_0^2 = \sigma_1^2/n_1$ gives a posterior distribution after the second sample which is

$$N[(w_1 x_1 + w_2 x_2)/(w_1 + w_2), (w_1 + w_2)^{-1}],$$

the precisions of the first sample (now the prior precision) and the second sample being added, and the means weighted with them. Continuing in this way to r samples we have the general result. This second proof is interesting because it shows how one's beliefs change in a sequence as the new samples become available, and how the posterior distribution after one set of observations, or samples, becomes the prior distribution for the next set of observations.

Case of unknown variances

Notice that in order to make the best determination, namely the weighted mean, it is only necessary to know the ratios of the weights, or the precisions, or the variances σ_i^2. (Their absolute values are needed for the precision of the best determination.) Theorem 2 is available when these ratios are known but the final precision has to be estimated because the absolute values are not known. It is often useful when σ_i^2 is known to be constant, but there are other situations of frequent occurrence where the relative precisions are known but the absolute values are not. An example is where two instruments have been used, one of which is known to have twice the precision of the other when used on the same material, but where the actual precision depends on the material.

The result is best appreciated by noting that it is an extension of theorem 1 in the same way that Student's result (theorem 5.4.1) is an extension of theorem 5.2.1: namely the variance is estimated by a mean square, and the t-distribution replaces the normal. Thus if the result of theorem 1 is written, $(\Sigma w_i)^{\frac{1}{2}}(\theta - x_{..})$ is $N(0, 1)$, or, in the notation of theorem 2, $(\Sigma \tilde{w}_i)^{\frac{1}{2}}(\theta - \tilde{x}_{..})/\phi^{\frac{1}{2}}$ is $N(0, 1)$, the comparison is obvious except for the apparently rather unusual estimate, $\tilde{s}$, that replaces $\phi^{\frac{1}{2}}$. The form of $\tilde{s}$ can be easily understood by using analysis of variance terminology of the last section. We have seen in §6.1 that several estimates of variance should be combined by adding the sums of squares and adding the degrees of freedom, and dividing the former by the latter. In the situation of theorem 2 each sample provides an estimate of variance, the sum of squares for the ith sample being $\sum_j (x_{ij} - x_{i.})^2$. But this sample has variance $\phi\tau_i^2$, so $\sum_j (x_{ij} - x_{i.})^2/\tau_i^2$ is clearly the sum of squares appropriate to ϕ. Summing these over the samples gives the within sum of squares, the first term in the first form for $\tilde{s}^2$ in (5). But some more information is available about ϕ. The sample means $x_{i.}$ have variances $\phi\tau_i^2/n_i = \phi\tilde{w}_i^{-1}$, and hence $x_{i.}\tilde{w}_i^{\frac{1}{2}}$ has variance ϕ. Taking the sum of squares of the $x_{i.}$ about the weighted mean $\tilde{x}_{..}$ with weights $\tilde{w}_i$ gives $\Sigma\tilde{w}_i(x_{i.} - \tilde{x}_{..})^2$, the remaining term in the first form for $\tilde{s}^2$ in (5). This latter expression is a generalization of the idea of the between sum of squares to the case where the means have different variances of known ratios. The two expressions may be added to give the second form for $\tilde{s}^2$ in (5), and the part in square brackets is therefore the total sum of squares. The degrees of freedom are $(n_i - 1)$ for each sample; giving $\Sigma(n_i - 1) = \Sigma n_i - r$, for the within sum, and $(r-1)$ for the between sum; giving $\Sigma n_i - r + (r-1) = \Sigma n_i - 1 = \nu$, in all. This last result is also intuitively reasonable since the Σn_i readings have lost one degree of freedom for the single unknown θ.

It would be useful to have an extension of theorem 1 to the case where the σ_i^2 are unknown, but no simple result is available. The situation is complex because r integrations are needed (to remove each of the σ_i^2), and the distribution depends on many parameters. The reader can compare the difficulties with those encountered in Behrens's problem (§6.3). There are two things one can do: if the σ_i^2 are not likely to be too unequal then one can assume them all equal to an unknown σ^2, and use theorem 2; alternatively, one can estimate σ_i^2 in the usual way by

$$s_i^2 = \sum_j (x_{ij} - x_{i.})^2/(n_i - 1),$$

replace σ_i^2 by s_i^2 and ignore any distributional changes that might result. The latter is only likely to be satisfactory if all the sample sizes are large so that the error in replacing σ_i^2 by s_i^2 is small.

Multivariate problems

The alternative proof of theorem 1, using the corollary to theorem 5.1.1, demonstrates a connexion between the combination of observations and the combination of prior information and likelihood, and therefore between theorem 1 and theorem 3, which is an extension of theorem 5.1.1 from one variable to several. (In the univariate case we can take $\mathbf{A}$ to be the unit matrix.) It says that a multivariate normal prior distribution combines with a multivariate normal likelihood to give a multivariate normal posterior distribution. If $\mathbf{A} = \mathbf{I}$, the unit matrix (necessarily $p = n$), then the inverses of the dispersion matrices add (equation (9)) and the posterior mean is an average of the two means with weights equal to these inverses.

Example 1. We consider two applications of theorem 3. The first is to the weighing example of §3.3. In the discussion of that example in §3.3 we were concerned only with the variances of certain quantities (equations 3.3.2), suggested as determinations of the weights.† We now use theorem 3 to show that these determinations are the best possible. It is clear that the x_i of §3.3 have expectations (equation 3.3.1)

$$\mathscr{E}(x_1) = \theta_1 + \theta_2 - \theta_3 - \theta_4, \tag{12}$$

etc., and that, granted the normality, they satisfy the conditions of the theorem with

$$\mathbf{A} = \begin{pmatrix} 1 & 1 & -1 & -1 \\ 1 & -1 & 1 & -1 \\ 1 & -1 & -1 & 1 \\ 1 & 1 & 1 & 1 \end{pmatrix} \tag{13}$$

and $\mathbf{C}$ equal to the unit matrix times σ^2. If the prior knowledge of the true weights, θ_i, is vague compared with σ^2, then $\mathbf{C}_0^{-1}$ may be ignored and the posterior distribution of the true weights has mean (equation (8)) equal to $(\mathbf{A}'\mathbf{A})^{-1}\mathbf{A}'\mathbf{x}$. An easy calculation shows that $\mathbf{A}'\mathbf{A}$ is four times the unit matrix so that the mean is $\frac{1}{4}\mathbf{A}'\mathbf{x}$, the values given in equation 3.3.2. The dispersion matrix of the posterior distribution is $(\mathbf{A}'\mathbf{A})^{-1}\sigma^2$: so that the variances are all $\frac{1}{4}\sigma^2$ and the covariances zero. Note that, apart from a factor 4, $\mathbf{A}$ is an orthogonal matrix ($\mathbf{A}'\mathbf{A} = 4\mathbf{I}$), hence the name, *orthogonal* experiment, used in §3.2.

Example 2. The second example applies to the analysis of variance between and within samples discussed in §§6.4, 6.5. Suppose the situation to be that described in the statement of theorem 6.4.2 with ϕ known, equal to σ^2, say. Without loss of generality we can suppose $n = 1$ and denote the observations by x_i, since, with $n > 1$, $\{x_{i.}\}$ is sufficient. Then the likelihood is as in theorem 3 with $\mathbf{A}$ equal to the unit matrix of size $r \times r$ and $\mathbf{C}$ equal to the unit matrix times σ^2. In §6.5 we

† The reader should be careful to avoid confusing the weights put on the balance and the statistical weights of the observations.

made the point that in the prior distribution the θ_i were not always independent. Suppose that $\boldsymbol{\theta}$ is $N(\boldsymbol{\mu}_0, \mathbf{C}_0)$ with all the elements of $\boldsymbol{\mu}_0$ equal to μ_0, say, and with $\mathbf{C}_0$ having the special form of a matrix with σ_0^2 down the leading diagonal and $\rho\sigma_0^2$ everywhere else: denote such a matrix by $\|\sigma_0^2, \rho\sigma_0^2\|$. Then the θ_i have the same prior means μ_0 and variances σ_0^2, and the correlation between any pair is ρ. In particular, for any unequal i, j,

$$\mathscr{D}^2(\theta_i - \theta_j) = 2\sigma_0^2(1-\rho), \tag{14}$$

by theorem 3.3.2. This prior distribution is one in which the θ_i's are all alike in means and variances, and are equally correlated. Some algebra is necessary to obtain the mean of the posterior distribution, but this is considerably lightened by using the fact (which is easily verified) that if the matrices are of size $r \times r$

$$\|a, b\|^{-1} = \|[a+(r-2)b]c, -bc\| \quad \text{with} \quad c^{-1} = [a+(r-1)b](a-b). \tag{15}$$

It immediately follows from this result that

$$\mathbf{C}_0^{-1} = \|[1+(r-2)\rho]\lambda, -\rho\lambda\| \quad \text{with} \quad \lambda^{-1} = [1+(r-1)\rho](1-\rho)\sigma_0^2.$$

Now we are interested (as explained in §6.5) in cases where any one θ_i is vaguely known but any two are thought to be close together. We therefore consider the case where $\sigma_0^2 \to \infty$ but $\sigma_0^2(1-\rho) \to \tau^2$, say, which is finite: this will mean that $\rho \to 1$. From (14) it follows that the difference between any two θ_i's has finite variance, $2\tau^2$, but the variance of any θ_i tends to infinity. In this limit

$$\mathbf{C}_0^{-1} = \left\| \frac{r-1}{r\tau^2}, \quad -\frac{1}{r\tau^2} \right\|$$

and

$$\mathbf{C}_0^{-1} + \mathbf{C}^{-1} = \left\| \frac{r-1}{r\tau^2} + \frac{1}{\sigma^2}, \quad -\frac{1}{r\tau^2} \right\|.$$

Using (15) again, we obtain

$$(\mathbf{C}_0^{-1} + \mathbf{C}^{-1})^{-1} = \left\| \left(\frac{1}{r\tau^2} + \frac{1}{\sigma^2}\right)\xi, \frac{1}{r\tau^2}\xi \right\| \quad \text{with} \quad \xi^{-1} = \frac{\sigma^2+\tau^2}{\sigma^4\tau^2}.$$

Since $\mathbf{C}_0^{-1}\boldsymbol{\mu}_0 = \mathbf{0}$ and $\mathbf{C}^{-1}\mathbf{x}$ has typical element x_i/σ^2, we have for a typical element of $\boldsymbol{\mu}_1$, the value

$$\left\{\left(\frac{1}{r\tau^2}+\frac{1}{\sigma^2}\right)x_i+\frac{1}{r\tau^2}\sum_{j\neq i}x_j\right\}\Big/\left(\frac{\sigma^2+\tau^2}{\sigma^2\tau^2}\right)=\frac{x_i/\sigma^2+x_{.}/\tau^2}{1/\sigma^2+1/\tau^2}. \quad (16)$$

In words, the expected posterior value of θ_i is a weighted mean of x_i and the mean of the x's, $x_{.}$, with weights equal to the precision of each x_i and twice the precision of the prior difference of any two θ_i's. The effect is to replace the natural value, x_i, for the expectation of θ_i by a value shifted toward the common mean. The values of θ_i with x_i near x will be little affected, but those with the extreme values of x_i will be shifted to a greater degree towards $x_{.}$. This is in satisfactory agreement with common-sense ideas for it is usual to feel that the extreme values should be a little discounted just because they are extreme values. For example, in the numerical illustration of §6.5 the second method had the lowest mean and was, on the evidence of the experiment, the best; but one might not feel happy at saying its expected value was $157/6 = 26{\cdot}2$ because it may just have done rather well in this experiment: in another experiment method 3 might be lower.

General inference problems

Theorem 3 does not easily extend to deal with the case of unknown σ^2 and τ^2 in example 2 (evidence on these is available from the within and between sums of squares respectively). But it does show how prior knowledge should be incorporated into the data. This is perhaps a convenient point to make the general remark that every inference situation should be judged on its own. In this book we give recipes that fit a wide class of problems but it should not be thought that these are the only cases that arise. The best approach to any inference problem is to consider the likelihood and the prior distributions separately and then to form their product to obtain the posterior distribution. If they fit into the framework of a standard method so much the better, but if not, one can always try to put the posterior distribution into a convenient form and, if all else fails, it can always be calculated numerically in any particular case.

Suggestions for further reading

The suggestions given in chapter 5 are relevant. Two more advanced books on mathematical statistics are Kendall and Stuart (1958, 1961) of which a third volume is promised, and Wilks (1962). Another important work is the older book on probability and statistics by Cramér (1946).

Exercises

1. The table gives the yields of two varieties A and B of wheat grown in strictly comparable conditions at each of twelve Canadian testing stations. Obtain the posterior distribution of the difference in average yield of the two varieties:

Station...	1	2	3	4	5	6	7	8	9	10	11	12
A	25	28	28	20	23	23	26	29	26	26	37	31
B	17	15	21	18	22	14	25	28	25	19	34	25

(Camb. N.S.)

2. The following results relate to the yields of two varieties of tomato grown in pairs in nine different positions in a greenhouse:

Position, i	Variety X, x_i	Variety Y, y_i	Position, i	Variety X, x_i	Variety Y, y_i
1	1·39	1·17	6	1·63	1·22
2	1·41	1·22	7	0·88	0·84
3	1·07	1·08	8	1·14	0·94
4	1·65	1·42	9	1·32	1·25
5	1·57	1·39			

($\Sigma x_i = 12{\cdot}06, \Sigma y_i = 10{\cdot}53, \Sigma x_i^2 = 16{\cdot}7258, \Sigma y_i^2 = 12{\cdot}6123, \Sigma x_i y_i = 14{\cdot}4768.$)

Give 99 % confidence limits for the difference between the average yields of X and Y.

Discuss whether the arrangement of the experiment in pairs has been advantageous. (Camb. N.S.)

3. A new wages structure is introduced throughout a certain industry. The following values of output per man-hour are obtained from a number of factories just before and just after its introduction:

Output per man-hour	
Old wages structure	New wages structure
122·3	119·0
141·0	131·8
120·1	118·7
109·6	106·0
119·1	118·2
128·0	126·6
132·8	124·1

If ξ is the average change in output per man-hour over all factories in the industry, consequent upon the change in wages structure, calculate, on conventional assumptions, 95 % confidence limits for ξ in each of the following cases:

Case I. The data refer to seven different factories, corresponding figures in the two series coming from the same factory.

Case II. The data refer to fourteen different factories.

Suppose in case II that further values were obtained, making up n factories in each column instead of 7. Roughly how large would n have to be for one to be reasonably sure of detecting that ξ was not zero, if in fact ξ was equal to $-2\cdot0$ units? (Manch. Dip.)

4. Two normal populations of known equal variance σ_0^2 have means θ_1, θ_2. It is required to calculate, from random samples of size n taken from each population, a 100α % confidence interval for $\Delta = \theta_2 - \theta_1$. The following procedure is suggested: 100α % confidence intervals are calculated separately for θ_1 and θ_2 and are, say, (l_1, u_1) and (l_2, u_2). Any value in (l_1, u_1) is, at this probability level, a possible value for θ_1, and similarly for θ_2. Hence any value in the interval $(l_2 - u_1, u_2 - l_1)$ is a possible value for Δ at confidence level 100α % and is the confidence interval required. Explain why this argument is false and compare the answer it gives with that obtained from the correct procedure. (Lond. B.Sc.)

5. It is expected that a certain procedure for estimating the percentage amount of a substance X in a sample of material will give a standard error approximately equal to 0·12. Two chemists take samples from a given bulk of material and each performs five independent determinations of the percentage amount of X present in the material. Their readings are as follows:

First chemist: 2·43, 2·42, 2·22, 2·29, 2·06

Second chemist: 2·48, 2·72, 2·43, 2·40, 2.58

Do these figures support the assumption that the standard error of a determination is 0·12? Is there any evidence that one chemist's readings are biased relative to the other? (Camb. N.S.)

6. Two workers carried out an experiment to compare the repeatabilities with which they used a certain chemical technique. The first worker made eight determinations on the test substance and estimated the standard deviation of a single measurement as 0·74; the second worker made fifteen determinations on the same substance and obtained a standard deviation of 1·28. Compare the repeatabilities of the two workers' measurements. (Camb. N.S.)

7. Two analysts each made six micro-analytical determinations of the carbon content of a chemical, giving the values:

Analyst I: 59·09, 59·17, 59·27, 59·13, 59·10, 59·14

Analyst II: 59·06, 59·40, 59·00, 59·12, 59·01, 59·25

Discuss whether these results provide evidence that the two analysts differ in their accuracy of determination. (Camb. N.S.)

8. Two methods, I and II, are available for determining the sulphur dioxide content of a soft drink. The variances associated with single determinations are σ_1^2 and σ_2^2 respectively, but it is usual to use several independent determinations with one of the two methods. Show that if I costs c times as much to carry out as II per determination, then I is only to be preferred if

$$\sigma_2^2/\sigma_1^2 > c.$$

Thirteen determinations were made using method I on one drink and thirteen using method II with another drink. The sample variances were equal. Show that, on the basis of this information, in order to be able to make a fairly confident decision about which method to use, one method would have to be at least 3·28 times as costly as the other.

[Interpret 'fairly confident' as a 95 % level of probability.]

(Wales Maths.)

9. A random sample $x_1, x_2, \ldots, x_m$ is available from $N(\theta_1, \phi)$ and a second independent random sample $y_1, y_2, \ldots, y_n$ from $N(\theta_2, 2\phi)$. Obtain, under the usual assumptions, the posterior distribution of $\theta_1 - \theta_2$, and of ϕ.

10. Use the result of exercise 5 in chapter 2 to show that tables of the F-distribution can be used to obtain tables of the binomial distribution. Specifically, in an obvious notation show that if s is $B(n, p)$ then

$$p(s > r) = p\left[F(2(n-r+1), 2r) > \frac{qr}{p(n-r+1)}\right].$$

11. The following table gives the values of the cephalic index found in two random samples of skulls, one consisting of fifteen and the other of thirteen individuals:

Sample I: 74·1, 77·7, 74·4, 74·0, 73·8, 79·3, 75·8, 82·8, 72·2, 75·2, 78·2, 77·1, 78·4, 76·3, 76·8

Sample II: 70·8, 74·9, 74·2, 70·4, 69·2, 72·2, 76·8, 72·4, 77·4, 78·1, 72·8, 74·3, 74·7

If it is known that the distribution of cephalic indices for a homogeneous population is normal, test the following points:

(*a*) Is the observed variation in the first sample consistent with the hypothesis that the standard deviation in the population from which it has been drawn is 3·0?

(*b*) Is it probable that the second sample has come from a population in which the mean cephalic index is 72·0?

(*c*) Use a more sensitive test for (*b*) if it is known that the two samples are obtained from populations having the same but unknown variance.

(*d*) Obtain the 90 % confidence limits for the ratio of the variances of the populations from which the two samples are derived. (Leic. Gen.)

12. The table gives the time in seconds for each of six rats ($A, B, \ldots, F$) to run a maze successfully in each of four trials (1, 2, 3, 4). Perform an analysis of variance to determine whether the rats give evidence of differing in their ability.

	A	*B*	*C*	*D*	*E*	*F*
1	15	18	11	18	19	14
2	10	15	11	22	14	16
3	13	16	13	17	16	16
4	14	16	11	16	14	15

(Camb. N.S.)

13. Twelve tomato plants are planted in similar pots and are grown under similar conditions except that they are divided into three groups A, B, C of four pots each, the mixture of soil and fertilizer being the same within each group but the groups differing in the type of fertilizer. The yields (in fruit weight given to the nearest pound) are tabulated below. Decide by carrying out an analysis of variance (or otherwise) whether the evidence that the fertilizers are of different value is conclusive or not.

Groups	Yield of plants			
A	3	3	4	6
B	6	8	8	10
C	9	12	13	14

(Camb. N.S.)

14. An experiment was carried out to investigate the efficacies of four compounds; two, P_1, P_2, based on one antibiotic and two, S_1, S_2, based on another antibiotic. A control C was also used. The compounds were each used five times in a suitable arrangement to ensure independence, with the following results:

C	P_1	P_2	S_1	S_2
5·48	4·79	7·16	4·28	5·74
4·42	7·32	6·78	2·92	6·80
4·97	5·80	7·05	3·97	6·04
3·28	6·52	9·21	5·07	6·93
4·50	6·11	7·36	3·72	6·50

Find the posterior distributions of the following quantities:

(1) the difference between the mean of P_1 and P_2 and the control;

(2) the difference between the mean of S_1 and S_2 and the control.

Is there any evidence that the P compounds differ from the S compounds? Or that the two P compounds are more different amongst themselves than the two S compounds?

15. Four independent determinations are made of a physical constant using apparatuses of known standard deviations. The means and standard deviations of the posterior distributions associated with the four determinations are, in the form $\mu \pm \sigma$:

$$1{\cdot}00 \pm 0{\cdot}10, \quad 1{\cdot}21 \pm 0{\cdot}08, \quad 1{\cdot}27 \pm 0{\cdot}11, \quad 1{\cdot}11 \pm 0{\cdot}06.$$

Find the most probable value for the constant, and its standard deviation. (Camb. N.S.)

16. An apparatus for determining g, the acceleration due to gravity, has standard deviation σ_1 when used at one place and σ_2 at another. It is proposed to make N determinations—some at one place, some at the other—to find the difference between g at the two places. Determine the optimum allocation of the determinations between the two places in order to minimize the posterior variance of the difference.

If $\sigma_1^2 = 0{\cdot}0625$, $\sigma_2^2 = 0{\cdot}1250$ find how large the observed mean difference will have to be before a test of the hypothesis that the real difference is zero is significant at the 5 % level. How large must N be if the real difference is 0·25 and we wish to have a 90 % chance of the above test yielding a significant result?

17. In an experiment to determine an unknown value θ for a substance, independent measurements are made on the substance at times $t_1, t_2, \ldots, t_n$. A measurement made at time t has mean $\theta e^{-\kappa t}$, where $\kappa > 0$ is a known constant, and has known constant coefficient of variation v. Determine the posterior distribution of θ under the assumption that the measurements are normally distributed.

18. In the experiment discussed in detail in §6.5 a possible prior distribution might be one in which the four methods were thought unlikely to differ much among themselves but might differ substantially from the control (in either direction). On the lines of the example discussed in §6.6 put this prior distribution into a suitable multivariate normal form and, assuming the common sample variance known, obtain algebraic expressions for the means of the posterior distributions for the five methods analogous to those of equation 6.6.16.

19. A random sample $x_1, x_2, \ldots, x_m$ from $N(\theta_1, \theta_2)$ has sufficient statistics

$$\bar{x} = \sum_{i=1}^{m} x_i/m \quad \text{and} \quad S_x^2 = \sum_{i=1}^{m} (x_i - \bar{x})^2.$$

It is proposed to take a second independent random sample of size n from the same distribution and to calculate from the values $y_1, y_2, \ldots, y_n$ the two statistics

$$\bar{y} = \sum_{i=1}^{n} y_i/n \quad \text{and} \quad S_y^2 = \sum_{i=1}^{n} (y_i - \bar{y})^2.$$

Show that $$p(\bar{y}, S_y^2 | \bar{x}, S_x^2) = \int p(\bar{y} | \bar{x}, \theta_2) p(S_y^2 | \theta_2) \pi(\theta_2 | S_x^2) \, d\theta_2,$$

and hence that

$$p(S_y^2 | \bar{x}, S_x^2) = p(S_y^2 | S_x^2) = \int p(S_y^2 | \theta_2) \pi(\theta_2 | S_x^2) \, d\theta_2.$$

Using the known expressions for $p(S_y^2|\theta_2)$ (exercise 5.20) and $\pi(\theta_2|S_x^2)$ (equation 5.3.4) evaluate this integral and show that $p(S_y^2|\bar{x}, S_x^2)$ is such that

$$\frac{(m-1)S_y^2}{(n-1)S_x^2}$$

has an F-distribution on $(n-1)$ and $(m-1)$ degrees of freedom.

The posterior distribution of θ_1 based on the combined sample of $m+n$ observations will, if $m+n$ is large (so that the t-distribution is almost normal) and the little information about θ_2 provided by the difference between $\bar{x}$ and $\bar{y}$ is ignored, be approximately normal with mean $(m\bar{x}+n\bar{y})/(m+n)$ and variance $(S_x^2+S_y^2)/(m+n-2)(m+n)$. Show how to use the above result and tables of the F-distribution to determine the value of n to be taken for the second sample in order to be 90 % certain that the 95 % confidence interval for θ_1 obtained by this approximation has width less than some pre-assigned number c.

Obtain also $p(\bar{y}|\bar{x}, S_x^2)$.

20. Two independent Poisson processes are each observed for a unit time. m and n incidents are observed respectively and the unknown rates (the Poisson means) are θ_1 and θ_2 respectively. If θ_1 and θ_2 have independent uniform prior distributions over the positive real line show that the posterior distribution of $\psi = \theta_1/(\theta_1+\theta_2)$ has density

$$\frac{(m+n+1)!}{m!n!}\psi^m(1-\psi)^n. \qquad (*)$$

Show that the conditional distribution of m, given $m+n$, depends only on ψ.

A statistician who wishes to make inferences about ψ (but not about θ_1 and θ_2 separately) decides to use this last fact to avoid consideration of θ_1 and θ_2. Show that if he assumes ψ to have a uniform prior distribution in (0, 1) then he will obtain a posterior distribution identical to (*). Does this mean that $m+n$ gives no information about ψ?

21. A complex organic compound contains a chemical which decays when the compound is removed from living tissue. If p is the proportion of the chemical present at the time of removal ($t = 0$), the proportion present at time t is $pe^{-\lambda t}$ where $\lambda\ (> 0)$ is known. To determine p a quantity of the compound is removed and independent, unbiased estimates are made of the proportions present at times $t = r\tau\ (r = 0, 1, 2, ..., N)$ in individual experiments each of which is of duration τ, the estimates having variances $\kappa e^{-\lambda r\tau}$ (i.e. proportional to the proportion present at the beginning of each experiment). Determine the best estimate of p and its variance when N is large.

A second method of determination is suggested in which each experiment is twice as quick (i.e. τ is reduced to $\frac{1}{2}\tau$) but with each variance increased by 50 %. Show that this second method is only better if $\lambda < \tau^{-1}\ln 4$. (Camb. N.S.)

22. The random variables $Y_1, ..., Y_n$ are independently distributed with $\mathscr{E}(Y_i) = \beta x_i$, where $x_1, ..., x_n$ are known.

How would you estimate β if the errors $Y_i - \beta x_i$ are independently rectangularly distributed over $(-\theta, \theta)$, where θ is unknown?

(Lond. M.Sc.)

7

APPROXIMATE METHODS

The results obtained for normal distributions in the two previous chapters were exact for the specified prior distribution. In dealing with samples for distributions other than normal it is often necessary to resort to approximations to the posterior density even when the prior distribution is exact. In this chapter we shall mainly be concerned with binomial, multinomial and Poisson distributions, but begin by describing an approximate method of wide applicability.

7.1. The method of maximum likelihood

Let $\mathbf{x}$ be any observation with likelihood function $p(\mathbf{x}|\theta)$ depending on a single real parameter θ. The value of θ, denoted by $\hat{\theta}(\mathbf{x})$, or simply $\hat{\theta}$, for which the likelihood for that observation is a maximum is called the *maximum likelihood estimate* of θ. Notice that $\hat{\theta}$ is a function of the sample values only: it is an example of what we have previously called a statistic (§5.5). The definition generalizes to a likelihood depending on several parameters $p(\mathbf{x}|\theta_1, \theta_2, ..., \theta_s)$: the set $(\hat{\theta}_1, \hat{\theta}_2, ..., \hat{\theta}_s)$ for which the likelihood is a maximum form the set of maximum likelihood estimates, $\hat{\theta}_i$, of θ_i $(i = 1, 2, ..., s)$. The estimate is particularly important in the special case where $\mathbf{x} = (x_1, x_2, ..., x_n)$ is a random sample of size n from a distribution with density $f(x_i|\theta)$. Then

$$p(\mathbf{x}|\theta) = \prod_{i=1}^{n} f(x_i|\theta). \tag{1}$$

In this case the logarithm of the likelihood can be written as a sum:

$$L(\mathbf{x}|\theta) = \ln p(\mathbf{x}|\theta) = \sum_{i=1}^{n} \ln f(x_i|\theta). \tag{2}$$

Important properties of the log-likelihood, $L(\mathbf{x}|\theta)$, can be deduced from the strong law of large numbers (theorem 3.6.3) in the following way. In saying that $\mathbf{x}$ is a random sample we imply that the sample values have the same density $f(x_i|\theta)$ for some θ fixed, for all i. Denote this value of θ by θ_0. We refer to it as the true value of θ. It is, of course, unknown. Then, for each θ, the quantities, $\ln f(x_i|\theta)$, are independent random variables with a common distribution, depending on θ_0, and, by the strong law, their mean converges strongly to their common expectation. By definition this expectation is

$$\mathscr{E}_0\{\ln f(x_i|\theta)\} = \int \ln f(x_i|\theta) . f(x_i|\theta_0)\, dx_i, \tag{3}$$

where the suffix has been added to the expectation sign to indicate the true value, θ_0, of θ. Hence the law says that with probability one

$$\lim_{n\to\infty} \{n^{-1}L(\mathbf{x}|\theta)\} = \mathscr{E}_0\{\ln f(x_i|\theta)\}. \tag{4}$$

Similarly, provided the derivatives and expectations exist,

$$\lim_{n\to\infty} \{n^{-1}\partial^r L(\mathbf{x}|\theta)/\partial\theta^r\} = \mathscr{E}_0\{\partial^r \ln f(x_i|\theta)/\partial\theta^r\}. \tag{5}$$

Equation (4) may be expressed in words by saying that, for large n, the log-likelihood behaves like a constant times n, where the constant is the expectation in (3). Similar results apply in the case of several parameters.

Theorem 1. *If a random sample of size n is taken from $f(x_i|\theta)$ then, provided the prior density, $\pi(\theta)$, nowhere vanishes, the posterior density of θ is, for large n, approximately normal with mean equal to the maximum likelihood estimate and variance, σ_n^2, given by†*

$$\sigma_n^{-2} = -\partial^2 L(\mathbf{x}|\hat{\theta})/\partial\theta^2. \tag{6}$$

It is not possible to give a rigorous proof of this theorem at the mathematical level of this book. The following 'proof' should convince most readers of the reasonableness of the result.

The posterior density is proportional to

$$\exp\{L(\mathbf{x}|\theta) + \ln \pi(\theta)\},$$

and since we have seen that $L(\mathbf{x}|\theta)$ increases like n, it will ultimately, as $n \to \infty$, dwarf $\ln \pi(\theta)$ which does not change with n. Hence the density is, apart from a constant, approximately

$$\exp\{L(\mathbf{x}|\theta)\} = \exp\{L(\mathbf{x}|\hat{\theta}) + \tfrac{1}{2}(\theta-\hat{\theta})^2 \partial^2 L(\mathbf{x}|\hat{\theta})/\partial\theta^2 + R\},$$

† $\partial^2 L(\mathbf{x}|\hat{\theta})/\partial\theta^2$ denotes the second derivative with respect to θ evaluated at $\hat{\theta}$.

on expanding $L(\mathbf{x}|\theta)$ by Taylor's theorem about $\hat{\theta}$, where R is a remainder term. Since the likelihood, and hence the log-likelihood, has a maximum at $\hat{\theta}$ the first derivative vanishes there. Also the second derivative will be negative there and may therefore be written $-\sigma_n^{-2}$. Furthermore, since it does not involve θ, the first term may be incorporated into the omitted constant of proportionality and we are left with

$$\exp\{-\tfrac{1}{2}(\theta-\hat{\theta})^2/\sigma_n^2+R\}. \tag{7}$$

From the discussions of the normal density in §2.5 it is clear that the term $\exp\{-\frac{1}{2}(\theta-\hat{\theta})^2/\sigma_n^2\}$ is negligible if $|\theta-\hat{\theta}|>3\sigma_n$; so that since σ_n^{-2} is, by (5), of order n, this term is only appreciable if θ differs from $\hat{\theta}$ by something of the order of $n^{-\frac{1}{2}}$. In that case the remainder term, R, which may be written

$$\tfrac{1}{6}(\theta-\hat{\theta})^3\,\partial^3L(\mathbf{x}|\theta_1)/\partial\theta^3$$

for some θ_1, is of order $n^{-\frac{3}{2}}$ times n (by (5)). Hence it is of order $n^{-\frac{1}{2}}$ and is negligible compared with the other term in (7). Hence, inserting the constant of proportionality, the posterior density is approximately

$$(2\pi\sigma_n)^{-\frac{1}{2}}\exp\{-\tfrac{1}{2}(\theta-\hat{\theta})^2/\sigma_n^2\}, \tag{8}$$

which establishes the result. Notice that σ_n^2 is, under the assumptions made here, of order n^{-1}.

Theorem 2. *If a random sample of size n is taken from*

$$f(x_i|\theta_1,\theta_2,\ldots,\theta_s)$$

then, provided the joint prior density, $\pi(\theta_1,\theta_2,\ldots,\theta_s)$ nowhere vanishes, the joint posterior density is, for large n, approximately multivariate normal with means equal to the maximum likelihood estimates $\hat{\theta}_i$ and a dispersion matrix whose inverse has typical element

$$-\partial^2L(\mathbf{x}|\hat{\theta}_1,\hat{\theta}_2,\ldots,\hat{\theta}_s)/\partial\theta_i\partial\theta_j. \tag{9}$$

This is the extension of theorem 1 to the case of several parameters. The proof proceeds as in that case. The important terms in the Taylor series expansion of $L(\mathbf{x}|\theta_1,\theta_2,\ldots,\theta_s)$ are

$$\frac{1}{2}\sum_{i,j=1}^{s}(\theta_i-\hat{\theta}_i)(\theta_j-\hat{\theta}_j)\,\partial^2L(\mathbf{x}|\hat{\theta}_1,\hat{\theta}_2,\ldots,\hat{\theta}_s)/\partial\theta_i\partial\theta_j, \tag{10}$$

and a comparison with the multivariate normal density (equation 3.5.17) establishes the result.

The matrix, whose typical element is given by (9), will be called the *information matrix*. It is the inverse of the dispersion matrix of the posterior density (compare the definition of precision in §5.1). Similarly (6) is called the *information*.

General remarks

Although known to earlier writers, the method of maximum likelihood has only become widely used through the work of R. A. Fisher, who obtained its principal properties. The main advantages of the method are that it produces a description of the posterior distribution which, because it is normal, is easy to handle, and which has a particularly simple mean and variance. (We shall see below that these are easy to compute.) Fisher used the method to provide a point estimate of θ. We shall not have much to say in this book about the problem of point estimation; by which is usually meant the problem of finding a single statistic which is, in some sense, near to the true value of a parameter (see §5.2); our reason for not doing so is that posterior distributions cannot be adequately described by one statistic.† But the problem is much discussed by some statisticians and, in large samples, is adequately solved by the maximum likelihood estimate, though other approximations are available. There is, as we shall see below, a close relationship between $\mathscr{D}^2(\hat{\theta})$ and σ_n^2 above: so that $\hat{\theta}$ and its variance do provide, because of the normality, an adequate description, in large samples, of the posterior density.

In the definition of $\hat{\theta}$ the word 'maximum' is used in the sense of 'largest value': that is, $L(\mathbf{x}|\hat{\theta}) \geqslant L(\mathbf{x}|\theta)$ for all θ. The estimate is therefore not necessarily determined by equating the first derivative to zero. This latter process will only yield the local maxima (or minima). An example where this process is inapplicable is provided by the uniform distribution discussed below. (In the 'proof' it has been assumed that the first derivative is zero at the maximum.)

† If we wished to use a single statistic we could take the mean of the posterior distribution. But this would not be helpful without, in addition, at least the variance.

Notice that the method has little or nothing to recommend it in small samples. There are two reasons for this. First, the posterior distribution is not necessarily normal. Secondly, the prior distribution is relevant in small samples because the information provided by it may be comparable with the information provided by the sample, and any method based on the likelihood alone may be misleading. We have mentioned the diminishing importance of the prior distribution as the sample size increases in connexion with the normal distribution (§§5.1, 5.2) but the point is quite general. Of course, if the prior knowledge is very vague (as in theorem 5.2.1), even a small sample may contain virtually all the information. Notice that, in the statement of the theorems, it has been necessary to assume that the prior density nowhere vanishes. If it did vanish near θ_1, say, then no amount of evidence would ever convince one that θ was near θ_1 (cf. §5.2) and the posterior density would vanish near θ_1 even if $\hat{\theta} = \theta_1$. (In the proof $\ln \pi(\theta)$ would be minus infinity, and certainly not negligible.)

Example: normal distribution

We begin with situations already studied, in order to see how the approximation compares with the exact result. In the case of the normal mean (§§5.1, 5.2) with known variance the likelihood is given by equation 5.1.9. The log-likelihood is therefore

$$L(\mathbf{x}\,|\,\theta) = C - \tfrac{1}{2}(\bar{x}-\theta)^2\,(n/\sigma^2),$$

where C is a constant. Differentiating and equating to zero gives the result $(\bar{x}-\hat{\theta})\,(n/\sigma^2) = 0$, so that $\hat{\theta} = \bar{x}$. A second differentiation gives

$$\sigma_n^{-2} = -\partial^2 L(\mathbf{x}\,|\,\hat{\theta})/\partial\theta^2 = (n/\sigma^2),$$

so that the posterior density is approximately $N(\bar{x}, \sigma^2/n)$, which agrees with theorem 5.2.1, and is exact (corollary to theorem 5.1.1) if the prior distribution of θ is uniform over the real line.

If the variance is also unknown (§5.4) then the logarithm of the likelihood is, from equation 5.4.3 rearranged in the same way as the posterior distribution was rearranged to obtain 5.4.4,

$$C - \tfrac{1}{2}n \ln \theta_2 - \{\nu s^2 + n(\bar{x}-\theta_1)^2\}/2\theta_2.$$

To obtain the maximum likelihood estimates we differentiate partially with respect to θ_1 and θ_2 and equate to zero. The results are

$$\left.\begin{aligned}\frac{\partial L}{\partial \theta_1} &= n(\bar{x}-\hat{\theta}_1)/\hat{\theta}_2 = 0,\\ \frac{\partial L}{\partial \theta_2} &= -\tfrac{1}{2}n/\hat{\theta}_2 + \{\nu s^2 + n(\bar{x}-\hat{\theta}_1)^2\}/2\hat{\theta}_2^2 = 0,\end{aligned}\right\} \quad (11)$$

so that $\hat{\theta}_1 = \bar{x}$ and $\hat{\theta}_2 = \nu s^2/n = \Sigma(x_i - \bar{x})^2/n$. (12)

The matrix whose elements are minus the second derivatives of the log-likelihood at the maximum (the information matrix of (9)) is easily seen to be

$$\begin{pmatrix} n\hat{\theta}_2^{-1} & 0 \\ 0 & \tfrac{1}{2}n/\hat{\theta}_2^2 \end{pmatrix}, \quad (13)$$

with inverse

$$\begin{pmatrix} \hat{\theta}_2/n & 0 \\ 0 & 2\hat{\theta}_2^2/n \end{pmatrix}. \quad (14)$$

The posterior distribution of θ_1 is thus approximately $N(\hat{\theta}_1, \hat{\theta}_2/n)$, or, from (12), $n^{\frac{1}{2}}(\theta_1 - \bar{x})/s(\nu/n)^{\frac{1}{2}}$ is approximately $N(0, 1)$. This is in agreement with the exact result (theorem 5.4.1) that $n^{\frac{1}{2}}(\theta_1 - \bar{x})/s$ has Student's distribution, since this distribution tends to normality as $n \to \infty$ (§5.4), and $\nu/n \to 1$. The distribution of θ_2 is approximately $N(\hat{\theta}_2, 2\hat{\theta}_2^2/n)$. This agrees with the exact result (theorem 5.4.2) that $\nu s^2/\theta_2$ is χ^2 with ν degrees of freedom, because the mean and variance of θ_2 are s^2 and $2s^4/\nu$ (equations 5.3.5 and 6) in large samples and the distribution of θ_2 tends to normality. This last result was proved in §5.3. Finally we note that the covariance between θ_1 and θ_2 is zero, which is in exact agreement with the result obtained in §5.4.

Example: binomial distribution

Consider a random sequence of n trials with constant probability θ of success. If r of the trials result in a success the likelihood is (cf. equation 5.5.9)

$$\theta^r(1-\theta)^{n-r}. \quad (15)$$

The derivative of the log-likelihood is therefore

$$r/\theta - (n-r)/(1-\theta),$$

so that $\hat{\theta} = r/n$, the proportion of successes in the n trials. The second derivative is

$$-r/\theta^2-(n-r)/(1-\theta)^2,$$

which gives $\sigma_n^2 = r(n-r)/n^3$. These results agree with the exact results of §7.2.

Example: exponential family

The method of maximum likelihood is easily applied to any member of the exponential family. In the case of a single sufficient statistic for a single parameter the density is (equation 5.5.5)

$$f(x_i \mid \theta) = F(x_i) G(\theta) e^{u(x_i)\phi(\theta)}$$

and the log-likelihood is, apart from a constant,

$$ng(\theta)+t(\mathbf{x})\phi(\theta),$$

where $g(\theta) = \ln G(\theta)$ and $t(\mathbf{x}) = \sum_{i=1}^{n} u(x_i)$. The posterior density of θ is therefore approximately normal with mean equal to the root of

$$ng'(\theta)+t(\mathbf{x})\phi'(\theta) = 0 \tag{16}$$

and variance equal to

$$\{-ng''(\theta)-t(\mathbf{x})\phi''(\theta)\}^{-1}$$

evaluated at that root. Similar results apply in the case of several sufficient statistics and parameters.

Solution of maximum likelihood equation

The equation for the maximum of the likelihood,

$$\partial L(\mathbf{x} \mid \theta)/\partial\theta = 0,$$

or, in the case of the exponential family, (16) above, may not be solvable in terms of elementary functions. However, there is an elegant numerical method of solving the equation, in the course of which σ_n^2 is also obtained. This is Newton's method of solving an equation. A reference to fig. 7.1.1 will explain the idea. On a graph of the derivative of the log-likelihood against θ a tangent to the graph is drawn at a first approximation, $\theta^{(1)}$, to the value of $\hat{\theta}$. The tangent intersects the θ-axis at a second approximation $\theta^{(2)}$ which is typically nearer to $\hat{\theta}$ than $\theta^{(1)}$ is, and, in any case, may be used in place of $\theta^{(1)}$ to repeat the process, obtaining $\theta^{(3)}$, and so on. The sequence $\{\theta^{(i)}\}$ usually converges to $\hat{\theta}$. Algebraically the method may be expressed by expanding in a Taylor series

$$\partial L(\mathbf{x}\,|\,\hat{\theta})/\partial\theta = 0 = \partial L(\mathbf{x}\,|\,\theta^{(1)})/\partial\theta + (\hat{\theta}-\theta^{(1)})\,\partial^2 L(\mathbf{x}\,|\,\theta^{(1)})/\partial\theta^2 + \ldots$$

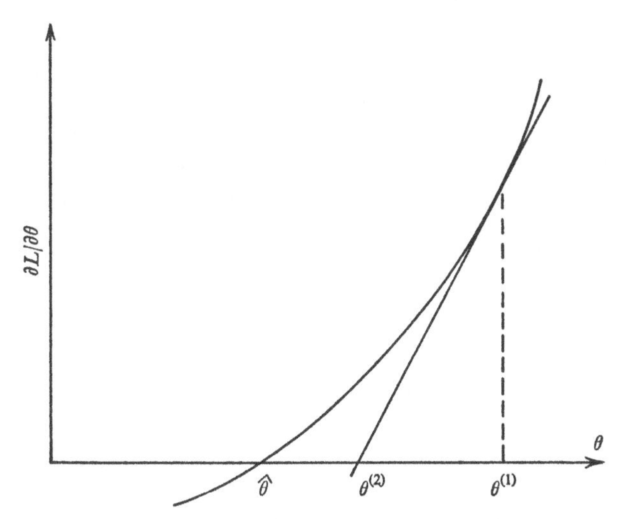

Fig. 7.1.1. Newton's method for solution of the maximum likelihood equation.

and retaining only the first two terms. The root of the equation for $\hat{\theta}$ so obtained is $\theta^{(2)}$, that is

$$\theta^{(2)}-\theta^{(1)} = \{\partial L(\mathbf{x}\,|\,\theta^{(1)})/\partial\theta\}/\{-\partial^2 L(\mathbf{x}\,|\,\theta^{(1)})/\partial\theta^2\}. \qquad (17)$$

It is not necessary to recalculate the second derivative at each approximation: the method will still work with a single value retained throughout. A final recalculation may be advisable at the termination of the process when $\hat{\theta}$ has been obtained to sufficient accuracy (that is when $\theta^{(r)}-\theta^{(r-1)}$ is negligible) in order to obtain a better value for $\sigma_n^2 = \{-\partial^2 L(\mathbf{x}\,|\,\theta^{(r)})/\partial\theta^2\}$.

The method is equally convenient when several parameters are involved. The Taylor series expansion gives

$$\partial L(\mathbf{x}\,|\,\boldsymbol{\theta}^{(1)})/\partial\theta_i = \sum_j (\theta_j^{(2)}-\theta_j^{(1)})\,\{-\partial^2 L(\mathbf{x}\,|\,\boldsymbol{\theta}^{(1)})/\partial\theta_i\partial\theta_j\}, \qquad (18)$$

where $\boldsymbol{\theta}^{(1)} = (\theta_1^{(1)}, \theta_2^{(1)}, ..., \theta_s^{(1)})$, a set of linear equations for $\boldsymbol{\theta}^{(2)} - \boldsymbol{\theta}^{(1)}$. The matrix which has to be inverted is the information matrix at argument $\boldsymbol{\theta}^{(1)}$ instead of at argument $\hat{\boldsymbol{\theta}}$. At the final approximation, $\boldsymbol{\theta}^{(r)}$, this has to be inverted in any case in order to obtain the dispersion matrix. Thus the method is well suited, not only to the evaluation of the means, but also to the evaluation of the dispersion matrix, of the posterior distribution. Numerical methods for the inversion of matrices are given in §8.4.

Example

As an illustration of Newton's method consider random samples from a Γ-distribution, or equivalently a χ^2-distribution, with both the index and the parameter unknown. The density for a single observation is (equation 2.3.7)

$$f(x_i \mid \theta_1, \theta_2) = \frac{\theta_1^{\theta_2}}{(\theta_2 - 1)!} e^{-x_i\theta_1} x_i^{\theta_2 - 1}, \tag{19}$$

where we have written θ_1 for λ and θ_2 for n. The likelihood for a random sample of size n is thus

$$\left\{\frac{\theta_1^{\theta_2}}{(\theta_2 - 1)!}\right\}^n \exp\left\{-\theta_1 \sum_{i=1}^n x_i + (\theta_2 - 1) \sum_{i=1}^n \ln x_i\right\}. \tag{20}$$

This shows that the distribution belongs to the exponential family and that $\bar{x} = \Sigma x_i/n$ and $\bar{y} = \Sigma \ln x_i/n$ are jointly sufficient statistics. Differentiation of the log-likelihood gives

$$\left.\begin{aligned} \frac{\partial}{\partial\theta_1}&: \quad n(-\bar{x} + \hat{\theta}_2/\hat{\theta}_1) = 0, \\ \frac{\partial}{\partial\theta_2}&: \quad n\left\{\ln\hat{\theta}_1 + \bar{y} - \frac{d}{d\theta_2}\ln(\hat{\theta}_2 - 1)!\right\} = 0. \end{aligned}\right\} \tag{21}$$

The first of these equations is sufficiently simple to enable $\hat{\theta}_1$ to be eliminated and a single equation,

$$\ln\hat{\theta}_2 - \ln\bar{x} + \bar{y} - \frac{d}{d\theta_2}\ln(\hat{\theta}_2 - 1)! = 0, \tag{22}$$

for $\hat{\theta}_2$ to be solved by Newton's method. The derivative of the left-hand side of (22) is $\hat{\theta}_2^{-1} - (d^2/d\theta_2^2)\ln(\hat{\theta}_2 - 1)!$ and tables of the derivatives of the logarithm of the factorial function (see, for example, Davis (1933)) enable the calculations to be carried out. It is necessary, however, to start with a first approximation. Here this is most easily obtained by using the approximation† to $d\ln(\hat{\theta}_2-1)!/d\theta_2$ of $\ln\hat{\theta}_2 - 1/2\hat{\theta}_2$, which, on insertion in (22), gives a value of θ_2 equal to $\{2(\ln\bar{x}-\bar{y})\}^{-1}$ to use as $\theta_2^{(1)}$ in the iteration. The approximation is remarkably good for all values of θ_2 except those near zero, so that except in that case, a single stage of Newton's procedure should be sufficient. We leave the reader to verify that the dispersion matrix is the matrix

$$\begin{pmatrix} \dfrac{d^2}{d\theta_2^2}\ln(\hat{\theta}_2-1)! & \hat{\theta}_1^{-1} \\ \hat{\theta}_1^{-1} & \hat{\theta}_2/\hat{\theta}_1^2 \end{pmatrix} \tag{23}$$

with each element divided by $n\hat{\theta}_1^{-2}\{\hat{\theta}_2 d^2\ln(\hat{\theta}_2-1)!/d\theta_2^2-1\}$.

These results might be of value if one wished to investigate whether observed incidents were occurring in a Poisson process (§2.3). It might be reasonable to suppose the intervals between successive incidents to be independent (for example if the incidents were failures of a component which was immediately replaced by a new one when it failed, §4.4), with a distribution of Γ-type. The Poisson process is the case $\theta_2 = 1$ (theorem 2.3.2), so one could perform an approximate significance test of the null hypothesis that $\theta_2 = 1$ by remarking that the posterior distribution of θ_2 is approximately normal with mean $\hat{\theta}_2$ and variance $\{n(d^2\ln(\hat{\theta}_2-1)!/d\theta_2^2-\hat{\theta}_2^{-1})\}^{-1} = \sigma_n^2$, say. The approximation to $d^2\ln(\theta_2-1)!/d\theta_2^2$ of $(1/\theta_2)+(1/2\theta_2^2)$, obtained from the above approximation to the first derivative by another differentiation of Stirling's formula, shows that σ_n^2 is approximately $2\hat{\theta}_2^2/n$. The result will therefore be significant at the 5 % level if $\hat{\theta}_2$ exceeds $1+2\sigma_n$. Notice that, in agreement with the general result, σ_n^2 is of the order n^{-1}.

Choice of parameter

In the method of maximum likelihood there is no distinction between the estimation of θ and the estimation of a function of θ, $\phi(\theta)$. We have the obvious relation that $\hat{\phi} = \phi(\hat{\theta})$. The

† This may be obtained by taking logarithms and differentiating both sides of Stirling's formula, equation 4.4.15.

variance of ϕ, $\{-\partial^2 L(\mathbf{x}|\hat{\phi})/\partial\phi^2\}^{-1}$, may also be related to the variance of θ in the following way. Write L for the log-likelihood in order to simplify the notation. Then

$$\frac{\partial L}{\partial \phi} = \frac{\partial L}{\partial \theta}\frac{d\theta}{d\phi} \quad \text{and} \quad \frac{\partial^2 L}{\partial \phi^2} = \frac{\partial^2 L}{\partial \theta^2}\left(\frac{d\theta}{d\phi}\right)^2 + \frac{\partial L}{\partial \theta}\frac{d^2\theta}{d\phi^2},$$

so that, since $\partial L/\partial\theta = 0$ at $\hat{\theta}$, the second equation gives

$$\mathscr{D}^2(\phi|\mathbf{x}) = \left(\frac{d\phi}{d\theta}\right)^2 \mathscr{D}^2(\theta|\mathbf{x}), \tag{24}$$

where the derivative is evaluated at the maximum likelihood value. These results may also be obtained from theorem 3.4.1 since the variances are small, being of order n^{-1}. Thus in changing from θ to ϕ the means and variances change in the usual approximate way. Since the method does not distinguish between θ and ϕ, both parameters have an approximately normal distribution. At first glance this appears incorrect since if θ is normal then, in general, ϕ will not be normal. But it must be remembered that these results are only limiting ones as $n \to \infty$ and both the distributions of θ and ϕ can, and indeed do, tend to normality. What will distinguish θ from ϕ in this respect will be the rapidity of approach to normality: ϕ may be normal to an adequate approximation for smaller n than is the case with θ. It often pays, therefore, to consider whether some transform of θ is likely to be more nearly normal than θ itself and, if so, to work in terms of it. Of course, there is some transform of θ which is exactly normal since any (sufficiently respectable) distribution can be transformed into any other (compare the argument used in §3.5 for obtaining random samples from any distribution), but this transform will involve the exact posterior distribution and since the point of the approximation is to provide a simple result this is not useful. What is useful is to take a simple transformation which results in a more nearly normal distribution than is obtained with the untransformed parameter. No general results seem available here, but an example is provided by the variance θ_2 of a normal distribution just discussed. The distribution of θ_2, as we saw in §5.3, has a longer tail to the right (large θ_2) than to the left (small θ_2). This suggests considering $\ln\theta_2$ which might remove the effect of the long tail.

Detailed calculations show that the posterior distribution of $\ln \theta_2$ is more nearly normal than that of θ_2, though even better transformations are available. The approximate mean and variance of the distribution of $\ln \theta_2$ may be found either by maximum likelihood directly, equation (9), or from the results for θ_2, equation (14), combined with equation (24). Other examples will occur in later sections.

Distribution of the maximum likelihood estimate

We saw in §5.1 that when making inferences that involved using the mean of a normal distribution there were two distinct results that could be confused (statements (*a*) and (*b*) of that section). A similar situation obtains here because of the approximate normality of the posterior distribution. The two statements are:

(*a*) the maximum likelihood estimate, $\hat{\theta}$, is approximately normally distributed about θ_0 with variance the inverse of

$$I_n(\theta_0) = \mathscr{E}_0\{-\partial^2 L(\mathbf{x} \mid \theta_0)/\partial\theta^2\}; \tag{25}$$

(*b*) the parameter θ is approximately normally distributed about $\hat{\theta}$ with variance σ_n^2.

(θ_0 is the true, unknown, fixed value of θ as explained before, equation (3).) Statement (*b*) is the result of theorem 1, statement (*a*) can be proved in essentially the same way as that theorem was proved. Statement (*a*) is a result, in frequency probability, about a statistic, $\hat{\theta}$: (*b*) is a result, in degrees of belief, about a parameter θ. In practice (*a*) and (*b*) can be confused, as explained in §5.1, without harm. Actually (*a*) is rarely used in the form given, since θ_0 is unknown and yet occurs in the variance (equation (25)). Consequently, (25) is usually replaced by $I_n(\hat{\theta})$. This still differs from σ_n^{-2} because of the expectation† used in (25) but not in the expression for σ_n^{-2}. It is interesting to note that the use of the expectation makes no difference in random samples of fixed size from an exponential family: there $I_n(\hat{\theta}) = \sigma_n^{-2}$. (See equation (16) and the one immediately following.)

† Those who have read the relevant paragraph in §5.6 will notice that the use of the expectation violates the likelihood principle, and is, on this score, unsatisfactory.

Exceptional cases

It is necessary to say a few words about the question of rigour in the proofs of the theorems. A complete proof with all the necessary details is only possible when certain assumptions are made about the likelihood: for example, assumptions about the existence and continuity of derivatives and their expectations. These assumptions are not always satisfied and the theorem is not always true; the most common difficulty arises when the range of possible values of x_i depends on θ. The difficulty is the same as that encountered in discussing sufficiency in §5.5 and the same example as was used there suffices to demonstrate the point here. If

$$f(x_i|\theta) = \theta^{-1} \quad (0 \leqslant x_i \leqslant \theta),$$

and is otherwise zero; then the likelihood is θ^{-n} provided $\theta \geqslant \max_i x_i = X$, say, and is otherwise zero. Hence the posterior density is approximately proportional to θ^{-n} and clearly this does not tend to normality as $n \to \infty$. Indeed, the maximum value is at $\theta = X$, so that $\hat{\theta} = X$, at the limit of the range of values of θ with non-zero probability. If the prior distribution of θ is uniform over the positive half of the real line, a simple evaluation of the constant shows that

$$\pi(\theta|\mathbf{x}) = (n-1)X^{n-1}\theta^{-n} \quad (\theta \geqslant X, n > 1),$$

with mean $(n-1)X/(n-2)$ and variance $(n-1)X^2/(n-3)(n-2)^2$ if $n > 3$. As $n \to \infty$ the variance is approximately X^2/n^2, whereas the theorem, if applicable here, would give a result that is of order n^{-1}, not n^{-2}. The estimation of θ is much more accurate than in the cases covered by the theorem. The practical reason for the great accuracy is essentially that any observation, x_i, immediately implies that $\theta < x_i$ has zero posterior probability; since, if $\theta < x_i$, x_i has zero probability. This is a much stronger result than can usually be obtained. The mathematical reason is the discontinuity of the density and its differential with respect to θ at the upper extreme of the range of x. Difficulties can also occur with estimates having smaller accuracy than suggested by the theorem, when dealing with several parameters. Examples of this phenomenon will not arise in this book.

7.2. Random sequences of trials

In this section we consider the simple probability situation of a random sequence of trials with constant probability, θ, of success, and discuss the inferences about θ that can be made. If a random variable, x, has a density

$$\frac{(a+b+1)!}{a!\,b!}x^a(1-x)^b, \tag{1}$$

for $0 \leqslant x \leqslant 1$ and $a, b > -1$, it is said to have a *Beta-distribution* with parameters a and b. We write it $B_0(a, b)$, the suffix distinguishing it from the binomial distribution $B(n, p)$ with index n and parameter p (§2.1).

Theorem 1. *If, with a random sequence of n trials with constant probability, θ, of success, the prior distribution of θ is $B_0(a, b)$, then the posterior distribution of θ is $B_0(a+r, b+n-r)$ where r is the number of successes.*

The proof is straightforward:

$$\pi(\theta) \propto \theta^a(1-\theta)^b, \quad \text{from (1)}, \tag{2}$$

the likelihood is $\quad p(\mathbf{x}\,|\,\theta) = \theta^r(1-\theta)^{n-r},$

so that

$$\pi(\theta\,|\,\mathbf{x}) \propto \theta^{a+r}(1-\theta)^{b+n-r}, \tag{3}$$

proving the result. (Here $\mathbf{x}$ denotes the results of the sequence of trials. Since r is sufficient $\pi(\theta\,|\,\mathbf{x})$ may be written $\pi(\theta\,|\,r)$.)

Corollary 1. *Under the conditions of the theorem the posterior distribution of*

$$F = \left(\frac{b+n-r+1}{a+r+1}\right)\left(\frac{\theta}{1-\theta}\right) \tag{4}$$

is $F[2(a+r+1), 2(b+n-r+1)]$.

From (4) $\quad dF/d\theta \propto (1-\theta)^{-2}$

and also $\quad \theta = a'F/(b'+a'F),$

where $a' = a+r+1$, $b' = b+n-r+1$. Substitution in (3), not forgetting the derivative (theorem 3.5.1), gives

$$\begin{aligned}\pi(F\,|\,\mathbf{x}) &\propto F^{a'-1}/(b'+a'F)^{a'+b'} \\ &\propto F^{a'-1}/(2b'+2a'F)^{a'+b'}.\end{aligned}$$

A comparison with the density of the F-distribution (equation 6.2.1) establishes the result.

Corollary 2. *The posterior distribution of* $\ln\{\theta/(1-\theta)\}$ *has, for large r and* $(n-r)$, *approximate mean*

$$\ln\{r/(n-r)\} + \frac{a+\frac{1}{2}}{r} - \frac{b+\frac{1}{2}}{n-r}, \tag{5}$$

and variance

$$\frac{1}{r} + \frac{1}{(n-r)}. \tag{6}$$

In §6.2 the mean and variance of the F-distribution were obtained. From those results and corollary 1, the mean of F, given by equation (4), is

$$\frac{b+n-r+1}{b+n-r} = 1 + \frac{1}{b+n-r} \doteq 1 + \frac{1}{n-r}, \tag{7}$$

and the variance is approximately $n/r(n-r)$. Since the variance of F is small when r and $(n-r)$ are large, the mean and variance of $u = \ln F$ can be found to the same order of approximation as in (7) by use of theorem 3.4.1.

$$\begin{aligned} \mathscr{E}(u) &\doteq \ln\left(1+\frac{1}{n-r}\right) - \frac{n}{2r(n-r)}\left(1+\frac{1}{n-r}\right)^{-2} \\ &\doteq \frac{1}{n-r} - \frac{n}{2r(n-r)} = \frac{1}{2}\left(\frac{1}{n-r} - \frac{1}{r}\right), \end{aligned} \tag{8}$$

and
$$\mathscr{D}^2(u) \doteq \left(1+\frac{1}{n-r}\right)^{-2} \frac{n}{r(n-r)} \doteq \frac{1}{r} + \frac{1}{n-r}. \tag{9}$$

Since $u = \ln\{(b+n-r+1)/(a+r+1)\} + \ln\{\theta/(1-\theta)\}$ the expression, (6), for the variance immediately follows. That for the mean, (5), is easily obtained since

$$\begin{aligned} \mathscr{E}\{\ln[\theta/(1-\theta)]\} &\doteq \mathscr{E}(u) + \ln\{(a+r+1)/(b+n-r+1)\} \\ &\doteq \frac{1}{2}\left(\frac{1}{n-r} - \frac{1}{r}\right) + \ln\frac{r}{n-r} + \frac{a+1}{r} - \frac{b+1}{n-r} \\ &= \ln\frac{r}{n-r} + \frac{a+\frac{1}{2}}{r} - \frac{b+\frac{1}{2}}{n-r}. \end{aligned}$$

Beta-distribution

The family of Beta-distributions is convenient for random variables which lie between 0 and 1. An important property of the family is that $y = 1-x$ also has a Beta-distribution, but with a and b interchanged, that is $B_0(b, a)$. This is obvious from the density. This makes it particularly useful if x is the probability of success, for y is then the probability of failure and has a distribution of the same class. That the integral of (1) is unity follows from the Beta-integral (equation 5.4.7). The same integral shows that the mean is $(a+1)/(a+b+2)$ and the variance (from equation 2.4.5) is

$$\frac{(a+1)(a+2)}{(a+b+2)(a+b+3)} - \frac{(a+1)^2}{(a+b+2)^2} = \frac{(a+1)(b+1)}{(a+b+2)^2(a+b+3)}. \quad (10)$$

If a and b are positive the density increases from zero at $x = 0$ to a maximum at $x = a/(a+b)$ and falls again to zero at $x = 1$. If $-1 < a < 0$ or $-1 < b < 0$ the density increases without limit as x approaches 0 or 1, respectively. The case of small a and large b corresponds to a distribution in which x is usually nearer 0 than 1, because the mode and mean are then both near zero. The relation between the Beta- and F-distributions has already been seen in deriving equation 6.2.5.

Change in Beta-distribution with sampling

Apart from the reasons just given, the family of Beta-distributions is the natural one to consider as prior distributions for the probability θ of success. We saw in §5.5 that any member of the exponential family had associated with it a family of conjugate prior distributions (cf. equation 5.5.18). The likelihood here belongs to the exponential family (equation 5.5.12) and the conjugate family is easily seen to be the Beta-family of distributions. Theorem 1 merely expresses the result discussed after equation 5.5.18 that the posterior distribution also belongs to the same family but with different parameters. In fact a and b change into $a+r$ and $b+(n-r)$. It is often convenient to represent the situation on a diagram (fig. 7.2.1). The axes are of a and b. The prior distribution corresponds to the point (a, b) with, say, integer co-ordinates. Every time a success occurs 1 is added to the value of a; every time a failure occurs 1 is added to the value of b. In this way the random sequence of trials can be represented by a path in the diagram. The figure shows the path corresponding to a sequence of trials where the prior distribution has $a = 1, b = 2$ and beginning *SFFS* (*S*—success: *F*—failure). Notice

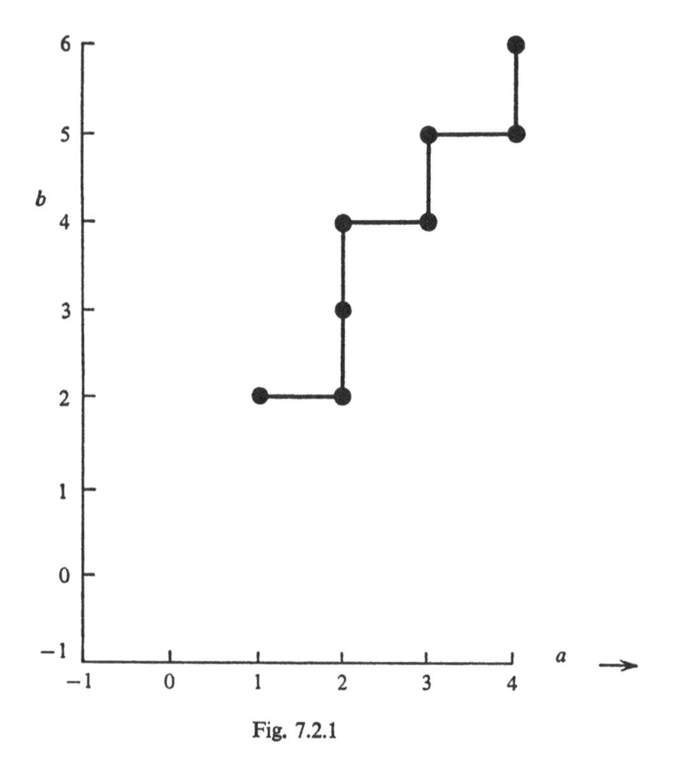

Fig. 7.2.1

that at each stage $(a+b)$ always changes by one, and hence after n trials has increased by n, so that the posterior variance of θ (equation (10)) is small for large n, corresponding to increased knowledge about the value of θ.

Vague prior knowledge

In any application it is necessary to choose values of a and b for the prior distribution. As discussed in connexion with the normal distribution, we usually require a prior distribution corresponding to a fair amount of ignorance about the parameter. The situation where the prior knowledge of θ is not vague is discussed below. Since any observation always increases either a or b it corresponds to the greatest possible ignorance to take a and b as small as possible. For the prior density to have total integral 1 it is necessary and sufficient that both a and b exceed -1. Therefore the usual prior distribution to take is that with $a = b = -1$. It is true that the resulting density can only be defined as a conditional density, but we have already used this type of prior density in supposing the mean or log-variance of the normal distribution to be uniformly distributed. In fact the density just suggested is proportional to $\{\theta(1-\theta)\}^{-1}$: this is equivalent to saying that $\phi = \ln\{\theta/(1-\theta)\}$ is uniformly distributed over the whole real line, since $d\phi/d\theta = \{\theta(1-\theta)\}^{-1}$. Hence the convenient parameter to consider is ϕ, the logarithm of the odds in favour of the event of success in a trial. Notice that it is this parameter which occurs in the natural representation of the likelihood as a member of the exponential family (equation 5.5.12). With $a = b = -1$ the posterior distribution is $B_0(r-1, n-r-1)$, the two parameters being one less than the number of successes and failures respectively. Bayes himself suggested supposing θ to be uniformly distributed over the interval (0, 1) and this distribution has often been suggested by later writers; a common argument in its favour being that if we are ignorant of θ then any value of θ is as likely as any other. Unfortunately this argument also applies to ϕ. The practical difference between these two apparently very different densities is, however, slight because two observations, one a success and one a failure, are sufficient to change one (ϕ uniform) into the other (θ uniform). This is clearly seen from fig. 7.2.1.† So in arguing in favour of one against the other we are only arguing over the equivalent of two observations, usually a small fraction of the total number, n, of observations on which the posterior distribution is based. If the uniform distribution of ϕ is used then the posterior distribution can only be defined as a conditional density until both a success and a failure have been observed. It is not unreasonable to say that reliable inferences

† Notice that in the figure the axes pass through $a = b = -1$.

cannot be made about the ratio of successes to failures until an example of each has occurred.

Use of F-tables

In order to reduce the number of tables of distribution functions that is needed, it is convenient in practice to use the relation between the Beta-distribution and the F-distribution and use tables of the latter (compare the discussion at the end of §5.4). This is the purpose of corollary 1. With $a = b = -1$ the quantity having the F-distribution is

$$F = \left(\frac{n-r}{r}\right)\left(\frac{\theta}{1-\theta}\right) = \left(\frac{\theta}{1-\theta}\right)\Big/\left(\frac{r}{n-r}\right), \tag{11}$$

the ratio of the odds in favour of success to the empirical odds in favour of success; the empirical odds being the ratio of observed number of successes to number of failures. The degrees of freedom are $2r$ and $2(n-r)$; twice the numbers of successes and failures respectively. Confidence intervals for F are given in the usual way (§6.2) with the F-distribution. A confidence interval for F of coefficient β is given by

$$\underline{F} = \underline{F}_{\frac{1}{2}\alpha}[2r, 2(n-r)] < F < \bar{F}_{\frac{1}{2}\alpha}[2r, 2(n-r)] = \bar{F},$$

with $\alpha = 1-\beta$. In terms of θ this is equivalent to

$$\frac{r\underline{F}/(n-r)}{1+r\underline{F}/(n-r)} < \theta < \frac{r\bar{F}/(n-r)}{1+r\bar{F}/(n-r)}. \tag{12}$$

Since $\underline{F}$ is not directly tabulated it is more convenient to use equation 6.2.7 and to write (12) as

$$\frac{1}{1+(n-r)\bar{F}'/r} < \theta < \frac{1}{1+(n-r)/r\bar{F}}, \tag{13}$$

where

$$\bar{F}' = \bar{F}_{\frac{1}{2}\alpha}[2(n-r), 2r] = \{\underline{F}_{\frac{1}{2}\alpha}[2r, 2(n-r)]\}^{-1} = \underline{F}^{-1}.$$

The limits given by (13) are wider than people usually expect and a numerical example may prove instructive. In 17 trials, 5 successes (and 12 failures) were observed. The degrees of freedom for F are therefore 10 and 24, so that $\bar{F} = 2{\cdot}64$ with $\alpha = 0{\cdot}05$, and $\bar{F}' = 3{\cdot}37$, from the tables. Hence $(n-r)/r\bar{F} = 0{\cdot}91$

and $(n-r)\bar{F}'/r = 8{\cdot}09$, so that the 95 % confidence interval for θ is, from (13), (0·11, 0·52). The mean value of the posterior distribution is r/n, namely 0·29. Notice that the interval is not symmetric about the mean, the upper value differing from it by 0·23, but the lower only by 0·18. The interval is only symmetric when $r = n-r$, the mean then being $\frac{1}{2}$.

Approximate methods

The Beta- and F-distributions are a little awkward to handle, especially when inferences are to be made on the basis of several samples, and it is desirable to develop suitable approximations. The maximum likelihood method of §7.1 provides one such approximation, and we saw in that section that the posterior distribution of θ was approximately $N(r/n, r(n-r)/n^3)$. The mean and variance agree, in large samples, with the exact results, from equation (10) above, of $(a+r+1)/(a+b+n+2)$ and

$$(a+r+1)\,(b+n-r+1)/(a+b+n+2)^2\,(a+b+n+3),$$

but the approach to normality is slow unless r is about $\frac{1}{2}n$. Thus in the numerical example the posterior distribution is, according to the approximation, and with $a = b = -1$, $N(0{\cdot}29, 0{\cdot}0122)$, so that the 95 % confidence limits are (0·07, 0·51), not in too good agreement with the exact result, especially at the lower limit. Notice that this interval, unlike the exact one, is necessarily symmetrical about the mean. As explained in §7.1 some parametric functions will have a more nearly normal distribution than others and so we look for a better approximation using a parameter different from θ. The F-distribution is no more nearly normal than is the Beta-distribution, but since (§6.2), for all except very small values of the degrees of freedom, the density increases rapidly from zero at $F = 0$ to a maximum near $F = 1$ and then decreases slowly to zero as $F \to \infty$, it looks as though $\ln F$ will be much more nearly normal. The logarithm will have a stronger effect on the sharp left-hand tail, between $F = 0$ and $F = 1$, than on the right-hand tail, above $F = 1$. This transformation was first suggested by Fisher† and $\ln F$ may be shown to have a normal distribution to a much better approxi-

† Fisher took $\frac{1}{2}\ln F$, but the $\frac{1}{2}$ is more conveniently omitted in this context.

mation than does F itself. The transformation also commends itself in the present situation because $\ln\{\theta/(1-\theta)\}$, which differs from $\ln F$ by a constant, is the natural parameter to consider, both in the representation of the original distribution in exponential form (equation 5.5.12) and as the parameter having a uniform prior distribution.

Corollary 2 gives the mean and variance of the posterior distribution of $\ln\{\theta/(1-\theta)\}$, as far as the terms in $1/r$ and $1/(n-r)$, which, in conjunction with the assumption of normality, are sufficient to define the posterior distribution. The logarithm is the dominant term in the mean, the next term is of smaller order and may be ignored for large r and $(n-r)$: in any case it is of smaller order than the standard deviation. Notice that the prior distribution only enters into this second term, and not into the dominant terms of mean and variance. This is another example of the fact that for large samples the prior distribution can be ignored. Notice that, to the same order (5) may be written

$$\ln\left\{\frac{r+a+\frac{1}{2}}{n-r+b+\frac{1}{2}}\right\}, \tag{14}$$

which is more convenient for calculation. Consider this approximation in the numerical case just cited with $n-r = 12$ and $r = 5$, and $a = b = -1$. The logarithm of the odds is approximately normal with mean $-0{\cdot}939$ and standard deviation $0{\cdot}532$, and hence 95 % confidence limits are $-1{\cdot}982$ and $+0{\cdot}104$, giving 95 % confidence limits for θ of $(0{\cdot}12, 0{\cdot}53)$, agreeing very well with the exact values of $(0{\cdot}11, 0{\cdot}52)$ obtained above using the F-distribution.

Approximation to the likelihood

Another method of obtaining a useful approximate result is to approximate to the likelihood instead of to the posterior distribution directly. In particular, if the likelihood can be changed into an approximately normal one, then the inferences appropriate to a normal distribution can be made. Now one reason why inferences for the binomial distribution are not as simple as for the normal is that the parameter of interest affects both mean and variance. It is therefore sometimes convenient to find a transformation which will remove the parameter from the variance. We saw how this could be done in certain cases in §3.4. In particular for the binomial distribution (example 3 of that section), if r is $B(n, p)$ then $\sin^{-1}\sqrt{(r/n)}$ has approximate variance $1/4n$, not dependent on p, about a mean of $\sin^{-1}\sqrt{p}$. Furthermore, it may be shown that the distribution of the inverse-sine is more nearly normal than is that of r itself. Consequently, it is now possible to make inferences using the approximate normality of $\sin^{-1}\sqrt{(r/n)}$. The transformation is tabulated in Lindley and Miller (1961) and in the numerical example above, $\sin^{-1}\sqrt{(5/17)} = 0{\cdot}573$ (in radians) with a standard deviation of $1/\sqrt{68} = 0{\cdot}121$. Hence 95 % confidence limits for the inverse sine are approximately $0{\cdot}573 \pm 1{\cdot}96 \times 0{\cdot}121$, that is $(0{\cdot}336, 0{\cdot}810)$. Applying the inverse transformation the limits for θ are $(0{\cdot}11, 0{\cdot}52)$ in agreement with the exact values.

The inverse-sine transformation is particularly valuable because it enables all the inference methods appropriate to the normal distribution to be used with the binomial. For example, suppose we have k independent sets of random sequences each of n trials, with constant probabilities of success θ_i, and giving r_i successes $(i = 1, 2, \ldots, k)$. Then $x_i = \sin^{-1}\sqrt{(r_i/n)}$ will be approximately normal with means $\sin^{-1}\sqrt{\theta_i}$ and common variance $1/4n$. Consequently the hypothesis that the θ_i's are all equal can be tested by the analysis of variance method of §6.4, using theorem 2 of that section, since the variance is known. The test statistic will be $4n\Sigma(x_i - x_.)^2$, which may be approximately compared with a χ^2-distribution with $(k-1)$ degrees of freedom.

A disadvantage of the inverse-sine transformation is that the transform is not a useful parameter to consider. Thus, if we carry out an analysis of variance as described in §8.5 on the inverse-sines and learn that an interaction is probably zero, that result is not usually of much value to the experimenter to whom the inverse-sine is not a physically meaningful quantity.

The reader who has studied the final part of §5.6 may object to the above argument since it uses the distribution of r to derive the transformation and not just the likelihood. This may be avoided, however, by replacing the argument of example 3 of §3.4, by the comparable argument of §7.1 in connexion with maximum likelihood. Equation (24) of that section replaces equation 3.4.2 and the purpose of the transformation is to find a parameter about which the information is constant. This appears to have been Fisher's intention in producing the transformation originally.

Several samples

The true worth of any of these approximations is only appreciated when considering several samples. Suppose, for example, that we have two independent random sequences of n_i trials, with constant probability, θ_i, of success, yielding r_i successes ($i = 1, 2$). Then we may wish to inquire about differences between θ_1 and θ_2 and, in particular, to make a significance test of the hypothesis that $\theta_1 = \theta_2$. This hypothesis may be conveniently rephrased in terms of the logarithms of odds by investigating $\phi_1 - \phi_2 = \ln\{\theta_1(1-\theta_2)/\theta_2(1-\theta_1)\}$ and, in particular, testing whether $\phi_1 - \phi_2 = 0$. The posterior distributions of ϕ_1 and ϕ_2 will be independent and both approximately normal, so that the posterior distribution of $\phi_1 - \phi_2$ is approximately normal with mean

$$\ln\left(\frac{r_1}{n_1-r_1}\right) - \ln\left(\frac{r_2}{n_2-r_2}\right) = \ln\left\{\frac{r_1(n_2-r_2)}{(n_1-r_1)r_2}\right\} \tag{15}$$

and variance
$$\frac{1}{r_1}+\frac{1}{n_1-r_1}+\frac{1}{r_2}+\frac{1}{n_2-r_2}. \tag{16}$$

(In (15) the second term of (5) has been ignored, or equivalently we have supposed $a = b = -\frac{1}{2}$.) Hence the null hypothesis that $\theta_1 = \theta_2$, or equivalently $\phi_1 = \phi_2$, will be judged significant at the 5% level if (15) exceeds 1·96 times the standard deviation obtained from (16): that is, if

$$\frac{\left[\ln\left\{\frac{r_1(n_2-r_2)}{(n_1-r_1)r_2}\right\}\right]^2}{\left(\frac{1}{r_1}+\frac{1}{n_1-r_1}+\frac{1}{r_2}+\frac{1}{n_2-r_2}\right)} \geqslant (1{\cdot}96)^2 = 3{\cdot}84. \tag{17}$$

This method may be extended to cover numerous other situations but we shall not pursue it any further here since another method, using a different approximation, is available and is more commonly used. This will be studied in §7.6.

Inference with appreciable prior knowledge

We now consider the analysis suitable when the prior knowledge of θ is comparable in its importance to the knowledge to be obtained from the likelihood function. The situation we have in mind is exemplified by a problem concerning a defect in a mass-produced model of a car. The parameter in question is the proportion of cars of this particular model exhibiting the defect by the time they are one year old. After the model had been in production about a year the manufacturer realized, from complaints received, that this defect was of reasonably common occurrence. He naturally wished to find out how common it was and proposed to carry out a survey. For our purposes it will suffice to suppose that the survey consists in taking a random sample of size n of one-year-old cars and observing the number, r, with the defect. The likelihood will then be that considered in this section. However, there is already, before the sample is taken, some knowledge of θ. It is difficult to be precise about this because some customers will not notice the defect, others will not complain even if they do notice it, and not all complaints will reach the group in the manufacturing company conducting the investigation. Nevertheless, some information is available and should be incorporated into the analysis.

The question is: how should the prior knowledge be obtained and expressed? It will be most convenient if it is expressed in terms of the Beta-distribution, should this be possible. In order to do this it is necessary to obtain the values of a and b: when this has been done theorem 1 may be applied. In the example the company felt that the defect most probably occurred in about 15 % of their cars, and that it must certainly occur in about 5 % in order to explain the complaints already received. They also felt it to be most unlikely that more than 1 in 3 of the cars would exhibit the defect. This information can be turned into a probability density by supposing that the upper and lower limits quoted correspond roughly to the upper and lower 5 % points of a Beta-distribution. From tables of the percentage points of this distribution (for example those in Pearson and Hartley (1958), Table 16) the values $a = 2$, $b = 14$ provide upper and lower 5% points of θ at 0·050 and 0·326 respectively.† The mean value of θ is then $(a+1)/(a+b+2) = 3/18 = 0{\cdot}167$ and the mode (the most likely value) is $a/(a+b) = 2/16 = 0{\cdot}125$. These three figures agree tolerably well with those suggested above and the distribution $B_0(2, 14)$ might therefore be used to express the prior knowledge. Notice that the extent of the prior knowledge assumed is equivalent to observing the defect in about 3 cars out of a randomly chosen 18 when initially vague about θ: that is, with $a = b = -1$ (cf. §§5.2, 5.5). This comment suggests that if the survey is to consider a random sample of the order of a hundred or more cars then the prior knowledge is small compared with that to be obtained from the survey. Thus, if 100 cars are inspected and 32 have the defect the posterior distribution is $B_0(34, 82)$ which is not very different from $B_0(31, 67)$ which would have been obtained without the prior knowledge. Approximations in terms of the log-odds can, of course, still be used.

Methods of this type are always available whenever the likelihood belongs to the exponential family. Consequently when dealing with this family, it is no real restriction to confine attention to the prior distribution corresponding to substantial ignorance of θ provided, in any special case, the actual prior knowledge is capable of being represented by a distribution of the conjugate family. (Cf. the discussion in §5.5.)

Binomial distribution

The methods of this section are often said to be appropriate to the binomial distribution. This is certainly true, for if a fixed number, n, of trials is observed to result in r successes, the distribution of r is $B(n, p)$, where p is the probability of success, and the methods of this section can be used. But it is not advisable to associate the methods only with the binomial distribution; that is, only with fixed n. The methods do not assume n fixed. They are valid, for example, when r is fixed and the experimenter continues sampling until some random number, n, of trials yield r successes: or again, if n was the random

† Notice that the tables referred to use $a-1$ and $b-1$ for what we here call a and b.

number that the experimenter had time to investigate (§5.5). This is an example of the use of the likelihood principle (§5.6) and readers who have read the latter part of that section will appreciate that this is another example of the irrelevance of the sample space to the inference being made.

7.3. The Poisson distribution

In this section we discuss exact methods of making inferences with the Poisson distribution before proceeding to general approximations in §7.4.

Theorem 1. *If $(r_1, r_2, \ldots, r_n)$ is a random sample of size n from a Poisson distribution $P(\theta)$, and if the prior distribution of θ is $\Gamma(s, m)$; then the posterior distribution of θ is $\Gamma(s+n\bar{r}, m+n)$, where $\bar{r} = n^{-1}\sum_{i=1}^{n} r_i$.*

From the definition of the density of the Γ-distribution (equation 2.3.7) the prior density is

$$\pi(\theta) \propto e^{-m\theta}\theta^{s-1} \tag{1}$$

for $\theta \geqslant 0$, and otherwise zero. The likelihood is

$$p(\mathbf{x}|\theta) = e^{-n\theta}\theta^{\sum_{i=1}^{n} r_i} \Big/ \prod_{i=1}^{n}(r_i!) \propto e^{-n\theta}\theta^{n\bar{r}}; \tag{2}$$

where $\mathbf{x} = (r_1, r_2, \ldots, r_n)$. This shows that $\bar{r}$ is sufficient. Hence, multiplying (1) by (2), we have

$$\pi(\theta|\mathbf{x}) \propto e^{-(m+n)\theta}\theta^{s+n\bar{r}-1}, \tag{3}$$

and a further comparison with the density of the Γ-distribution establishes the result.

Corollary. *The posterior density of $2(m+n)\theta$ is χ^2 with $2(s+n\bar{r})$ degrees of freedom.*

This follows immediately from the relationship between the Γ- and χ^2-distributions (§5.3) that if y is $\Gamma(n, \lambda)$ then $2\lambda y$ is χ^2 with $2n$ degrees of freedom.

Theorem 2. *If two independent random samples of sizes n_1 and n_2 are available from Poisson distributions $P(\theta_1)$ and $P(\theta_2)$, and if the* *prior distributions of* θ_i *are independent and* $\Gamma(s_i, m_i)$ $(i = 1, 2)$*; then the posterior distribution of*

$$(\theta_1/\theta_2)\,\{(m_1+n_1)\,(s_2+n_2\bar{r}_2)/(m_2+n_2)\,(s_1+n_1\bar{r}_1)\}$$

is F with $[2(s_1+n_1\bar{r}_1), 2(s_2+n_2\bar{r}_2)]$ *degrees of freedom, where* $\bar{r}_i$ *is the mean of the ith sample.*

From the corollary the posterior densities of $2(m_i+n_i)\theta_i$ are χ^2 on $2(s_i+n_i\bar{r}_i)$ degrees of freedom $(i = 1, 2)$; and because of the independence, both of the prior distributions and the samples, they will be independent. Now in §6.2 we remarked that if χ_i^2 were independent and had χ^2-distributions with ν_i degrees of freedom $(i = 1, 2)$ then $(\chi_1^2/\nu_1)/(\chi_2^2/\nu_2)$ had an F-distribution with (ν_1, ν_2) degrees of freedom. (The proof was given in the course of proving theorem 6.2.1.) Applying this result here we immediately have that

$$\frac{2(m_1+n_1)\,\theta_1/2(s_1+n_1\bar{r}_1)}{2(m_2+n_2)\,\theta_2/2(s_2+n_2\bar{r}_2)} \tag{4}$$

has the F-distribution with $[2(s_1+n_1\bar{r}_1), 2(s_2+n_2\bar{r}_2)]$ degrees of freedom, as required.

Exact methods

The Poisson distribution is a member of the exponential family (§5.5), as is immediately seen by writing the Poisson density in the form

$$\left(\frac{1}{r_i!}\right) e^{-\theta}\, e^{r_i \ln \theta} \tag{5}$$

and comparing it with equation 5.5.5 with the r of that equation equal to 1. The conjugate prior distribution which fits naturally with it, in the way described in §5.5, is easily seen to be the Γ-distribution with density (of θ) proportional to $e^{-m\theta}\, e^{(s-1)\ln \theta}$. Theorem 1 expresses the known general form of the change in this distribution (§5.5) with observations. The parameter m changes deterministically by the addition of 1 for each sample taken from the Poisson distribution: the parameter s changes randomly by the addition of the sample value, r_i. The known connexion between the Γ- and χ^2-distributions enables tables of the latter (depending on only one parameter, the degrees of freedom) to be used for inferences.

Vague prior knowledge

Since both parameters of the Γ-distribution increase with any observation, the greatest possible ignorance is reflected in a prior distribution with these parameters as small as possible. For convergence of the Γ-distribution it is necessary that both parameters be positive (§2.3); hence the usual prior distribution to take to represent considerable ignorance about θ is that with both parameters zero. The prior density, (1), is then proportional to θ^{-1}. Consequently, in order to represent vague knowledge, we assume $\ln\theta$ to have a uniform prior distribution over the whole real line. This also agrees with the representation of the Poisson distribution as a member of the exponential family, equation (5), where the natural parameter is $\phi = \ln\theta$. With this prior distribution the posterior density of $2n\theta$ is χ^2 with $2n\bar{r}$ degrees of freedom. Notice that, since the sum of independent Poisson variables is itself a Poisson variable (§3.5), the sufficient statistic $\sum_{i=1}^{n} r_i$ has a Poisson distribution with parameter $n\theta$: so that inferences from a sample of size n from $P(\theta)$ are equivalent to inferences from a single sample from $P(n\theta)$.

Two samples

Theorem 2 gives an exact method of making inferences about the ratio of two Poisson means. The ratio is again the natural quantity to consider (rather than, say, the difference) because it is a function of the difference of the logarithms, which are, as we have just seen, the natural parameters for each distribution separately. Using the prior distribution of ignorance with $s_i = m_i = 0$ the posterior distribution of $(\theta_1/\bar{r}_1)/(\theta_2/\bar{r}_2)$ is F with $(2n_1\bar{r}_1, 2n_2\bar{r}_2)$ degrees of freedom.

Consider a numerical illustration. Suppose two independent Poisson processes had been observed for $\frac{1}{2}$ and 1 hour respectively, and gave 12 and 30 incidents respectively over the periods. (Notice that since Σr_i is sufficient no additional information would be obtained by considering the numbers of incidents in, say, separate 5-minute periods.) Then 12 is the value of a Poisson variable with mean $\frac{1}{2}\theta_1$ and 30 is the value of a

Poisson variable with mean θ_2, where θ_1 and θ_2 are the rates of occurrence per hour of incidents in the two processes. It follows that the posterior distribution of $(\frac{1}{2}\theta_1/12)/(\theta_2/30)$ is F with 24 and 60 degrees of freedom ($n_1 = n_2 = 1$). The upper and lower $2\frac{1}{2}$ % points of the F-distribution with those degrees of freedom are 1·88 and 0·48 respectively, so that a 95 % confidence interval for θ_1/θ_2 is

$$(\tfrac{24}{30})\, 0{\cdot}48 < \theta_1/\theta_2 < (\tfrac{24}{30})\, 1{\cdot}88,$$

that is

$$0{\cdot}38 < \theta_1/\theta_2 < 1{\cdot}50.$$

Approximations

The logarithmic transformation of F can be used, if the F-tables are not available, with the approximate normality and mean and variance obtained as in the previous section.

The logarithmic transformation can be useful when handling several Poisson distributions, though the method of § 7.4 (especially equation 7.4.16) is also available. We remarked in § 3.4 that if a random variable was $\Gamma(n, \lambda)$ then its logarithm had an approximate standard deviation $n^{-\frac{1}{2}}$, independent of λ. Also the logarithm is more nearly normally distributed than is the original Γ-variable. Consequently the use of $\ln \theta$ in the present case gives an approximately normal posterior distribution in the same way that the log-odds, in the binomial situation, had an approximately normal distribution (corollary 2 to theorem 7.2.1). Indeed, since the F-distribution is the ratio of two independent Γ-distributions (§ 6.2) the earlier result follows from this remark.

Another approximate method for the Poisson distribution is to transform the distribution to one more nearly normal by arranging that the variance (or the information) is approximately constant, whatever be the mean. In § 3.4, example 2, we saw that this could be done by taking the square root of the Poisson variable, with approximately constant variance of 1/4. As with the inverse-sine transformation for the binomial, the analysis of variance techniques are then available. For example, the equality of θ_1 and θ_2 in the above numerical example could be tested by this method: $\surd 12$ is approximately $N(\surd(\frac{1}{2}\theta_1), \frac{1}{4})$, or $\surd 24$ is approximately $N(\surd(\theta_1), \frac{1}{2})$; and $\surd 30$ is approximately

$N(\sqrt{(\theta_2)}, \frac{1}{4})$; so that by the methods of §6.1 the posterior distribution of $\sqrt{\theta_1}-\sqrt{\theta_2}$ is approximately $N(\sqrt{24}-\sqrt{30}, \frac{1}{2}+\frac{1}{4})$. This gives confidence limits for $\sqrt{\theta_1}-\sqrt{\theta_2}$ and a test of significance of the hypothesis $\theta_1 = \theta_2$ in the usual way. However, the square root of θ is not a useful parameter to consider so that other techniques are usually to be preferred.

7.4. Goodness-of-fit tests

In this section we discuss an important approximation, of wide application, to the posterior distribution when, in each of a random sequence of trials, one of a number of exclusive and exhaustive events of constant probability is observed to occur. The events will be denoted by $A_1, A_2, \ldots, A_k$ and their probabilities by $\theta_1, \theta_2, \ldots, \theta_k$, so that $\sum_{i=1}^{k} \theta_i = 1$. The case, $k = 2$, studied in §7.2, is a special case.

Theorem 1. *If, in a random sequence of n trials, the exclusive and exhaustive events A_i have constant probabilities of success θ_i and occur r_i times ($\Sigma r_i = n$); then the posterior distribution of*

$$\sum_{i=1}^{k} (r_i - n\theta_i)^2/n\theta_i \tag{1}$$

is, for large n, approximately χ^2 with $(k-1)$ degrees of freedom.

Suppose that the prior distribution of the θ's is uniform over the region $\theta_i \geqslant 0$, $\Sigma\theta_i = 1$. Then the posterior distribution of the θ's, $\pi(\boldsymbol{\theta}\,|\,\mathbf{r})$, where $\boldsymbol{\theta} = (\theta_1, \theta_2, \ldots, \theta_k)$ and $\mathbf{r} = (r_1, r_2, \ldots, r_k)$, is proportional to the likelihood; that is, to

$$\theta_1^{r_1}\,\theta_2^{r_2}\ldots\theta_k^{r_k}. \tag{2}$$

Hence
$$\ln \pi(\boldsymbol{\theta}\,|\,\mathbf{r}) = C + \sum_{i=1}^{k} r_i \ln \theta_i, \tag{3}$$

where C denotes a constant (that is, a quantity which does not depend on $\boldsymbol{\theta}$) and will not necessarily be the same constant throughout the argument. Now (3) only obtains if $\Sigma\theta_i = 1$, so that the value of $\boldsymbol{\theta}$ for which (3) is a maximum can be found by differentiating $\Sigma r_i \ln\theta_i - \lambda\Sigma\theta_i$ partially with respect to each θ_i and equating to zero, where λ is a Lagrange undetermined multiplier. The result is obviously that (3) has its maximum at $\theta_i = \hat{\theta}_i = r_i/n$ which, because of the uniform prior distribution, is also the maximum likelihood estimate of θ_i. Let us therefore write $\delta_i = \theta_i - r_i/n$, with $\Sigma\delta_i = 0$, and obtain, the Jacobian being unity,

$$\ln \pi(\boldsymbol{\delta} \mid \mathbf{r}) = C + \Sigma r_i \ln(r_i/n + \delta_i). \tag{4}$$

We saw in §7.1 (after equation 7.1.7) that the posterior distribution was only appreciable in an interval about the maximum likelihood estimate of width of order $n^{-\frac{1}{2}}$. Hence δ_i may be taken to be of this order and (4) may be expanded in a power series in the δ's, terms higher than the third degree being ignored. Then

$$\begin{aligned}\ln \pi(\boldsymbol{\delta} \mid \mathbf{r}) &= C + \Sigma r_i \ln(r_i/n) + \Sigma r_i \ln(1 + n\delta_i/r_i) \\ &= C + \Sigma r_i(n\delta_i/r_i) - \tfrac{1}{2}\Sigma r_i(n\delta_i/r_i)^2 \\ &\quad + \tfrac{1}{3}\Sigma r_i(n\delta_i/r_i)^3 \\ &= C - \tfrac{1}{2}\Sigma n^2\delta_i^2/r_i + \tfrac{1}{3}\Sigma n^3\delta_i^3/r_i^2,\end{aligned} \tag{5}$$

the terms in δ_i vanishing since $\Sigma\delta_i = 0$. Since δ_i is of order $n^{-\frac{1}{2}}$ and r_i is of order n, the first term after the constant is of order one and the next of order $n^{-\frac{1}{2}}$. For the moment, then, ignore the last term and write

$$\lambda_i = n\delta_i/\sqrt{(r_i)} \quad \text{with} \quad \Sigma\sqrt{(r_i)}\lambda_i = 0.$$

The Jacobian may be absorbed into the constant and approximately

$$\ln \pi(\boldsymbol{\lambda} \mid \mathbf{r}) = C - \tfrac{1}{2}\Sigma\lambda_i^2. \tag{6}$$

Hence the joint posterior density of the λ's is constant over spheres with centres at the origin, in the space of $(k-1)$ dimensions formed from the k-dimensional space of the λ's constrained by $\Sigma\sqrt{(r_i)}\lambda_i = 0$, a plane passing through the origin. Hence we may argue that confidence sets must be based on $\Sigma\lambda_i^2$, exactly as in the proof of theorem 6.4.2, which was there shown to have a χ^2-distribution with $(k-1)$ degrees of freedom, $(k-1)$ being the dimension of the spheres.

In terms of the θ's the result just established is that the posterior distribution of $\Sigma(r_i - n\theta_i)^2/r_i$ is approximately χ^2 with $(k-1)$ degrees of freedom. This differs from the result required in that r_i replaces $n\theta_i$ in the denominator. In order to show that this replacement does not affect the result we have to show

(*a*) that (5) may be written $C - \frac{1}{2}\Sigma n^2\delta_i^2/n\theta_i$, plus terms of the same order as the final term in (5),

(*b*) that the Jacobian of the transformation from θ_i (or equivalently δ_i) to

$$\mu_i = (n\theta_i - r_i)/(n\theta_i)^{\frac{1}{2}} \tag{7}$$

instead of to

$$\lambda_i = (n\theta_i - r_i)/r_i^{\frac{1}{2}}$$

introduces a term of higher order than those retained in (6). If (*a*) and (*b*) are established we shall be able to obtain $\ln \pi(\boldsymbol{\mu}\,|\,\mathbf{r}) = C - \frac{1}{2}\Sigma\mu_i^2$ instead of (6) and the χ^2 result will be as before.

Proof of (*a*). We have

$$\begin{aligned} -\tfrac{1}{2}\Sigma n^2\delta_i^2/r_i &= -\tfrac{1}{2}\Sigma \frac{n^2\delta_i^2}{n\theta_i}\left(1 + \frac{n\delta_i}{r_i}\right) \\ &= -\tfrac{1}{2}\Sigma\frac{n^2\delta_i^2}{n\theta_i} - \tfrac{1}{2}\Sigma\frac{n^3\delta_i^3}{n\theta_i r_i} \\ &= -\tfrac{1}{2}\Sigma\frac{n^2\delta_i^2}{n\theta_i} - \tfrac{1}{2}\Sigma\frac{n^3\delta_i^3}{r_i^2}, \end{aligned}$$

ignoring terms of order $n\delta_i^4$. Equation (5) can therefore be written

$$\ln \pi(\boldsymbol{\delta}\,|\,\mathbf{r}) = C - \tfrac{1}{2}\Sigma n^2\delta_i^2/n\theta_i - \tfrac{1}{6}\Sigma n^3\delta_i^3/r_i^2. \tag{8}$$

Hence to order $n\delta_i^2$ it is immaterial whether r_i or $n\theta_i$ is in the denominator. In fact with $n\theta_i$ the next term in (8) is only half the magnitude that it is with r_i in (5): so the approximation is better with $n\theta_i$.

Proof of (*b*). For the purpose of this proof only, omit the suffixes. We easily obtain from (7) that

$$\frac{d\mu}{d\theta} = \frac{n\theta + r}{2n^{\frac{1}{2}}\theta^{\frac{3}{2}}}$$

and hence, with a little algebra,

$$\begin{aligned} \ln\frac{d\theta}{d\mu} &= \ln\frac{\sqrt{r}}{n} + \tfrac{3}{2}\ln\left(1 + \frac{n\delta}{r}\right) - \ln\left(1 + \frac{n\delta}{2r}\right) \\ &= \ln\frac{\sqrt{r}}{n} + \frac{n\delta}{r}, \end{aligned} \tag{9}$$

omitting terms of order n^{-1}. The logarithm of the Jacobian of the transformation from the θ's (or equally the δ's) to the μ's is obtained from the sum of k terms like (9) and gives

$$\ln\left|\frac{d\boldsymbol{\theta}}{d\boldsymbol{\mu}}\right| = C + n\Sigma\delta_i/r_i.$$

The summation term is of order $n^{-\frac{1}{2}}$ and therefore of the order of terms neglected. The sign of this term is variable and it is not a simple matter to see if the approximation is better with μ_i instead of λ_i. (This is discussed again below.)

These results, (*a*) and (*b*), show that $\boldsymbol{\mu}$ may replace $\boldsymbol{\lambda}$ in (6) and the proof is complete except for the remark that the effect of any other reasonable prior distribution besides the uniform one would have negligible effect as n increases.

Corollary. *An approximate significance test (valid as $n \to \infty$) of the hypothesis that $\theta_i = p_i$ $(i = 1, 2, \ldots, k)$ for assigned values of p_i, is obtained at level α by declaring the data significant when*

$$\chi^2 = \sum_{i=1}^{k} (r_i - np_i)^2/np_i \tag{10}$$

exceeds $\bar{\chi}^2_\alpha(k-1)$, the upper 100α % point of the χ^2-distribution with $(k-1)$ degrees of freedom (§5.3).

The result is significant if the point $(p_1, p_2, \ldots, p_k)$ lies outside the confidence set which, in terms of the λ's, is a sphere, centre the origin, and radius determined by the distribution of (1) (cf. theorem 6.4.1). This proves the result.

The quantity (10) is called *Pearson's chi-squared statistic*. The numbers, r_i, are called *observed* values and are often denoted by O_i: the numbers, np_i, are called *expected* values and are often denoted by E_i. (The reason for the latter term is that if $\theta_i = p_i$, that is if the null hypothesis is true, then $\mathscr{E}(r_i) = np_i$.) The statistic may then be written

$$\chi^2 = \sum_{i=1}^{k} (O_i - E_i)^2/E_i. \tag{11}$$

Pearson's test

The test of the corollary was one of the first significance tests to be developed. Pearson developed it in 1900 in a different way, but our approach, using Bayesian methods, gives essentially the same results as he obtained and subsequent writers have used. It is usually described as a goodness-of-fit test, for the following reason. Suppose one is interested in a particular null hypothesis for the probabilities of the events A_i, namely that $\theta_i = p_i$ ($i = 1, 2, \ldots, k$), and wishes to see how well the observations agree with it. The null hypothesis may be thought of as a model to describe the sampling behaviour of the observations and one wishes to know how good the model is in fitting the data: hence the terminology. The criterion, (10), may be justified on the following intuitive grounds: if $\theta_i = p_i$ we should expect E_i occurrences of the event A_i, whereas we actually obtained O_i. The discrepancy, for a single event A_i, is conveniently measured by $(O_i - E_i)^2$, just as the spread of a distribution is measured by squares of departures from expectation. But it would not be sensible to add these squares to provide an overall measure of discrepancy because, even if the null hypothesis is true, some will clearly be larger than others due to the fact that the E_i's will normally differ. In fact, since O_i has, for fixed n, a binomial distribution, the variance of O_i is $E_i(n - E_i)/n$, and therefore $(O_i - E_i)^2$ will be of this order. Additionally the O_i's are correlated and allowance for this shows that the order is more accurately measured by E_i. (This point will be elaborated below.) Hence a sensible criterion is $\Sigma(O_i - E_i)^2/E_i$, agreeing with the rigorous argument based on the posterior distribution. Note that if n is fixed the r_i will have a multinomial distribution (§3.1). The analysis does not, however, assume n fixed.

Example

An example of the use of Pearson's statistic is provided by the following genetical situation. If two recessive genes giving rise to phenotypes A and B are not linked, the proportions of the phenotypes AB, $A\bar{B}$, $\bar{A}B$, $\bar{A}\bar{B}$ (AB means exhibiting both phenotypes, etc., $\bar{A}$ denotes not-A) should be in the ratios 1:3:3:9. Hence if we take a random sample of n individuals and observe the numbers of each of the four phenotypes, they will have, on the null hypothesis of no linkage, a multinomial distribution of index n and four classes with probabilities 1/16, 3/16, 3/16 and 9/16. The hypothesis may be tested by

Pearson's method. A numerical example with $n = 96$ gave 6, 13, 16 and 61 in the four classes. These are the observed numbers, O_i. The expected numbers, E_i, are $96/16 = 6$, 18, 18 and 54 and χ^2 is

$$0+5^2/18+2^2/18+7^2/54 = 2{\cdot}52,$$

which is well within the confidence set for χ^2 with 3 degrees of freedom, the upper 5 % value being 7·81. Consequently there is no reason to doubt the null hypothesis on the evidence of these data.

Extension to any distribution

The range of applicability of Pearson's statistic is enormously increased by noticing that any case of random sampling can be converted to the multinomial situation by grouping (§2.4). To show this, suppose that each sample yields a single random variable, x (the extension to more than one variable is straightforward) and that the range of variation of x is divided into groups: for example, let $a_i \leqslant x < a_{i+1}$ be such a group for two fixed numbers a_i and a_{i+1}, and let A_i denote the event that x falls in the group. Then if there are k exclusive and exhaustive groups with corresponding exclusive and exhaustive events $A_1, A_2, \ldots, A_k$, the distribution of the number of occurrences, r_i, of the event A_i ($i = 1, 2, \ldots, k$) will (because of the random sampling) be multinomial with index n and parameters

$$p_i = p(A_i) = \int_{a_i}^{a_{i+1}} p(x)\,dx, \tag{12}$$

where $p(x)$ is the density of x. Hence, if it is desired to test the hypothesis that the density of x is of some particular form, $p(x)$, this can be done, at the loss of some information, by testing, by Pearson's method, the hypothesis that the grouped results have probabilities given by (12). This equation is in the form for continuous distributions but the idea is equally applicable to discrete distributions which may already be thought of as grouped, or may be put into larger groups with

$$p(A_i) = \sum_{a_i \leqslant s < a_{i+1}} q_s,$$

where q_s is the density for the discrete variable. As an example consider the case where the null hypothesis is that the random variable is uniformly distributed in the interval (0, 1). If this interval is divided into k equal intervals each of length k^{-1} the probability associated with each of them under the null hypothesis is k^{-1}, so that the expected numbers in a sample of size n are n/k.

Relevance of prior knowledge

It might be thought that this method could be used in a situation illustrated by the following example, and it is important to understand why it would not be appropriate. In §5.6, using results developed in §5.1, we discussed the case of a random sample of size n from a normal distribution of known variance σ^2, and developed a test of the null hypothesis that the mean, θ, was a specified value, say zero. This was done by declaring the result significant at 5% if $|\bar{x}|$ exceeded $1{\cdot}96\sigma/\sqrt{n}$. The same null hypothesis could be tested by grouping the observations in the way just described and using Pearson's statistic with the density of $N(0, \sigma^2)$; the probabilities in (12) being obtained from tables of the normal distribution function. The difference† between these two tests for the same null hypothesis lies in the form of the prior distribution. In the first test, using normal theory, the prior information consisted of three parts: (i) knowledge that the distribution was normal, (ii) knowledge of the variance of the distribution, (iii) ignorance of the mean of the distribution. In the second test, using the χ^2-statistic, there was considerable prior ignorance of the values of the p's, contrasting markedly with the precise knowledge contained in (i) and (ii). The second test would therefore not be appropriate if the knowledge of normality and variance was available. Pearson's is a test which is appropriate when the prior knowledge is much vaguer than in the tests described in the last two chapters; though notice that it does assume some considerable knowledge, namely that the trials are random with constant probabilities. There are tests designed for situations where even this prior knowledge is not

† There is also a difference due to the grouping used in one method. But this is slight if the grouping interval is small. (See Sheppard's corrections in §7.5.)

available, but they will not be discussed in this book. They include tests for randomness.

Confidence sets

Notice that the theorem, as distinct from the corollary, does give a confidence set for the θ_i. This is not often useful in practice, at least when k is not small, because of its complexity. The test is preferred for much the same reasons as the F-test was in §6.4. When $k = 2$, the binomial situation, the result does provide approximate confidence intervals for the single parameter $\theta = \theta_1 = 1-\theta_2$. If the result is expressed in the binomial notation of §7.2, that is $r_1 = r$, $r_2 = n-r$, it says that

$$\begin{aligned}\frac{(r_1-n\theta_1)^2}{n\theta_1}+\frac{(r_2-n\theta_2)^2}{n\theta_2} &= (r-n\theta)^2\left(\frac{1}{n\theta}+\frac{1}{n(1-\theta)}\right)\\ &= \frac{(r-n\theta)^2}{n\theta(1-\theta)}\end{aligned} \tag{13}$$

has approximately a χ^2-distribution with one degree of freedom. Thus a 95 % confidence interval for θ is given by (13) being less than $\lambda = 3{\cdot}84$, the upper 5 % point of the χ^2-distribution with one degree of freedom. This yields a quadratic inequality for θ

$$(n^2+\lambda n)\theta^2-(2rn+\lambda n)\theta+r^2 < 0. \tag{14}$$

Clearly the roots of the quadratic are real and lie between 0 and 1, and the coefficient of θ^2 is positive; so that the confidence set for θ is the interval between the two roots formed by equating the quadratic to zero. Although this result is only a large sample one, so that the exact form of the prior distribution is irrelevant, it was obtained with a uniform prior distribution for θ. If the distribution of ignorance suggested in §7.2 had been used, with density proportional to $\{\theta(1-\theta)\}^{-1}$ the effect would have been to have replaced (3) by $C+\Sigma(r_i-1)\ln\theta_i$. Hence in order to compare (14) with the exact result and approximations of §7.2, one should be subtracted from the values of r and $(n-r)$ before using (13). In the numerical case of §7.2 with $n = 17$, $r = 5$, the reduction to $n = 15$, $r = 4$ will give, for the quadratic

$$(225+15\lambda)\theta^2-(120+15\lambda)\theta+16 = 0$$

($\lambda = 3{\cdot}84$), with roots $\theta = 0{\cdot}11, 0{\cdot}52$. The values agree with the exact ones obtained in §7.2.

Tests for Poisson means

The test can be used in a rather unexpected context. Suppose that we have k independent random variables $r_1, r_2, \ldots, r_k$ with Poisson distributions of parameters $\theta_1, \theta_2, \ldots, \theta_k$. (Notice that, as explained in §7.3, several random samples from the same Poisson distribution are equivalent to a single random sample from a Poisson distribution, so that each r_i may be obtained from several samples from one distribution.) The joint density of the r_i is then

$$p(r_1, r_2, \ldots, r_k \mid \theta_1, \theta_2, \ldots, \theta_k) = \exp\left[-\sum_{i=1}^{k} \theta_i\right] \prod_{i=1}^{k} (\theta_i^{r_i}/r_i!).$$

This can be written as the product of the density of $n = \Sigma r_i$ and the joint conditional density of the r_i, given n. But n, being the sum of independent Poisson variables, will have a Poisson distribution of parameter $\theta = \Sigma\theta_i$ (§3.5), so that

$$p(n \mid \theta_1, \theta_2, \ldots, \theta_k) = e^{-\theta}\theta^n/n!.$$

Now we may write

$$p(r_1, r_2, \ldots, r_k \mid \theta_1, \theta_2, \ldots, \theta_k) = \left\{\frac{e^{-\theta}\theta^n}{n!}\right\}\left\{\frac{n!}{\Pi(r_i!)}\,\Pi\left(\frac{\theta_i}{\theta}\right)^{r_i}\right\}, \quad (15)$$

the expression in the first set of braces being $p(n \mid \theta_1, \theta_2, \ldots, \theta_k)$; that in the second set is therefore $p(r_1, r_2, \ldots, r_k \mid n: \theta_1, \theta_2, \ldots, \theta_k)$. If we write $\phi_i = \theta_i/\theta$, so that $\phi_i \geqslant 0$ and $\Sigma\phi_i = 1$ the latter distribution is easily seen to be multinomial with index n and parameters ϕ_i. So (15) may be written

$$p(\{r_i\} \mid \theta, \{\phi_i\}) = p(n \mid \theta)\, p(\{r_i\} \mid n, \{\phi_i\})$$

and a comparison with equation 5.5.21 shows that if θ is known, or has a prior distribution independent of the ϕ's, n is an ancillary statistic and may be supposed fixed for the purposes of inferences about the ϕ's. When n is fixed we return to the multinomial situation. If, therefore, we wish to test the null hypothesis that the θ's are all equal, or, more generally, are in assigned ratios, and we are prepared to make the prior judgement of the independence of θ and the ϕ's, we can perform the test by expressing the null hypothesis in terms of the ϕ's and use only the likelihood of the ϕ's, and hence Pearson's statistic. If we require to test that all the θ's are equal, or equivalently that all the ϕ's are k^{-1}, the statistic is

$$\frac{\Sigma(r_i - n/k)^2}{n/k} = \frac{\Sigma(r_i - \bar{r})^2}{\bar{r}}, \tag{16}$$

where $\bar{r} = n/k = k^{-1} \sum_{i=1}^{k} r_i$, the average of the r's, and may be compared with $\bar{\chi}^2_\alpha(k-1)$. The statistic (16) is often known as a *coefficient* (*or index*) *of dispersion*. Its form is interesting: if it were divided by $(k-1)$ the numerator would be the usual mean square estimate of the variance of the r_i, assuming them to have come from a common distribution (the null hypothesis), and the denominator would be the usual estimate of the mean. But for a Poisson distribution the mean and variance are equal so that the ratio should be about one, or (16) should be about $(k-1)$. The statistic is effectively a comparison of the variance and the mean. When $k = 2$, the test is an approximation to the exact method based on theorem 7.3.2.

There are two points of interest about the method just used. First, it shows that the multinomial distribution of random variables r_i, implying that the r_i are correlated (see §3.1), can be thought of as being obtained by taking *independent* r_i, each with a Poisson distribution, and investigating their distribution conditional on Σr_i being fixed. The substitution of correlated r_i by independent r_i is often convenient. For example, in the intuitive discussion of Pearson's statistic above we could think of the O_i as Poisson with means E_i, and therefore also variances E_i: hence weighting with the inverses of the variances in the usual way we obtain Pearson's criterion. Although a sum of k squares it has only $(k-1)$ degrees of freedom because of the constraint on Σr_i.

Small samples

We conclude by making a few remarks about the proof of theorem 1. The argument used in §7.2 to explain the use of the prior distribution $\pi(\theta) \propto \theta^{-1}(1-\theta)^{-1}$ in the case of $k = 2$ can easily be extended to the case of general k to justify the prior distribution with density

$$\pi(\theta_1, \theta_2, \ldots, \theta_k) \propto \left(\prod_{i=1}^{k} \theta_i\right)^{-1}.$$

The effect of using this instead of the uniform prior distribution would be to replace r_i everywhere, from (3) onwards, by $(r_i - 1)$. Hence, in applying the result it would probably be better to deduct one from each observed value (and k from n) before using it: but this is not general practice. This was done in the numerical application to the binomial case $k = 2$, above, with successful results. Actually, although the theorem is only a limiting result as $n \to \infty$, the exact distribution of Pearson's statistic is very nearly χ^2 for quite small samples. The general recommendation, based on considerable investigations of particular cases, is that the statistic and test may safely be used provided each expected value is not less than two. In these investigations the comparison has been made with the sampling distribution of χ^2, that is, for fixed θ_i and varying r_i, rather than with the posterior distribution, but similar conclusions probably apply to the latter.

The basis of the approximation is the observation that the logarithm of the posterior density has a maximum at $\hat{\boldsymbol{\theta}}$ and that the density falls away fairly sharply from this maximum. The logarithmic density can therefore be replaced by a second-degree surface about this maximum value: this corresponds to the retention of only the second-degree (in δ_i) terms in (5). This is essentially the same type of approximation that was used with the maximum likelihood method (§7.1). It leads naturally to the quantities λ_i and the statistic

$$\chi'^2 = \sum_{i=1}^{k} (O_i - E_i)^2/O_i \tag{17}$$

to replace (11). This statistic is often called '*chi-dash-squared*'. The use of χ'^2 would be similar to saying that θ_i has approximately a normal posterior distribution with mean r_i/n and variance r_i/n^2; so that χ'^2 is the sum of squares of k standard normal variables with the single constraint $\Sigma\theta_i = 1$. But this is not too good an approximation as we saw in the case $k = 2$ in §7.2, where we took θ to be approximately normal on the basis of maximum likelihood theory: the normal approximation gave symmetric confidence limits, whereas the true ones are asymmetric. Some of this asymmetry can be introduced by replacing

O_i in the denominator of (17) by E_i, and we saw above, in one numerical example, that this introduces the required degree of asymmetry. In general, the improvement in the approximation obtained by using E_i is reflected in the smaller size of the highest-order term neglected in the expansion of the logarithm of the posterior density. The term of order $n^{-\frac{1}{2}}$ in (8) is only half the magnitude of the corresponding term in (5), and the Jacobian of the transformation introduces a term of order $n^{-\frac{1}{2}}$, equation (9), which, on the average is zero, since $\mathscr{E}(\delta_i | \theta_i) = 0$.

One great limitation of Pearson's statistic as developed in this section is that it is only available for testing a completely specified hypothesis, namely that all the θ_i's are equal to assigned values p_i. There are many situations where one wishes to test a hypothesis which only partially specifies the θ_i's. For example, we have described how to test the hypothesis that a random sample comes from a normal distribution of specified mean and variance, but we cannot, using the methods of this section, test the less specific hypothesis that it comes from some normal distribution. It is to the type of problems of which this is an example that we turn in the next section.

7.5. Goodness-of-fit tests (continued)

The notation is the same as that of the previous section.

Theorem 1. *If, in each of a random sequence of n trials, the exclusive and exhaustive events A_i $(i = 1, 2, \ldots, k)$ have constant probabilities of success θ_i and occur r_i times $(\Sigma r_i = n)$; then an approximate significance test (valid as $n \to \infty$) of the hypothesis that l $(< k)$ functions, functionally independent of each other and of $\Sigma\theta_i$, are all zero,*

$$\phi_1(\theta_1, \theta_2, \ldots, \theta_k) = \phi_2(\theta_1, \theta_2, \ldots, \theta_k) = \ldots = \phi_l(\theta_1, \theta_2, \ldots, \theta_k) = 0, \tag{1}$$

is obtained at level α by declaring the data significant when

$$\chi^2 = \sum_{i=1}^{k} (r_i - n\tilde{\theta}_i)^2 / n\tilde{\theta}_i \tag{2}$$

exceeds $\overline{\chi}^2_\alpha(l)$, the upper $100\alpha\%$ point of the χ^2-distribution *with l degrees of freedom, where $\tilde{\theta}_i$ is the maximum likelihood estimate of θ_i assuming that* (1) *obtains.*

We shall use the asymptotic result of theorem 7.4.1. That result was based on the remark that the posterior distribution had a maximum at $\theta_i = r_i/n$ and that $\delta_i = \theta_i - r_i/n$ would be small, in fact of order $n^{-\frac{1}{2}}$, so that only a small region in the neighbourhood of the maximum need be considered. In that region it will be assumed that the functions ϕ_i of (1) are sufficiently smooth for them to be replaced by linear functions of the θ's: that is, we suppose that the constraints $\phi_i = 0\,(i = 1, 2, ..., l)$ can be approximately written

$$\sum_{j=1}^{k} a_{ij}\theta_j = c_i \tag{3}$$

for $i = 1, 2, ..., l$. These may alternatively be written

$$\sum_{j=1}^{k} a'_{ij}\lambda_j = c'_i, \tag{4}$$

where $a'_{ij} = a_{ij}\sqrt{(r_j)}/n$ and $c'_i = c_i - \sum_{j=1}^{k} a_{ij}r_j/n$, and, as in §7.4,

$$\lambda_i = n\delta_i/\sqrt{r_i} = n(\theta_i - r_i/n)/\sqrt{r_i}. \tag{5}$$

Now it is well known (see, for example, most books on linear algebra) that an orthogonal $k \times k$ matrix with elements b_{ij} can be found such that the restrictions imposed on the λ's by (4) are equivalent to the restrictions

$$\sum_{j=1}^{k} b_{ij}\lambda_j = d_i, \tag{6}$$

for $i = 1, 2, ..., l$. (For this to be so it is necessary that the ϕ's, and hence the left-hand sides of (3), be independent functions of the θ's.) Define new parameters ψ_i by the equations

$$\psi_i = \sum_{j=1}^{k} b_{ij}\lambda_j, \tag{7}$$

for *all* i, so that the constraints (6) are $\psi_i = d_i$ for $i \leqslant l$. Then the hypothesis we are interested in testing is that $\psi_i = d_i$ for $i \leqslant l$, and the parameters ψ_i for $i > l$ are not of interest to us. In accord with the general principles of inference we have to find the posterior distribution of the ψ's involved in the hypothesis; that is, ψ_i for $i \leqslant l$.

Now from equation 7.4.6

$$\ln \pi(\boldsymbol{\lambda} | \mathbf{r}) = C - \frac{1}{2} \sum_{i=1}^{k} \lambda_i^2 \tag{8}$$

to the order of approximation being used. It immediately follows that to the same order

$$\ln \pi(\boldsymbol{\psi} | \mathbf{r}) = C - \frac{1}{2} \sum_{i=1}^{k} \psi_i^2 \tag{9}$$

since the Jacobian is unity. This result follows because the matrix of elements b_{ij} is orthogonal. Consequently, integrating the posterior probability density with respect to $\psi_{l+1}, \psi_{l+2}, \ldots, \psi_k$

$$\ln \pi(\psi_1, \psi_2, \ldots, \psi_l | \mathbf{r}) = C - \frac{1}{2} \sum_{i=1}^{l} \psi_i^2. \tag{10}$$

Also since the ϕ's are independent of $\Sigma\theta_i$, so are the ψ_i's for $i \leqslant l$ and hence there is no linear relation between the ψ_i's in (10). It follows as in theorem 6.4.2 that the distribution of $\sum_{i=1}^{l} \psi_i^2$ is χ^2 with l degrees of freedom and hence that a significance test is obtained by declaring the data significant if $\sum_{i=1}^{l} d_i^2$ exceeds $\bar{\chi}_\alpha^2(l)$. It remains to calculate $\sum_{i=1}^{l} d_i^2$.

Since, in deriving (8), and hence (9), the prior distribution of the θ's was supposed uniform, the right-hand side of (9) is equally the logarithm of the likelihood of the ψ's, the ψ's being linear functions of the θ's. If (1) obtains this logarithmic likelihood is

$$C - \frac{1}{2} \sum_{i=1}^{l} d_i^2 - \frac{1}{2} \sum_{i=l+1}^{k} \psi_i^2,$$

with maximum obtained by putting $\psi_i = 0$ for $i > l$, giving

$$C - \frac{1}{2} \sum_{i=1}^{l} d_i^2.$$

Consequently $\frac{1}{2} \sum_{i=1}^{l} d_i^2$ is equal to the value of $\frac{1}{2} \sum_{i=1}^{k} \lambda_i^2$ at the maximum likelihood values of the θ's when (1) obtains. This proves the result with r_i instead of $n\tilde{\theta}_i$ in the denominator of (2). We may argue, as in the proof of theorem 7.4.1, that the replacement will not affect the result as $n \to \infty$. The theorem is therefore proved.

The quantity (2) may be written in Pearson's form

$$\Sigma(O_i - E_i)^2/E_i, \tag{11}$$

provided the E_i are suitably interpreted. E_i is the expected number of occurrences of A_i if the maximum likelihood value under the null hypothesis were the true value.

Special case: $l = k-1$

The corollary to theorem 7.4.1 is a special case of the above. In that corollary the null hypothesis being tested was completely specified so that the number of functions in (1) must be $l = k-1$, which, together with $\Sigma\theta_i = 1$, completely determine the θ's. Since the θ's are completely determined at values p_i it follows that $\tilde{\theta}_i = p_i$ and (2) reduces to (7.4.10), the degrees of freedom being $l = k-1$.

Test for a binomial distribution

In the last section we saw how to test the hypothesis that a random sample comes from a completely specified distribution; for example, binomial with index† s and parameter p, where s and p are given. The new result enables a test to be made of the hypothesis that a random sample comes from a family of distributions; for example, binomial with index s, the parameter being unspecified. To do this suppose that we have n observations, all integer and between 0 and s, of independent random variables from a fixed, but unknown distribution which may be a binomial distribution with index s. Let the number of observations equal to i be r_i $(0 \leqslant i \leqslant s)$, so that $\Sigma r_i = n$. Then the r's have a multinomial distribution, with index $s+1$ (since the original random variables are supposed independent and from a fixed distribution) and if the further assumption is made

† We use s, instead of the usual n, for the index in order not to confuse it with the value n of the theorem.

that the distribution is binomial, the parameters of the multinomial distribution will be, for $0 \leqslant i \leqslant s$,

$$\theta_i = \binom{s}{i} \theta^i (1-\theta)^{s-i}, \tag{12}$$

where θ is the unknown binomial parameter. If θ were specified, say equal to p, the θ_i would be known by putting $\theta = p$ in (12); the E_i would be $n\theta_i$ and Pearson's statistic could be used with s degrees of freedom. But if θ is unknown it has to be estimated by maximum likelihood. The logarithm of the likelihood is, apart from a constant, $\Sigma r_i \ln \theta_i$ (equation 7.4.3) or, by (12),

$$\sum_{i=0}^{s} i r_i \ln \theta + \sum_{i=0}^{s} (s-i) r_i \ln (1-\theta). \tag{13}$$

The maximum is easily seen (cf. §7.1) to be given by

$$\tilde{\theta} = \sum_{i=0}^{s} i r_i / ns. \tag{14}$$

Consequently if θ, in (12), is put equal to $\tilde{\theta}$ the resulting values $\tilde{\theta}_i$ may be used to compute expected values $n\tilde{\theta}_i$ and Pearson's statistic compared with χ^2.

Degrees of freedom

It only remains to determine the degrees of freedom. According to the theorem the number of degrees of freedom is equal to the number of constraints on the θ_i that the null hypothesis (12) imposes. The latter number is most easily determined by remarking that the number of θ_i's (the number of groups) is $s+1$, the k of the theorem, and hence the values of $s\ (=k-1)$ constraints (apart from $\Sigma\theta_i = 1$) are needed to specify the θ_i's completely: but the binomial hypothesis leaves only one value, θ, unspecified so that it must implicitly specify $(s-1)$ constraints. Hence there are $(s-1)$ constraints in (1): that is $l = s-1$ and the degrees of freedom are therefore $s-1$.

This method of obtaining the degrees of freedom is always available and is often useful. In essence, one considers the number of constraints, additional to those in (1), that would have to be specified in order to determine all the θ_i's. If this number is m, then $l+m = k-1$ and the degrees of freedom are

$l = (k-1)-m$. There is a loss of one degree of freedom for each unspecified constraint in the null hypothesis. Each such constraint corresponds to a single unspecified parameter in the null hypothesis (here θ) and the rule is that there is a loss of one degree of freedom for each parameter that has to be estimated by maximum likelihood. Rather roughly, the ability to vary each of these parameters enables one to make the E_i agree more closely with the O_i, so that the statistic is reduced, and this is reflected in a loss of degrees of freedom—the mean value of χ^2 being equal to the degrees of freedom (§5.3). The result on the reduction in the degrees of freedom is due to Fisher; Pearson having been under the mistaken impression that no correction for estimation was necessary.

Test for Poisson distribution

Similar ideas may be applied to any discrete distribution. Another example would be the Poisson distribution; but there the number of possible values would be infinite, namely all non-negative integers, and some grouping would be necessary, as explained in §7.4. Usually it will be enough to group all observations not less than some value c, and suppose them all equal to c, the value of c being chosen so that the expectation is not below about two. This introduces a slight error into the maximum likelihood estimation, but this is not likely to be serious. It is important to notice the difference between this type of test and the test for Poisson distributions in §7.4. The latter was for the differences between Poisson means; that is, the distributions were assumed, because of prior knowledge, to be Poisson and only their means were compared. The present test supposes it known only that the variables come from some distribution and asks if that distribution could be Poisson. There is obviously less prior knowledge available here and consequently a larger sample is needed to produce a useful test.

Continuous distributions

The same ideas may be applied to fitting a continuous distribution: for example, we might ask if the variables come from an unspecified normal distribution. The observations would be divided into k convenient groups and the numbers r_i in the different groups counted. It only remains to find the maximum likelihood estimates of the θ_i when the null hypothesis is true. If the null hypothesis distribution depends on parameters $\alpha_1, \alpha_2, \ldots, \alpha_s$ and has density $p(x \mid \alpha_1, \alpha_2, \ldots, \alpha_s)$, the θ_i's are given by (cf. equation 7.4.12)

$$\theta_i = \int_{a_i}^{a_{i+1}} p(x \mid \alpha_1, \alpha_2, \ldots, \alpha_s)\, dx, \tag{15}$$

and the logarithm of the likelihood is as usual. The maximum likelihood estimation presents some difficulties because the equations to be solved are

$$\sum_{i=1}^{k} r_i \theta_i^{-1} \partial\theta_i / \partial\alpha_j = 0 \quad (j = 1, 2, \ldots, s) \tag{16}$$

and the partial derivatives can be awkward. It is usual to approximate to (15) by assuming the density approximately constant in the interval when, if ξ_i is some point in the interval,

$$\theta_i = p(\xi_i \mid \alpha_1, \alpha_2, \ldots, \alpha_s)\,(a_{i+1} - a_i). \tag{17}$$

The maximum likelihood equations are then

$$\sum_{i=1}^{k} r_i \partial \ln p(\xi_i \mid \alpha_1, \alpha_2, \ldots, \alpha_s) / \partial\alpha_j = 0 \quad (j = 1, 2, \ldots, s). \tag{18}$$

If each observation x_t ($t = 1, 2, \ldots, n$) is replaced by the value ξ_i of the group to which it belongs (there then being r_i x's which yield a value ξ_i), equations (18) are the ordinary maximum likelihood equations for the α's when the sample values are the ξ's and the density is $p(x \mid \alpha_1, \alpha_2, \ldots, \alpha_s)$. These may be easier to solve than (16). Thus with the normal case where there are two parameters, the mean and the variance, this gives the usual estimates (equation 7.1.12—where the present α's are there θ's) in terms of the grouped values, ξ_i. These estimates can then be inserted in (17), or (15), to obtain the $\tilde{\theta}_i$. The procedure may be summarized as follows: (i) replace the observed values x_t by the grouped values ξ_i; (ii) estimate the parameters in the density by maximum likelihood applied to the ξ_i, (18); (iii) estimate the θ_i from (15) or (17); (iv) use these estimates in the χ^2 statistic (2), subtracting a degree of freedom for each parameter estimated. Thus with the normal distribution the degrees of freedom will be $(k-1)-2 = k-3$.

Sheppard's corrections

The replacement of (16) by (18) can lead to systematic errors. They can be reduced by a device which we illustrate for the case where there is a single unspecified parameter in the family of distributions, which will be denoted in the usual way by θ, and where the grouping intervals are equal (except for the end ones which will have to be semi-infinite). Equation (15) may be written

$$\theta_i = \int_{(i-\frac{1}{2})h}^{(i+\frac{1}{2})h} p(x|\theta)\,dx \tag{19}$$

for a convenient origin. An expansion in Taylor series gives an improvement on (17), namely

$$\theta_i = hp + \frac{h^3}{24}p'',$$

where the arguments of p are ih (formerly ξ_i) and θ, a remainder term has been ignored, and dashes denote differentiation with respect to x. Consequently

$$\ln\theta_i = \ln hp + \ln\left[1 + \frac{h^2}{24}\frac{p''}{p}\right] = \ln hp + \frac{h^2}{24}\frac{p''}{p}$$

to the same order, and

$$\theta_i^{-1}\,\partial\theta_i/\partial\theta = \frac{\partial}{\partial\theta}\ln p + \frac{h^2}{24}\frac{\partial}{\partial\theta}\left(\frac{p''}{p}\right). \tag{20}$$

We have to solve (16) with $\alpha_1 = \theta$. This is easily seen to be equivalent to summing the right-hand side of (20) over the observations at their grouped values and equating to zero. (The equation is the same as (18) with the addition of the term in h^2.) This equation can be solved by Newton's method (§7.1) with a first approximation given by $\theta^{(1)}$, the root of (18). The second approximation is seen, from (7.1.17), to be $\theta^{(2)} = \theta^{(1)} + \Delta$, where

$$\Delta = -\frac{h^2}{24}\left\{\sum_{t=1}^{n}\frac{\partial}{\partial\theta}\left(\frac{p''}{p}\right)\Big/\sum_{t=1}^{n}\frac{\partial^2}{\partial\theta^2}(\ln p)\right\}. \tag{21}$$

Here the x_t have been replaced by their grouped values and the other argument of p is $\theta^{(1)}$. One iteration is usually sufficient. The θ_i's can then be obtained from (19) with $\theta = \theta^{(2)}$. Generally $\theta^{(2)}$ will be nearer the true maximum likelihood estimate than $\theta^{(1)}$. The method extends without difficulty to several parameters. A correction related to Δ (equation (21)) was obtained by a different method by Sheppard and it is sometimes known as *Sheppard's correction.*

Prior knowledge

This method of testing goodness of fit of distributions has the disadvantage that the observations have to be grouped in an arbitrary way. Against this it has the advantage that a reasonable prior distribution for the grouped distribution is easy to specify. For example, it would be difficult to specify a parametric family, and an associated prior distribution of the parameters, appropriate for investigating the null hypothesis that a distribution was normal, unless some definite prior knowledge was available about possible alternatives to the normal distribution. In the absence of such prior knowledge the χ^2-test is a useful device.

The main application of theorem 1 will be described in the next section.

7.6. Contingency tables

Theorem 1. *If, in each of a random sequence of n trials, the exclusive and exhaustive events* $A_1, A_2, \ldots, A_s$ $(s > 1)$ *occur with unknown constant probabilities,* $\theta_{i.}$, *of success; and if, in the same sequence, the exclusive and exhaustive events* $B_1, B_2, \ldots, B_t$ $(t > 1)$ *also occur with unknown constant probabilities,* $\theta_{.j}$, *of success; then, if* θ_{ij} *is the probability that both* A_i *and* B_j *occur in a trial, an approximate significance test of the null hypothesis that for all i, j,*

$$\theta_{ij} = \theta_{i.}\theta_{.j} \tag{1}$$

for unspecified $\theta_{i.}$ *and* $\theta_{.j}$ *is obtained at level* α *by declaring the result significant if*

$$\sum_{i=1}^{s} \sum_{j=1}^{t} \frac{(r_{ij}-r_{i.}r_{.j}/n)^2}{r_{i.}r_{.j}/n} \tag{2}$$

exceeds $\bar{\chi}^2_\alpha(\nu)$, *where* $\nu = (s-1)(t-1)$. *In* (2), r_{ij} *is the number of times that both* A_i *and* B_j *occur in the n trials, so that* $\sum_{i,j} r_{ij} = n$, *and*†

$$r_{i.} = \sum_{j=1}^{t} r_{ij}, \quad r_{.j} = \sum_{i=1}^{s} r_{ij}, \tag{3}$$

so that these are respectively the number of times that A_i, *and the number of times that* B_j, *occur in the n trials.*

The situation here described is essentially the same as that of theorem 7.5.1 with a random sequence of n trials and $k = st$ exclusive and exhaustive events, namely the simultaneous occurrence of A_i and B_j, the events $A_i B_j$ in the notation of § 1.2. The null hypothesis (1) puts some constraints on the probabilities of these events and says that the events A_i and B_j are independent, that is

$$p(A_i B_j) = p(A_i) p(B_j),$$

without specifying the individual probabilities of the A's and B's. Theorem 7.5.1 can therefore be applied. We have only to determine (*a*) the maximum likelihood estimates of θ_{ij} under the null hypothesis, (1), and (*b*) the number, l, of constraints involved in (1).

(*a*) The logarithm of the likelihood is, if (1) obtains,

$$C + \sum_{i,j} r_{ij} \ln(\theta_{i.} \theta_{.j}) = C + \sum_i r_{i.} \ln \theta_{i.} + \sum_j r_{.j} \ln \theta_{.j}.$$

Remembering that $\sum_i \theta_{i.} = \sum_j \theta_{.j} = 1$ we see easily (compare the proof of theorem 7.4.1) that the maximum likelihood estimates of $\theta_{i.}$ and $\theta_{.j}$ are respectively $r_{i.}/n$ and $r_{.j}/n$. Hence

$$n\tilde{\theta}_{ij} = r_{i.} r_{.j}/n. \tag{4}$$

(*b*) As in the applications in the last section it is easiest to calculate l by finding the number of parameters unspecified in the null hypothesis. Here it is obviously the number of functionally independent $\theta_{i.}$ and $\theta_{.j}$, that is, $(s-1)+(t-1)$. Hence

$$l = (st-1)-(s-1)-(t-1) = st-s-t+1 = (s-1)(t-1) = \nu.$$

† Notice that the dot suffix is here used slightly differently from previously (§ 6.4). Thus $r_{i.}$ denotes the *sum* over the suffix j, not the *mean*.

Inserting the values of $\tilde{\theta}_{ij}$ given by (4) into equation 7.5.2 gives (2). This, with the value of l just obtained, establishes the result.

***Theorem* 2.** *Consider $s > 1$ independent random sequences of trials, with $r_{i.}$ trials in the ith sequence ($i = 1, 2, \ldots, s$) and $\Sigma r_{i.} = n$. Suppose that in each of these trials the exclusive and exhaustive events $B_1, B_2, \ldots, B_t$ occur with constant probabilities ϕ_{ij} ($j = 1, 2, \ldots, t$) in the ith sequence, so that $\sum_{j=1}^{t} \phi_{ij} = 1$ for all i. Then an approximate significance test of the null hypothesis that ϕ_{ij} does not depend on i: that is, that for all i, j,*

$$\phi_{ij} = \phi_j \tag{5}$$

for unspecified ϕ_j, is obtained by the test of theorem 1. *In* (2) *r_{ij} is the number of times B_j occurs in the ith sequences of $r_{i.}$ trials, so that $r_{i.} = \sum_{j=1}^{t} r_{ij}$, the number of trials in the ith sequence; and $r_{.j}$ is as defined in* (3), *the total number of times B_j occurs in all trials.*

Consider the situation described in theorem 1. The set of all r_{ij}, written $\{r_{ij}\}$, referring to the events $A_i B_j$, has probability density

$$p(\{r_{ij}\} \mid \{\theta_{ij}\}) \propto \prod_{i=1}^{s} \prod_{j=1}^{t} \theta_{ij}^{r_{ij}}, \tag{6}$$

and similarly the set of $r_{i.}$, written $\{r_{i.}\}$, referring to the events A_i, has probability density

$$p(\{r_{i.}\} \mid \{\theta_{ij}\}) \propto \prod_{i=1}^{s} \theta_{i.}^{r_{i.}} = \prod_{i=1}^{s} \prod_{j=1}^{t} \theta_{i.}^{r_{ij}}. \tag{7}$$

Hence, dividing (6) by (7),

$$p(\{r_{ij}\} \mid \{r_{i.}\}, \{\theta_{ij}\}) \propto \prod_{i=1}^{s} \prod_{j=1}^{t} (\theta_{ij}/\theta_{i.})^{r_{ij}}. \tag{8}$$

Let $\phi_{ij} = \theta_{ij}/\theta_{i.}$, so that $\sum_{j=1}^{t} \phi_{ij} = 1$ for all i, and change from a parametric description in terms of $\{\theta_{ij}\}$ to one in terms of $\{\theta_{i.}\}$ and $\{\phi_{ij}\}$. Then what has just been proved may be written

$$p(\{r_{ij}\} \mid \{\theta_{i.}\}, \{\phi_{ij}\}) = p(\{r_{i.}\} \mid \{\theta_{i.}\}) \, p(\{r_{ij}\} \mid \{r_{i.}\}, \{\phi_{ij}\}).$$

A comparison of this equation with equation 5.5.21 shows that, given $\{\phi_{ij}\}$, $\{r_{i.}\}$ is sufficient for $\{\theta_{i.}\}$; and, what is more relevant, given $\{\theta_{i.}\}$, $\{r_{i.}\}$ is ancillary for $\{\phi_{ij}\}$. By the argument leading to equation 5.5.22, if these two sets of parameters have independent prior distributions, inferences about $\{\phi_{ij}\}$ may be made with $\{r_{i.}\}$ either fixed or random. But the right-hand side of (8) is the likelihood in the situation of the present theorem, (6) is the likelihood in theorem 1, and in that theorem inferences were made about $\{\phi_{ij}\}$; namely that $\phi_{ij} = \theta_{.j}$, or ϕ_{ij} does not depend on i. This is equivalent to (5). Hence the same test may be used here as in theorem 1, with such a prior distribution. But the test is only asymptotic and the exact form of the prior distribution is almost irrelevant.

Variable margins

The situation described in theorem 1 arises when, in each trial, the outcome is classified in two ways into exclusive and exhaustive categories and we wish to know whether the classifications are related to each other, or, as we sometimes say, are contingent on each other. The data may be written in the form of an $s \times t$ *contingency table*, with s rows corresponding to the classification into the A classes, and t columns corresponding to the B's. The $r_{i.}$ and $r_{.j}$ are often referred to as the *margins* of the table (compare the definition of a marginal distribution in §3.1). For example, if we take a random sample of n school-children of a given age-group from a city, and classify them according to height and social class, where the range of heights has been divided into s convenient groups, and there are t social classes. The test would be applicable if we wanted to know whether the childrens' heights are influenced by the social class that they come from. If they are not influenced, then an event A_i, having a certain height, would be independent of an event B_j, belonging to a certain social class, and the null hypothesis would be true. The r_{ij} are the observed numbers, O_{ij}, say, of children of a given height and social class. The expected numbers of children of that height and social class, E_{ij}, say, are given by

$$E_{ij} = r_{i.}r_{.j}/n \tag{9}$$

and the test statistic (2) may be written

$$\sum_{i,j} (O_{ij} - E_{ij})^2/E_{ij}. \tag{10}$$

The choice of expected numbers can be justified intuitively in the following way. We do not know, and the null hypothesis does not say, what the probability, $\theta_{i.}$, of A_i, is; but, ignoring the other classification into the B's, our best estimate of it is about $r_{i.}/n$, the observed proportion of A_i's. If the null hypothesis is true (and remember the $n\tilde{\theta}_{ij} = E_{ij}$ are calculated on this assumption) then the proportion of A_i's should be the same, irrespective of the classification into the B's. So if we consider the $r_{.j}$ members of class B_j the expected proportion of A_i's amongst them would still be $r_{i.}/n$, and so the expected number in both A_i and B_j is $r_{i.}r_{.j}/n = E_{ij}$ agreeing with (9). The formula for the expected number is easily remembered by noting that E_{ij} corresponding to $O_{ij} = r_{ij}$ is the total $(r_{i.})$ for the row concerned, times the total $(r_{.j})$ for the column concerned, divided by the grand total.†

Example

In §7.4 we discussed a genetical example involving two genes. The individuals may be thought of as being classified in two ways, A or $\bar{A}$, B or $\bar{B}$, so that $s = t = 2$ and we have a 2×2 contingency table:

	B	$\bar{B}$	
A	6	13	19
$\bar{A}$	16	61	77
	22	74	96

Thus $r_{11} = 6$, $r_{1.} = 19$, $r_{.1} = 22$, $n = 96$, etc. The expected number in the class AB is $19 \times 22/96 = 4{\cdot}35$, the other classes being found similarly, and (10) is equal to $1{\cdot}02$. The upper 5 % point of χ^2 on one degree of freedom is $3{\cdot}84$; since the statistic is less than this there is no reason to think that the frequency of A's (against $\bar{A}$'s) depends on B. Notice the difference between this test and that of §7.4. The earlier test investigated whether $\theta_{11} = 1/16$, $\theta_{12} = 3/16$, $\theta_{21} = 3/16$, $\theta_{22} = 9/16$, in the notation of this section. The present test investigates whether $\theta_{ij} = \theta_{i.}\theta_{.j}$

† In an extension of our notation $n = r_{..}$.

for all i, j, and a little calculation shows that this is equivalent to $\theta_{11}/\theta_{12} = \theta_{21}/\theta_{22}$. The question asked here is less specific than that of §7.4.

The computational problem of dealing with a contingency table of general size may be reduced in the case of a 2×2 table by noting that (2) may, in that special situation, be written

$$\frac{(r_{11}r_{22} - r_{12}r_{21})^2 n}{r_{1.}r_{.1}r_{2.}r_{.2}}.$$

This can be proved by elementary algebra.

One margin fixed

The situation considered in theorem 2 is rather different. To change the language a little: in the first theorem we have a single sample of n individuals, each classified in two ways; in the second theorem we have s independent samples in which each individual is classified in only one way. However, the second situation may be related to the first by regarding the s samples as another classification, namely classifying the individual according to the number of the sample to which he belongs. The first stage in the proof of theorem 2 is to show that in the first situation with n individuals, the conditional probability distribution of the individuals with regard to the second (the B) classification given the first (the A) classification is the same as in the second situation, with the A-classification being into the different samples. Or, more mathematically, the probability distribution of the entries in the table (the r_{ij}) given one set of margins ($r_{i.}$) is the same as the distribution of s samples of sizes $r_{i.}$. Furthermore the distribution of the margin does not depend on the parameters of interest. The situation therefore is exactly that described in equation 5.5.21 with θ_2 there being the present ϕ's. Consequently inferences about the ϕ's may be made in the same way whether the margins are fixed or random. If they are random theorem 1 shows us how to make the significance test. Therefore the same test can be used in the situation of theorem 2 with the margin fixed. There is another way of looking at the same situation. As an example of an ancillary statistic in §5.5 we took the case of a random sample of size n, where n may have been chosen randomly according to a distribution which did not depend on the parameter of interest. We saw that n was ancillary and could therefore be supposed fixed. Here we have s samples of different sizes, possibly random, $r_{1.}, r_{2.}, \ldots, r_{s.}$; these sizes having a distribution which does not depend on the parameters, ϕ_{ij}, of interest. Exactly as before they may be supposed fixed.

The null hypothesis tested in theorem 2 is that the probabilities of the events B_j are the same for all samples: that is, the classification by the B's is not contingent on the sample (or equivalently the classification by the A's). It would arise, for example, if one took random samples of school-children of a given age-group in a city from each of the social classes, and asked if the distribution of height varied with the social class. The computations would be exactly the same as in the former example where the numbers in the social classes were also random.

Binomial distributions

The situation of theorem 2 with $s = t = 2$ is the same as that considered in §7.2 for the comparison of the probabilities of success in two random sequences of trials. The χ^2-test provides a much more widely used alternative to that using equation 7.2.17 based on the logarithmic transformation of the odds. The earlier test has the advantage that it can be extended to give confidence intervals for the ratio of odds in the two sequences. It is not common practice to quote confidence limits in dealing with contingency tables for much the same reasons as were discussed in §6.5 in connexion with the analysis of variance. But, where possible, this should certainly be done; or better still, the posterior distribution of the relevant parameter, or parameters, quoted.

The situation of theorem 2 with $t = 2$ and general s is the comparison of several probabilities of success in several random sequences of trials. The distribution of the r_{i1} will be $B(r_{i.}, \phi_{i1})$ and the null-hypothesis is that $\phi_{i1} = \phi$ for all i, with ϕ unspecified. It is easily verified that if $r_{i.} = n$, say,† the same for all i,

† We use n because it is the index of the binomial distribution. The n of the theorems is now sn.

so that the test is one for the equality of the parameters of binomial distributions of the same index, the test statistic (2) reduces to

$$\frac{\sum_{i=1}^{s} (r_i - \bar{r})^2}{n\frac{\bar{r}}{n}\frac{n-\bar{r}}{n}}, \tag{11}$$

where we have written $r_{i1} = r_i$, $r_{i2} = n - r_i$ and $\bar{r} = \sum_{i=1}^{s} r_i/s$. The form of this statistic is interesting and should be compared with that of equation 7.4.16. It is $(s-1)$ times the usual estimate of the variance of the r_i, divided by an estimate based on the assumption that the distributions are binomial with fixed ϕ, namely $\mathscr{D}^2(r_i) = n\phi(1-\phi)$ with ϕ replaced by $\bar{r}/n$. The statistic (11) is also known as a *coefficient* (*or index*) *of dispersion*. It can be referred to χ^2 with $(s-1)$ degrees of freedom. The test should not be confused with that of §7.5 for testing the goodness of fit of the binomial distribution: here the distributions are assumed binomial and the investigation concerns their means.

Example

It is sometimes as instructive to see when a result does not apply as to see when it does. We now describe a problem where a comparison of binomial probabilities seems appropriate but, in fact, is not. The papers of 200 candidates for an examination were marked separately by the examiners of two examining boards and were thereby classified by them as pass or fail. The results are shown below:

		Board B		
		Pass	Fail	Total
Board A	Pass	136	2	138
	Fail	16	46	62
Total		152	48	200

The percentage failure for board A (31 %) is higher than for board B (24 %) and one might contemplate comparing these by the method just described for the difference of two binomial proportions, or by the method of §7.2. But that would be quite false because the two samples are not independent: the same candidates are involved in both samples of 200. Again the mere look of the table may suggest carrying out a χ^2-test (theorem 1) on a 2×2 contingency table, but this would ask whether the proportion passing with board B was the same for those who failed with board A as for those who passed with board A, a question of no interest. What one wishes to know is whether one board has a higher failure rate than the other—the question that would have been answered by the first test had it been valid. To perform a test we note that the 182 candidates who were treated similarly by the two boards (both failed or both passed) can provide no information about the differences between the boards: so our interest must centre on the 18 other candidates. We can see this also by remarking that in the table as arranged the likelihood is, assuming the candidates to be a random sample,

$$\begin{aligned}
&\theta_{11}^{r_{11}}\theta_{12}^{r_{12}}\theta_{21}^{r_{21}}\theta_{22}^{r_{22}} \\
&= \{(\theta_{11}+\theta_{22})^{r_{11}+r_{22}}(\theta_{12}+\theta_{21})^{r_{12}+r_{21}}\} \\
&\qquad \times \left\{\frac{\theta_{11}^{r_{11}}\theta_{22}^{r_{22}}}{(\theta_{11}+\theta_{22})^{r_{11}+r_{22}}}\right\}\left\{\frac{\theta_{12}^{r_{12}}\theta_{21}^{r_{21}}}{(\theta_{12}+\theta_{21})^{r_{12}+r_{21}}}\right\} \\
&= \{\psi_1^{r_{11}+r_{22}}(1-\psi_1)^{r_{12}+r_{21}}\}\{\psi_2^{r_{11}}(1-\psi_2)^{r_{22}}\}\{\psi_3^{r_{12}}(1-\psi_3)^{r_{21}}\}, \quad (12)
\end{aligned}$$

where $\psi_1 = \theta_{11}+\theta_{22}$, $\psi_2 = \theta_{11}/(\theta_{11}+\theta_{22})$ and $\psi_3 = \theta_{12}/(\theta_{12}+\theta_{21})$. We are only interested in ψ_3 which is the probability, given that the candidate has been treated differently by the two boards, that he has been failed by B and passed by A. By the principle of ancillary statistics we can therefore consider only r_{12} and r_{21}, provided our prior judgment is that ψ_1 and ψ_2 are independent of ψ_3, which is reasonable. From the last set of braces in (12), the likelihood is seen to be of the form appropriate to $r_{12}+r_{21}$ independent trials with constant probability ψ_3. If the boards are acting similarly $\psi_3 = \frac{1}{2}$ and it is this hypothesis that we must test. The exact method is that of §7.2. From equation 7.2.13 a confidence interval for ψ_3 is

$$(1+8\bar{F}')^{-1} < \psi_3 < (1+8\bar{F}^{-1})^{-1}.$$

With 4 and 32 degrees of freedom and $\alpha = 0{\cdot}05$, $\bar{F} = 3{\cdot}22$ (using $\frac{1}{2}\alpha = 0{\cdot}025$) and $\bar{F}' = 8{\cdot}45$ so that $0{\cdot}015 < \psi_3 < 0{\cdot}29$ with probability 0·95. The value $\frac{1}{2}$ is outside the interval and there is evidence that board A is more severe than board B.

Suggestions for further reading

The suggestions given in chapter 5 are adequate.

Exercises

1. An individual taken from a certain very large biological population is of type A with probability $\frac{1}{2}(1+F)$ and of type B with probability $\frac{1}{2}(1-F)$. Give the probability that a random sample of n such individuals will consist of a of type A and b of type B. Find $\hat{F}$, the maximum likelihood estimate of F. Show that the expectation of $\hat{F}$ is F. Calculate the information about F. (Camb. N.S.)

2. A random sample is drawn from a population with density function

$$f(x|\theta) = \frac{\theta}{1-e^{-\theta}} e^{-\theta x} \quad (0 \leqslant x \leqslant 1).$$

Show that the mean of the sample is a sufficient statistic for the parameter θ and verify that the maximum likelihood estimate is a function of the sample mean. Let this estimate be denoted by θ_1.

Suppose that the only information available about each sample member concerns whether or not it is greater than a half. Derive the maximum likelihood estimate in this case and compare its posterior variance with that of θ_1. (Wales Dip.)

3. A continuous random variable x, defined in the range $0 \leqslant x \leqslant \frac{1}{2}\pi$, has distribution function proportional to $(1-e^{-\alpha \sin x})$ where $\alpha > 0$. Find the density function of x.

Given a random sample of n observations from this distribution, derive the maximum likelihood equation for $\hat{\alpha}$, the estimate of α. Indicate very briefly how this equation can be solved numerically.

Also prove that the posterior variance of α is

$$\frac{4\hat{\alpha}^2 \sinh^2 \frac{1}{2}\hat{\alpha}}{n(4\sinh^2\frac{1}{2}\hat{\alpha} - \hat{\alpha}^2)}.$$ (Leic. Gen.)

4. Calculate the maximum likelihood estimate, $\hat{\theta}$, of θ, the parameter of the Cauchy distribution

$$dF = \frac{1}{\pi} \frac{1}{1+(x-\theta)^2} dx \quad (-\infty < x < \infty),$$

given the sample of 7 observations,

3·2, 2·0, 2·3, 10·4, 1·9, 0·4, 2·6.

Describe how you could have used this calculation to determine approximate confidence limits for θ, if the sample had been a large one. (Manch. Dip.)

5. Independent observations are made on a Poisson variable of unknown mean θ. It is known only that of n observations, n_0 have the value 0, n_1 have the value 1, and the remaining $n-n_0-n_1$ observations have values

greater than one. Obtain the equation satisfied by the maximum likelihood estimate and suggest how the equation could be solved numerically.

(Lond. B.Sc.)

6. The random variables $X_1, \ldots, X_m, X_{m+1}, \ldots, X_{m+n}$ are independently, normally distributed with unknown mean θ and unit variance. After $X_1, \ldots, X_m$ have been observed, it is decided to record only the signs of $X_{m+1}, \ldots, X_n$. Obtain the equation satisfied by the maximum likelihood estimate $\hat{\theta}$, and calculate an expression for the asymptotic posterior variance of θ. (Lond. M.Sc.)

7. In an experiment on the time taken by mice to respond to a drug, the log dose x_i received by mouse i is controlled exactly, and the log response time t_i is measured individually for all mice responding before log time T. The results of the experiment consist of $(n-r)$ pairs (x_i, t_i) with $t_i \leqslant T$, and r pairs in which x_i is known exactly, but t_i is 'censored', and is known only to be greater than T. It is known that t_i is normally distributed with *known* uniform variance σ^2, about the regression line

$$\mathscr{E}(t_i) = \alpha + \beta x_i,$$

where α and β are unknown. Verify that the following iterative procedure will converge to the maximum likelihood estimates of α and β.

(i) Using preliminary estimates a_1 and b_1, calculate y for each 'censored' observation from the formula

$$y_i = a_1 + b_1 x_i + \sigma \nu \left(\frac{T - a_1 - b_1 x_i}{\sigma} \right),$$

where $\nu(u)$ is the ratio of the ordinate to the right-hand tail of the distribution $N(0, 1)$ at the value u.

(ii) For non-censored observations, take $y_i = t_i$.

From the n pairs (x_i, y_i) calculate the usual least squares estimates of regression parameters.† Take these as a_2, b_2, and repeat until stable values are reached. (Aberdeen Dip.)

8. Serum from each of a random sample of n individuals is mixed with a certain chemical compound and observed for a time T, in order to record the time at which a certain colour change occurs. It is observed that r individuals respond at times $t_1, t_2, \ldots, t_r$, and that the remaining $(n-r)$ have shown no response at the end of the period T. The situation is thought to be describable by a probability density function $\alpha e^{-\alpha t}$ $(0 \leqslant t)$ for a fraction β of the population $(0 \leqslant \beta \leqslant 1)$ and complete immunity to the reaction in the remaining fraction $(1-\beta)$.

Obtain equations for $\hat{\alpha}$, $\hat{\beta}$, the maximum likelihood estimates of α, β. If the data make clear that $\hat{\alpha}$ is substantially greater than $1/T$, indicate how you would solve the equations. By consideration of asymptotic variances of the posterior distribution, show that, if α is known and αT is small, taking a sample of $2n$ instead of n may be much less effective for improving the precision of the estimation of β than the alternative of observing for time $2T$ instead of for T. (Aberdeen Dip.)

† Defined in equation 8.1.21.

9. A new grocery product is introduced at time $t = 0$. Observations are made for a period T on a random sample of n households. It is observed that m of these buy the product for the first time at times $t_1, \ldots, t_m$ and that in the remaining $n-m$ households the product has not been bought by the end of the period T.

It is suggested that a fraction $1-\theta$ of the population are not susceptible and will never buy the product and that for the susceptible fraction the time of first buying has an exponential probability density function $\alpha e^{-\alpha t}$. Show that the likelihood of the observations is proportional to

$$(\theta\alpha)^m \exp[-m\alpha\bar{t}]\,(1-\theta+\theta\exp[-\alpha T])^{n-m},$$

where $\bar{t} = \Sigma t_i/m$. Hence obtain the equations satisfied by the maximum likelihood estimates $\hat{\theta}$, $\hat{\alpha}$ of the unknown parameters θ, α and show that in particular

$$\frac{\bar{t}}{T} = \frac{1}{\hat{\alpha}T} - \frac{1}{\exp[\hat{\alpha}T]-1}.$$

Indicate how you would determine $\hat{\alpha}$ numerically. (Lond. B.Sc.)

10. A household consists of two persons, either of whom is susceptible to a certain infectious disease. After one person has contracted the infection a period of time U elapses until symptoms appear, when he is immediately isolated, and during this period it is possible for him to transmit the infection to the other person. Suppose that U is a random variable with probability density $\beta e^{-\beta u}$ $(0 < u < \infty)$ and that for given U the probability that the other person has *not* been infected by a time t after the beginning of the period is $e^{-\alpha t}$ $(0 < t \leqslant U)$. What is the probability of both persons being infected, given that one person has contracted the infection from some external source, assuming that the possibility of the second person contracting the infection from an external source may be neglected?

Out of n households in which infection has occurred there are r in which only one person is infected and for the remaining s, the intervals between the occurrence of the symptoms for the two persons are $t_1, t_2, \ldots, t_s$. Assuming that the probabilities for these households are all mutually independent, write down the likelihood function of the data, and show that the maximum likelihood estimates of α and β are

$$\hat{\alpha} = s^2/rT, \quad \hat{\beta} = s/T, \quad \text{where} \quad T = \sum_{i=1}^{s} t_i.$$

Determine also the asymptotic posterior variances of α and β.

(Camb. Dip.)

11. A survey is carried out to investigate the incidence of a disease. The number of cases occurring in a fixed period of time in households of size k is recorded for those households having at least one case. Suppose that the disease is not infectious and that one attack confers immunity. Let $p = 1-q$ be the probability that any individual develops the disease during the given period independently of other individuals. Show that the probability distribution of the number of cases per household will be

$$p(r|p) = \binom{k}{r} p^r q^{k-r}(1-q^k)^{-1} \quad (r = 1, 2, ..., k).$$

Derive the equation for the maximum likelihood estimate, $\hat{p}$, of p and show that, if p is small,

$$\hat{p} \sim \frac{2(\bar{r}-1)}{k-1},$$

where $\bar{r}$ is the average number of cases per household. Fit this distribution to the following data for $k = 3$, and, if your results justify it, estimate the number of households with no cases.

Distribution of number of cases per household

($k = 3$)

r	1	2	3	Total
Frequency	390	28	7	425

(Wales Dip.)

12. A new type of electric light bulb has a length of life, whose probability density function is of the form

$$f(x) = \frac{x}{\lambda^2} e^{-x/\lambda} \quad (x > 0).$$

Forty bulbs were given a life test and failures occurred after the following times (in hours):

196, 327, 405, 537, 541, 660, 671, 710,
786, 940, 954, 1004, 1004, 1006, 1202, 1459,
1474, 1484, 1602, 1662, 1666, 1711, 1784, 1796,
1799.

The tests were discontinued after 1800 hours, the remaining bulbs not then having failed.

Determine the maximum likelihood estimate of λ, and obtain an approximate 95 % confidence interval for this parameter. Hence, or otherwise, determine whether these results justify the manufacturer's claim that the average life of his bulbs is 2000 hours. (Leic. Gen.)

13. The random variables Y, Z have the form

$$Y = X+U, \quad Z = X+V,$$

where X, U, V, are independently normally distributed with zero means and variances θ, 1, 1. Given n independent pairs $(y_1, z_1), ..., (y_n, z_n)$, obtain the maximum likelihood estimate $\hat{\theta}$. What is the posterior distribution of θ? (Lond. M.Sc.)

14. A random sequence of n trials with initially unknown constant probability of success gives r successes. What is the probability that the $(n+1)$st trial will be successful? Consider in particular the cases $r = 0$ and $r = n$.

15. A random sample of size n from $P(\theta)$ gave values $r_1, r_2, \ldots, r_n$, and the prior knowledge of θ was small. Determine the probability that a further random sample of size m from $P(2\theta)$ will have mean s.

16. A random sample from $B(20, \theta)$ gave the value $r = 7$. Determine 50, 95 and 99 % confidence limits for θ by the following methods and compare their values:

(i) the exact method of theorem 7.2.1 and its first corollary:
(ii) the approximation of the second corollary to that theorem;
(iii) the inverse-sine transformation (§ 7.2);
(iv) χ^2, with and without the finite correction for small prior knowledge.

17. You are going to play a pachinko machine at a fair and can either win or lose a penny on each play. If θ is the probability of winning, you know that $\theta \leqslant \frac{1}{2}$ since otherwise the fair owner would not continue to use the machine (though there is always the slight chance that it has gone wrong, for him, and he has not noticed it). On the other hand, θ cannot be too much less than $\frac{1}{2}$ since then people would not play pachinko. Suggest a possible prior distribution. You play it 20 times and win on 7 of them. Find 95 % confidence limits for θ and compare with the corresponding limits in the previous exercise.

18. Initially, at time $t = 0$, a Poisson process is set to produce on the average one incident per unit time. It is observed for a time T and incidents occur at $t_1, t_2, \ldots, t_n$. It is then learnt from other information that at time T the Poisson process is working at a rate of two incidents per unit time. If it is believed that the process doubled its rate at some instant of time, θ, between 0 and T and that θ is equally likely to be anywhere in $(0, T)$, discuss the form of the posterior distribution of θ given the observed pattern of incidents in $(0, T)$.

19. The table shows the number of motor vehicles passing a specified point between 2.00 p.m. and 2.15 p.m. on 6 days in 2 successive weeks.

	Mon.	Tues.	Wed.	Thurs.	Fri.	Sat.
Week 1	50	65	52	63	84	102
Week 2	56	49	60	45	112	90

The flow of vehicles on any day is believed to conform to a Poisson distribution, in such a way that the mean number of vehicles per 15 minutes is λ on Monday to Thursday, μ on Friday and Saturday. Indicate how you would test deviations from either Poisson distribution. Obtain 95 % confidence limits for λ and μ. Test the hypothesis $2\lambda = \mu$. (Aberdeen Dip.)

20. The table shows the numbers of births (rounded to nearest 100) and the numbers of sets of triplets born in Norway between 1911 and 1940.

Period	Total births	Set of triplets
1911–15	308,100	52
1916–20	319,800	52
1921–25	303,400	40
1926–30	253,000	30
1931–35	222,900	24
1936–40	227,700	20

Apply χ^2 to test the significance of differences between periods in the frequencies of triplets, explaining and taking advantage of any convenient approximations. Discuss the relation of your test to a test of homogeneity of Poisson distributions. Indicate briefly how any linear trend in the proportion of triplet births might be examined. (Aberdeen Dip.)

21. Cakes of a standard size are made from dough with currants randomly distributed throughout. A quarter is cut from a cake and found to contain 25 currants. Find a 95 % confidence interval for the mean number of currants per cake. (Camb. N.S.)

22. Past observations of a bacterial culture have shown that on the average $2\frac{1}{2}$ % of the cells are undergoing division at any instant. After treatment of the culture 200 cells are counted and only 1 is found to be dividing. Is this evidence that the treatment has decreased the ability of the cells to divide? (Camb. N.S.)

23. The following table shows the result of recording the telephone calls handled between 1 p.m. and 2 p.m. on each of 100 days, e.g. on 36 days no calls were made. Show that this distribution is consistent with calls arriving independently and at random. Obtain (on this assumption) a 99 % confidence limit for the probability that if the operator is absent for 10 minutes no harm will be done.

Calls	0	1	2	3	4 or more
Days	36	35	22	7	0

(Camb. N.S.)

24. Bacterial populations in a fluid suspension are being measured by a technique in which a given quantity of the fluid is placed on a microscope slide and the number of colonies on the slide is counted. If the laboratory technique is correct, replicate counts from the same culture should follow a Poisson distribution. Six slides prepared from a certain culture give the counts: 105, 92, 113, 90, 97, 102. Discuss whether there is any evidence here of faulty laboratory technique. (Camb. N.S.)

25. In an investigation of the incidence of death due to a particular cause it is required to find whether death is more likely at some times of the day than at others. In 96 cases the times of death are distributed as follows in 3-hourly intervals beginning at the stated times.

Mid-night	3 a.m.	6 a.m.	9 a.m.	Noon	3 p.m.	6 p.m.	9 p.m.	Total
19	15	8	8	12	14	9	11	96

Do these data provide adequate evidence for stating that the chance of death is not constant?

Indicate, without doing the calculations, what test you would have used had it been suggested *from information unconnected with these data*, that the chance of death is highest between midnight and 6 a.m. Why is the qualification in italics important? (Camb. N.S.)

26. In some classical experiments on pea-breeding Mendel obtained the following frequencies for different kinds of seed in crosses with round yellow seeds and wrinkled green seeds:

	Observed	Theoretical
Round and yellow	315	312·75
Wrinkled and yellow	101	104·25
Round and green	108	104·25
Wrinkled and green	32	34·75
	556	556·00

The column headed 'Theoretical' gives the numbers that would be expected on the Mendelian theory of inheritance, which predicts that the frequencies should be in proportions 9, 3, 3, 1. Can the difference between the observed and theoretical figures be ascribed to chance fluctuations? (Camb. N.S.)

27. In an experiment to investigate whether certain micro-organisms tended to move in groups or not, 20 micro-organisms were placed in the field of view A of a microscope and were free to move in a plane of which A was a part. After a lapse of time the number of micro-organisms remaining within A was counted. The experiment was repeated six times with the following results:

7, 12, 4, 3, 16, 17.

Perform a significance test designed to test the hypothesis that the micro-organisms move independently of each other, explaining carefully the appropriateness of your analysis to the practical problem described. (Camb. N.S.)

28. The independent random variables $X_1, \ldots, X_n$ follow Poisson distributions with means $\mu_1, \ldots, \mu_n$. The hypotheses H_1, H_2, H_3 are defined as follows:

$$H_1: \mu_1, \ldots, \mu_n \text{ are arbitrary;}$$
$$H_2: \mu_i = \alpha + i\beta \quad (i = 1, \ldots, n);$$
$$H_3: \mu_i = \gamma,$$

where a, β, γ are unknown. Develop tests for:

(i) the null hypothesis H_2 with H_1 as alternative;

(ii) the null hypothesis H_3 with H_2 as alternative.

(Lond. M.Sc.)

29. An observer records the arrival of vehicles at a road bridge during a period of just over an hour. In this period 54 vehicles reach the bridge, the intervals of time between their consecutive instants of arrival (in seconds) being given in numerical order in the table below.

It is postulated that the 53 intervals t between consecutive arrivals are distributed independently according to the exponential distribution

$$\frac{1}{\lambda}e^{-t/\lambda}$$

(or, equivalently, that the values of $2t/\lambda$ are independently distributed as χ^2 with two degrees of freedom). Test this hypothesis.

Observed values of t

2	11	29	44	77	148
4	15	31	47	97	158
6	17	33	50	114	163
7	18	34	51	114	165
9	19	35	53	116	180
9	19	37	61	121	203
9	28	38	68	124	340
10	29	38	73	135	393
10	29	43	74	146	

(Manch. Dip.)

30. A suspension of particles in water is thoroughly shaken. 1 c.c. of the suspension is removed and the number of particles in it is counted. The remainder of the suspension is diluted to half concentration by adding an equal volume of water, and after being thoroughly shaken another c.c. is removed from the suspension and a count of particles is made. This dilution process is repeated until four such counts in all have been made.

The following counts were obtained by two different experimenters A, B, using similar material.

Dilution	1	$\frac{1}{2}$	$\frac{1}{4}$	$\frac{1}{8}$
A	18	11	9	2
B	49	9	5	2

Deduce what you can about the adequacy of the experimental technique and the consistency of the experimenters. (Camb. N.S.)

31. The following table gives the results of 7776 sequences of throws of an ordinary die. For $s = 1, 2, 3$, n_s is the number of sequences in which a 6 was first thrown at the sth throw, and n_4 is the number of sequences in which a 6 was not thrown in any of the first 3 throws. Investigate if there is evidence of the die being biased. Suggest a simpler experimental procedure to test for bias.

s	1	2	3	4	Total
n_s	1216	1130	935	4495	7776

(Camb. N.S.)

32. Two fuses included in the same apparatus are assumed to have independent probabilities p and q of being blown during each repetition of a certain experiment. Show that the probabilities of neither, one only, or both being blown during any repetition are given respectively by

$$1-p-q+pq, \quad p+q-2pq, \quad pq.$$

In 154 repetitions, one only is blown 93 times, and both are blown 36 times. Assuming that $p = 0{\cdot}60$, $q = 0{\cdot}50$, test the assumption that the actions of the fuses are independent. What kind of departure from independence is in fact indicated? (Camb. N.S.)

33. A sequence is available of $n+1$ geological strata of four different types in all, denoted by a, b, c and d. The null hypothesis assumes that the occurrence of the four types occur randomly (with possibly different probabilities p_a, p_b, p_c and p_d), except that two consecutive strata of the same type cannot be separated and thus do not appear in the sequence. Show that the sequence is a Markov chain on the null hypothesis, and write down its transition probability matrix. What are the maximum likelihood estimates of p_a, p_b, p_c and p_d? It is suspected that the actual sequence may show some non-randomness, for example, some tendency to a repeated pattern of succession of states. How would you construct a χ^2 test to investigate this? Give some justification for your answer, indicating the formula you would use to calculate χ^2 and the number of degrees of freedom you would allocate to it. (Lond. Dip.)

34. The table given below, containing part of data collected by Parkes from herd-book records of Duroc-Jersey Pigs, shows the distribution of sex in litters of 4, 5, 6 and 7 pigs.

Examine whether these data are consistent with the hypothesis that the number of males within a litter of given size is a binomial variable, the sex ratio being independent of litter size. If you are not altogether satisfied with this hypothesis, in what direction or directions does it seem to fail?

No. of males in litter	Size of litter			
	4	5	6	7
0	1	22	3	—
1	14	20	16	21
2	23	41	53	63
3	14	35	78	117
4	1	14	53	104
5	—	4	18	46
6	—	—	—	21
7	—	—	—	2
Totals	53	116	221	374

(Lond. B.Sc.)

35. In connexion with a certain experiment an instrument was devised which, in each consecutive time period of $\frac{1}{10}$th of a second, either gave an impulse or did not give an impulse. It was hoped that the probability, p, of the impulse occurring would be the same for every time period and independent of when the previous impulses occurred.

To test the instrument it was switched on and a note was made of the number of periods between the start of the test and the first impulse and thereafter of the number of periods between consecutive impulses. The test was continued until 208 impulses had been noted and the data were then grouped into the number of impulses which occurred in the first period after the previous one, the number in the 2nd period after the previous one (i.e. after one blank period) and so on. These grouped data are given in the table.

Period after previous impulse (or start)	No. of impulses	Period after previous impulse (or start)	No. of impulses
1st	81	7th	4
2nd	44	8th	3
3rd	24	9th	5
4th	22	10th	3
5th	13		
6th	9	Total	208

Fit the appropriate probability distribution, based on the above assumptions, to these data and test for goodness of fit.

The makers of the instrument claim that the probability of an impulse occurring during any period is 0·35. Test whether the data are consistent with this claim. (Lond. B.Sc.)

36. If trees are distributed at random in a wood the distance r from a tree to its nearest neighbour is such that the probability that it exceeds a is $\exp[-\pi a^2/4\mu^2]$, where μ is the mean distance.

The following data (from E. C. Pielou, *J. Ecol.* **48**, 575, 1960) give the frequencies of the nearest neighbour distances for 158 specimens of *Pinus ponderosa*. Test whether the trees may be regarded as randomly distributed and comment on your results.

Distance to nearest neighbour (in cm)	Frequency	Distance to nearest neighbour (in cm)	Frequency
0–50	13	301–350	8
51–100	17	351–400	5
101–150	34	401–450	3
151–200	21	451–500	7
201–250	9	501 —	14
251–300	17	Total	158

(Leic. Stat.)

37. Antirrhinum flowers are either white, or various shades of red. Five seeds, taken from a seed-pod of one plant, were sown, and the number of seeds which produced plants with white flowers was noted. The process was repeated with five seeds from each of 100 seed-pods, and the following

table shows the number of seed-pods in which k seeds were produced with white flowers ($k = 0, 1, \ldots, 5$).

No. of plants with white flowers	0	1	2	3	4	5	Total
No. of seed pods	20	44	25	7	3	1	100

Test the hypotheses (*a*) that the probability of obtaining a white flower is constant for all 100 seed-pods, and (*b*) that the probability of obtaining a white flower for a seed chosen at random is 1/4. (Lond. Psychol.)

38. Birds of the same species, kept in separate cages, were observed at specified times and it was found that each bird had a probability of $\frac{1}{2}$ of being on its perch when observed. Nine of these birds were then put together into an aviary with a long perch. At each of the next 100 times of observation the number of birds on the perch was counted and the following results were found:

No. of birds	1	2	3	4	5
Frequency with which this number was observed	2	6	18	22	52

Show that there is some evidence for a change in perching habits. (Camb. N.S.)

39. Each engineering apprentice entering a certain large firm is given a written test on his basic technical knowledge, his work being graded *A*, *B*, *C*, or *D* (in descending order of merit). At the end of his third year with the firm his section head gives him a rating based on his current performance; the four possible classifications being 'excellent', 'very good', 'average', and 'needs to improve'. The results for 492 apprentices are given in the following two-way table. Analyse the data, and state clearly what conclusions you draw regarding the usefulness, or otherwise, of the written test as an indicator of an entrant's subsequent performance in the firm.

Section head's assessment	Written test result			
	A	*B*	*C*	*D*
Excellent	26	29	21	11
Very good	33	43	35	20
Average	47	71	72	45
Needs to improve	7	12	11	9

(Manch. Dip.)

40. The following data (from H. E. Wheeler, *Amer. J. Bot.* **46**, 361, 1959) give the numbers of fertile and infertile perithecia from cultures of *Glomerella* grown in different media:

Medium	Fertile	Infertile	Total
Oatmeal	179	56	235
Cornmeal	184	19	203
Potato dextrose	161	39	200
Synthetic	176	26	202

Determine whether there is a significant difference between these media in respect to fertility of the perithecia. If there is a significant difference obtain an approximate 95 % confidence interval for the proportion of fertile perithecia given by the best medium. Otherwise obtain a similar interval for the pooled results. (Leic. Stat.)

41. Each of n individuals is classified as A or not-A ($\bar{A}$) and also as B or $\bar{B}$. The probabilities are:

$$p(AB) = \theta_{11}, \quad p(A\bar{B}) = \theta_{12}, \quad p(\bar{A}B) = \theta_{21}, \quad p(\bar{A}\bar{B}) = \theta_{22},$$

with $\sum_{i,j} \theta_{ij} = 1$; and the individuals are independent. Show that the joint distribution of m, the number classified as A; r, the number classified as B; and a, the number classified as both A and B (that is, AB) is

$$\left(\frac{\theta_{11}\theta_{22}}{\theta_{12}\theta_{21}}\right)^a \theta_{12}^m \theta_{21}^r \theta_{22}^{n-m-r} \; N(m, r, a),$$

where $N(m, r, a)$ is some function of m, r and a, but not of the θ_{ij}. Hence show that the probability distribution of a, given r and m depends only on $\theta_{11}\theta_{22}/\theta_{12}\theta_{21}$.

Suggest a method of testing that the classifications into A and B are independent: that is $p(AB) = p(A)p(B)$. (Wales Maths.)

42. Test whether there is any evidence for supposing an association between the dominance of hand and of eye in the following contingency table (i.e. of the 60 people, 14 were left-handed and left-eyed):

	Left-eyed	Right-eyed	Totals
Left-handed	14	6	20
Right-handed	19	21	40
Totals	33	27	60

How would you test if the association was positive (i.e. left-handedness went with left-eyedness)? (Camb. N.S.)

43. One hundred plants were classified with respect to the two contrasts *large L* versus *small l*, *white W* versus *coloured w*, the numbers in the four resulting classes being as shown in the table:

	L	l	Totals
W	40	15	55
w	20	25	45
Totals	60	40	100

Investigate the following questions:

(*a*) Is there reason to believe that the four classes are not being produced in equal numbers?

(*b*) Is the apparent association between L and W statistically significant? (Camb. N.S.)

44. k samples S_i $(i = 1, 2, \ldots, k)$ of an insect population are taken at k different sites in a certain habitat. S_i consists of n_i insects of which a_i are black and $b_i(=n_i-a_i)$ are brown. When the samples are combined the total number of specimens is n of which a are black and $b(=n-a)$ are brown. It is desired to test whether the observed proportions of the two kinds of insect differ significantly from site to site. Calculate a χ^2 appropriate to test this hypothesis. State the number of degrees of freedom, and prove that the χ^2 can be put into any one of the following forms:

$$\chi^2 = \sum_{i=1}^{k} \frac{(na_i-n_ia)^2}{abn_i} = \sum_{i=1}^{k} \frac{(ba_i-ab_i)^2}{abn_i}$$

$$= \frac{n^2}{ab}\left\{\sum_{i=1}^{k} \frac{a_i^2}{n_i} - \frac{a^2}{n}\right\}$$

$$= \frac{n^2}{4ab}\left\{\sum_{i=1}^{k} \frac{(a_i-b_i)^2}{n_i} - \frac{(a-b)^2}{n}\right\}.$$

(Camb. N.S.)

45. A group of individuals is classified in two different ways as shown in the following table:

	X	Not-X	
Y	a	b	$a+b$
Not-Y	c	d	$c+d$
	$a+c$	$b+d$	n

Establish the formula

$$\chi^2 = n(ad-bc)^2/(a+c)(b+d)(a+b)(c+d)$$

for testing the independence of the two classifications.

An entomologist collects 1000 specimens of *Insecta corixida*, 500 from each of two lakes, and finds that 400 specimens from one lake and 375 from the other have long wings. Is he justified in reporting a significant difference in the proportions of long-winged corixids in the two lakes?

(Camb. N.S.)

46. A sample of 175 school-children was classified in two ways, the result being given in the table. P denotes that the child was particularly able and not-P that it was not so. Is the apparent association between ability and family prosperity large enough to be regarded as statistically significant?

	P	Not-P
Very well clothed	25	40
Well clothed	15	60
Poorly clothed	5	30

(Camb. N.S.)

47. In a routine eyesight examination of 8-year-old Glasgow school-children in 1955 the children were divided into two categories, those who wore spectacles (A), and those who did not (B). As a result of the test, visual acuity was classed as good, fair or bad. The children wearing spectacles were tested with and without them.

The following results are given:

Visual acuity of Glasgow school-children (1955)

Category	A, with spectacles: Boys	Girls	A, without spectacles: Boys	Girls	B: Boys	Girls
Good	157	175	90	81	5908	5630
Fair	322	289	232	222	1873	2010
Bad	62	50	219	211	576	612
Total	541	514	541	514	8357	8252

What conclusions can be drawn from these data, regarding (*a*) sex differences in eyesight, (*b*) the value of wearing spectacles?

The figures for 8-year-old boys and girls for the years 1953 and 1954 were not kept separately for the sexes. They are shown as follows:

Visual acuity of Glasgow school-children (1953, 1954)

Category	A, with spectacles: 1953	1954	A, without spectacles: 1953	1954	B: 1953	1954
Good	282	328	152	173	8,743	10,511
Fair	454	555	378	443	3,212	3,565
Bad	84	78	290	345	1,015	1,141
Total	820	961	820	961	12,970	15,217

Are there, in your opinion, any signs of changes in the 8-year-old population of visual acuity with time? Do you think it possible that your conclusions might be vitiated by the pooling of the frequencies for both sexes in 1953 and 1954? (Lond. B.Sc.)

48. A radioactive sample emits particles randomly at a rate which decays with time, the rate being $\lambda e^{-\kappa t}$ after time t. The first n particles emitted are observed at successive times $t_1, t_2, \ldots, t_n$. Set up equations for the maximum-likelihood estimates $\hat{\lambda}$ and $\hat{\kappa}$, and show that $\hat{\kappa}$ satisfies the equation

$$\frac{\hat{\kappa}t_n}{e^{\hat{\kappa}t_n}-1} = 1-\hat{\kappa}\bar{t},$$

where

$$\bar{t} = \frac{1}{n}\sum_i t_i.$$

Find a simple approximate expression for $\hat{\kappa}$ when t_n/t is a little greater than 2. (Camb. Dip.)

49. A man selected at random from a population has probabilities $(1-\theta)^2$, $2\theta(1-\theta)$, θ^2 of belonging to the categories AA, AB, BB respectively. The laws of inheritance are such that the probabilities of the six possible pairs of brothers, instead of being the terms of a trinomial expansion, are:

Brothers		No. of cases
AA, AA	$\frac{1}{4}(1-\theta)^2(2-\theta)^2$	n_1
AA, AB	$\theta(1-\theta)^2(2-\theta)$	n_2
AA, BB	$\frac{1}{2}\theta^2(1-\theta)^2$	n_3
AB, AB	$\theta(1-\theta)(1+\theta-\theta^2)$	n_4

Brothers		No. of cases
AB, BB	$\theta^2(1-\theta)^2$	n_5
BB, BB	$\frac{1}{4}\theta^2(1+\theta)^2$	n_6
Total	1	N

A random sample of N pairs of brothers is collected and is found to consist of $n_1, n_2, \ldots n_6$ instances of the six types. Obtain an equation for the maximum likelihood estimate of the parameter θ. Show that the posterior distribution of θ has approximate variance

$$\frac{g(1+g)(2+g)}{N(6+5g+4g^2)},$$

where $g = \theta(1-\theta)$.

An alternative and arithmetically simpler estimation procedure would be to score 0 for each AA individual, 1 for each AB, and 2 for each BB, and then to equate the total score of the whole sample to its expectation. Find the posterior variance of θ using this method and compare with the maximum likelihood value. (Camb. Dip.)

50. Bacteria in suspension form into clumps, the probability that a random clump will contain n bacteria is $\theta^{n-1}(1-\theta)$ $(n = 1, 2, \ldots)$. When subject to harmful radiation there is a probability $\lambda\delta t$ that a bacteria will be killed in any interval of length δt irrespective of the age of the bacteria in the same or different clumps. A clump is not killed until all the bacteria in it have been killed. Prove that the probability that a random clump will be alive after being exposed to radiation for a time t is

$$e^{-\lambda t}/(1-\theta+\theta e^{-\lambda t}).$$

In order to estimate the strength, λ, of the radiation a unit volume of the unradiated suspension is allowed to grow and the number n_1 of live clumps counted. The remainder of the suspension is irradiated for a time t and then a unit volume is allowed to grow free of radiation and the number r_1 of live clumps counted. It may be assumed that both before and after radiation the clumps are distributed randomly throughout the suspension. The experiment is repeated with new suspensions s times in all giving counts $(n_1, n_2, \ldots, n_s)$ and $(r_1, r_2, \ldots, r_s)$. Show that if θ is known the maximum likelihood estimate of λ is

$$t^{-1}\ln\left\{\frac{N-\theta R}{R(1-\theta)}\right\},$$

where $$N = \sum_{i=1}^{s} n_i \quad \text{and} \quad R = \sum_{i=1}^{s} r_i.$$ (Camb. Dip.)

51. There are two ways in which an item of equipment may fail. If failure has not occurred at time t, there is a chance $\lambda_1\delta t + o(\delta t)$ of failure of type I and a chance $\lambda_2\delta t + o(\delta t)$ of failure of type II in $(t, t+\delta t)$. A number n of items is placed on test and it is observed that r_1 of them fail from cause I at times $t'_1, \ldots, t'_{r_1}$, that r_2 of them fail from cause II at times $t''_1, \ldots, t''_{r_2}$ and that when the test is stopped at time T the remaining

$(n-r_1-r_2)$ items have not failed. (As soon as an item fails, it takes no further part in the test, and the separate items are independent.) Obtain maximum likelihood estimates of λ_1 and λ_2 and explain the intuitive justification of your formulae. Derive a test of the null hypothesis $\lambda_1 = \lambda_2$.
(Camb. Dip.)

52. In an experiment to measure the resistance of a crystal, independent pairs of observations (x_i, y_i) $(i = 1, 2, \ldots, n)$ of current x and voltage y are obtained. These are subject to errors (X_i, Y_i), so that

$$x_i = \xi_i + X_i, \quad y_i = \eta_i + Y_i,$$

where (ξ_i, η_i) are the true values of current and voltage on the ith occasion, and $\eta_i = \alpha\xi_i$, α being the resistance of the crystal. On the assumption that the errors are independently and normally distributed with zero means and variances $\mathscr{D}^2(X_i) = \sigma_1^2$, $\mathscr{D}^2(Y_i) = \sigma_2^2 = \lambda\sigma_1^2$, where λ is known, show that $\hat{\alpha}$, the maximum likelihood estimate of α, is a solution of the equation

$$\hat{\alpha}^2 S_{xy} + \hat{\alpha}\{\lambda S_{xx} - S_{yy}\} - \lambda S_{xy} = 0,$$

where $$S_{xy} = \Sigma x_i y_i/n, \quad S_{xx} = \Sigma x_i^2/n, \quad S_{yy} = \Sigma y_i^2/n.$$

Show also that if $\Sigma\xi_i^2/n$ tends to a limit as $n \to \infty$, then $\hat{\alpha}$ converges in probability to α.

Show that the method of maximum likelihood gives unsatisfactory results when λ is not assumed known. (Camb. Dip.)

53. Let $Z_1, Z_2, \ldots, Z_m$ be independent observations which are identically distributed with distribution function $1-\exp(-x^\beta)$, where β is an unknown parameter. Obtain an estimate of β useful as $m \to \infty$ and hence the approximate posterior distribution of β. If the calculation of your estimator involves iteration, explain how a first approximation to this is to be found. How would you test the hypothesis $\beta = 1$ against the two-sided alternative $\beta \neq 1$? (Camb. Dip.)

54. In an investigation into the toxicity of a certain drug an experiment using k groups of animals is performed. Each of the n_i animals in the ith group is given a dose x_i, and the number of resulting deaths r_i is recorded $(i = 1, 2, \ldots, k)$. The probability of death for an animal in the ith group is assumed to be

$$P_i(\alpha, \beta) = \frac{1}{1+e^{-(\alpha+\beta x_i)}},$$

the probabilities for different animals being mutually independent. Show that the maximum likelihood estimates $\hat{\alpha}, \hat{\beta}$ of α, β satisfy the equations

$$\Sigma r_i = \Sigma n_i P_i(\hat{\alpha}, \hat{\beta}),$$

$$\Sigma r_i x_i = \Sigma n_i x_i P_i(\hat{\alpha}, \hat{\beta}),$$

and indicate how these estimates may be determined by successive approximation.

Show also that the asymptotic variances and covariances of the posterior distribution of α and β are given by

$$\mathscr{D}^2(\alpha) = (1/\Sigma w_i)+(\bar{x}^2/S_{xx}), \quad \mathscr{D}^2(\beta) = 1/S_{xx},$$

$$\mathscr{C}(\alpha, \beta) = -\bar{x}/S_{xx},$$

where
$$w_i = n_i P_i(\hat{\alpha}, \hat{\beta})\,[1-P_i(\hat{\alpha}, \hat{\beta})],$$

$$\bar{x} = \Sigma w_i x_i/\Sigma w_i, \quad S_{xx} = \Sigma w_i(x_i-\bar{x})^2.$$

Hence obtain a large sample test of the hypothesis that $-\alpha/\beta$, the mean of the tolerance distribution associated with $P_i(\alpha, \beta)$, has a specified value μ_0, and derive a 95% large sample confidence interval for $-\alpha/\beta$.

(Camb. Dip.)

55. Show that the probability of needing exactly k independent trials to obtain a given number n of successes, when each trial can result in success or failure and the chance of success is p for all trials, is

$$\binom{k-1}{n-1} p^n q^{k-n},$$

where $q = 1-p$. Writing $k = n+s$, show that the mean and variance of s are nq/p and nq/p^2.

Two such sequences of trials were carried out in which the required numbers of successes were n_1 and n_2, and the numbers of trials needed were n_1+s_1 and n_2+s_2. Two estimates were proposed for the mean number of failures per success; the first was the total number of failures divided by the total number of successes, and the second was the average of the numbers of failures per success in the two sequences. Which estimate would you prefer? State your reasons carefully. (Camb. N.S.)

56. A subject answers n multiple-choice questions, each having k possible answers, and obtains R correct answers. Suppose that the subject really knows the answers to v of the questions, where v is an unknown parameter, and that his answers to the remaining $n-v$ questions are pure guesses, each having a probability $1/k$ of being correct. Write down the probability distribution of R, indicating its relation to the binomial distribution.

If θ is the probability that he knows the correct answer and has prior distribution $B_0(a, b)$, and if the questions are answered independently discuss the posterior distribution of θ given R. (Lond. B.Sc.)

57. An estimate is required of the number of individuals in a large population of size N, who possess a certain attribute A. Two methods are proposed. Method I would select a random sample of size αN and interview each member of the sample to determine whether he possesses A or not. Method II would send out the postal question 'Do you possess A?' to all N individuals and then, in order to avoid any possible bias due to misunderstanding of the question, etc., would interview a random sample of size αN_y of those N_y individuals who replied 'Yes' and a random sample of size αN_n of those N_n individuals who replied 'No'.

Assuming that, if method II were used, all N individuals would reply to the question, suggest estimates for the two methods in terms of the results of the interviews. If $p_y(=1-q_y)$ is the proportion of the yes-replying individuals who actually have A, and $p_n(=1-q_n)$ is the proportion of the no-replying individuals who actually have A, show that the variances of the estimates for methods I and II are

$$\frac{(N_y p_y + N_n p_n)(N_y q_y + N_n q_n)}{N\alpha} \quad \text{and} \quad \frac{N_y P_y q_y + N_n p_n q_n}{\alpha},$$

respectively.

Show that method II has the smaller variance. Discuss the relevance of these sample variances to the posterior distribution of the number in the population possessing A. (Wales Dip.)

8
LEAST SQUARES

The method of least squares is a method of investigating the dependence of a random variable on other quantities. Closely associated with the method is the technique of analysis of variance. We begin with a special case and later turn to more general theory. The reader is advised to re-read the section on the multivariate normal distribution in §3.5 before reading §8.3.

8.1. Linear homoscedastic normal regression

We are interested in the dependence of one random variable y on another random variable x. We saw in §3.2 that this could most easily be expressed in terms of the distribution of x and the conditional distribution of y, given x. These distributions will depend on parameters; suppose the dependence is such that

$$p(x, y \mid \theta, \phi) = p(x \mid \phi)\, p(y \mid x, \theta); \tag{1}$$

that is, the parameters in the two distributions are distinct. Then this is effectively equation 5.5.21 again with $\mathbf{x} = (x, y)$ and $t(\mathbf{x}) = x$. As explained in §5.5, and illustrated in proving theorem 7.6.2, if θ and ϕ have independent prior distributions, inferences about the conditional distribution may be made by supposing x to be fixed. We shall make these two assumptions ((1) and the prior independence of θ and ϕ) throughout this section, and use more general forms of them in later sections. Consequently all our results will be stated for fixed x. Provided the assumptions obtain, they are equally valid for random x. Linear homoscedastic normal regression has already been defined (equations 3.2.13 and 14).

Theorem 1. *If* $\mathbf{x} = (x_1, x_2, \ldots, x_n)$ *is a set of real numbers, and if, for fixed* $\mathbf{x}$, *the random variables* $y_1, y_2, \ldots, y_n$ *are independent normal random variables with*

$$\mathscr{E}(y_i \mid \mathbf{x}) = \alpha + \beta(x_i - \bar{x}) \tag{2}$$

and $$\mathscr{D}^2(y_i \mid \mathbf{x}) = \phi \tag{3}$$

($i = 1, 2, \ldots, n$); then if the prior distributions of α, β and $\ln\phi$ are independent and uniform:

(i) *the posterior distribution of*

$$(\beta - b)/\{S^2/S_{xx}(n-2)\}^{\frac{1}{2}} \tag{4}$$

is a t-distribution with $(n-2)$ degrees of freedom;

(ii) *the posterior distribution of*

$$(\alpha - a)/\{S^2/n(n-2)\}^{\frac{1}{2}} \tag{5}$$

is a t-distribution with $(n-2)$ degrees of freedom;

(iii) *the posterior distribution of*

$$S^2/\phi \tag{6}$$

is χ^2 with $(n-2)$ degrees of freedom.

The notation used in (4), (5) *and* (6) *is*

$$\left.\begin{aligned} S_{xx} = \Sigma(x_i - \bar{x})^2, \quad S_{xy} &= \Sigma(x_i - \bar{x})(y_i - \bar{y}), \\ S_{yy} &= \Sigma(y_i - \bar{y})^2, \end{aligned}\right\} \tag{7}$$

$$S^2 = S_{yy} - S_{xy}^2/S_{xx} \tag{8}$$

and $$a = \bar{y}, \quad b = S_{xy}/S_{xx}. \tag{9}$$

The likelihood of the y's, given the x's, is

$$p(\mathbf{y} \mid \mathbf{x}, \alpha, \beta, \phi) \propto \phi^{-\frac{1}{2}n} \exp\left[-\sum_{i=1}^{n} \{y_i - \alpha - \beta(x_i - \bar{x})\}^2/2\phi\right] \tag{10}$$

and hence the joint posterior distribution of α, β and ϕ is

$$\pi(\alpha, \beta, \phi \mid \mathbf{x}, \mathbf{y}) \propto \phi^{-\frac{1}{2}(n+2)} \exp\left[-\sum_{i=1}^{n} \{y_i - \alpha - \beta(x_i - \bar{x})\}^2/2\phi\right]. \tag{11}$$

The sum of squares in the exponential may be written as

$$\begin{aligned} &\Sigma\{(y_i - \bar{y}) + (\bar{y} - \alpha) - \beta(x_i - \bar{x})\}^2 \\ &\quad = S_{yy} + n(\bar{y} - \alpha)^2 + \beta^2 S_{xx} - 2\beta S_{xy} \\ &\quad = S_{yy} - S_{xy}^2/S_{xx} + n(\bar{y} - \alpha)^2 + S_{xx}(\beta - S_{xy}/S_{xx})^2 \\ &\quad = S^2 + n(\alpha - a)^2 + S_{xx}(\beta - b)^2, \end{aligned} \tag{12}$$

in the notation of (8) and (9). Hence

$$\pi(\alpha, \beta, \phi \mid \mathbf{x}, \mathbf{y}) \propto \phi^{-\frac{1}{2}(n+2)} \exp[-\{S^2 + n(\alpha - a)^2 + S_{xx}(\beta - b)^2\}/2\phi]. \quad (13)$$

Integration with respect to α gives

$$\pi(\beta, \phi \mid \mathbf{x}, \mathbf{y}) \propto \phi^{-\frac{1}{2}(n+1)} \exp[-\{S^2 + S_{xx}(\beta - b)^2\}/2\phi], \quad (14)$$

and then with respect to ϕ gives (using theorem 5.3.2)

$$\pi(\beta \mid \mathbf{x}, \mathbf{y}) \propto \{S^2 + S_{xx}(\beta - b)^2\}^{-\frac{1}{2}(n-1)}. \quad (15)$$

If t, defined by (4), replaces β; then, the Jacobian being constant,

$$\pi(t \mid \mathbf{x}, \mathbf{y}) \propto \{1 + t^2/(n-2)\}^{-\frac{1}{2}(n-1)}. \quad (16)$$

A comparison with the density of the t-distribution, equation 5.4.1, establishes (i).

(ii) follows similarly, integrating first with respect to β and then with respect to ϕ.

To prove (iii) we integrate (14) with respect to β, which gives

$$\pi(\phi \mid \mathbf{x}, \mathbf{y}) \propto \phi^{-\frac{1}{2}n} e^{-\frac{1}{2}S^2/\phi}, \quad (17)$$

and a comparison with equation 5.3.2 establishes the result.

Discussion of the assumptions

The basic assumptions, (1) and the prior independence of the separate parameters of the marginal and conditional distributions, are often satisfied in practice. In the language of § 5.5 we may say that x is ancillary for θ for fixed ϕ. As in the examples of sample size in § 5.5 and the margin of a contingency table in § 7.6, it does not matter how the x values were obtained.

One application of the results of the present section is to the situation where a random sample of n items is taken and two quantities, x and y, are measured on each item: for example, the heights and weights of individuals. Another application is to the situation where experiments are carried out independently at several different values $x_1, x_2, \ldots, x_n$ of one factor and measurements $y_1, y_2 \ldots, y_n$ are made on another factor: for example, a scientist may control the temperature, x_i, of each experiment and measure the pressure, y_i. The x value, usually called the *independent variable*, is different in the two cases: in the former it is a random variable, in the latter it is completely under the control of the experimenter. Nevertheless, the same analysis of the dependence of y on x applies in both cases. The y variable is usually called the *dependent variable*. The reader should be careful not to confuse the use of the word 'independent' here with its use in the phrase 'one random variable is independent of another'.

The form of the conditional distribution assumed in theorem 1 is that discussed in §3.2. The distributions are normal and therefore can be completely described by their means and variances. The means are supposed *linear* in the independent variable and the variances are constant (*homoscedastic*). (Notice that we have written the regression in (2) in a form which is slightly different from equation 3.2.13, the reason for which will appear below: essentially, α of 3.2.13 has been rewritten $\alpha - \beta\bar{x}$.) It is important that the assumptions of linearity and homoscedasticity should be remembered when making any application of the theorem. Too often regression lines are fitted by the simple methods of this section without regard to this point. An example of alternative methods of investigating the dependence of y on x without invoking them will be discussed below. Particular attention attaches to the case $\beta = 0$, when the two variables are independent and there is said to be no regression effect. A test of this null hypothesis will be developed below (table 8.1).

The prior distributions of α and ϕ are as usual: ϕ is a variance and $\alpha = n^{-1}\Sigma\mathscr{E}(y_i|\mathbf{x})$; that is, the expectation of y averaged over the values of x used in the experiment. We have chosen β, which is the only parameter that expresses any dependence of y on x, to be uniformly distributed: this is to be interpreted in the sense of §5.2, merely meaning that the prior knowledge of β is so diffuse that the prior density is sensibly constant over the effective range of the likelihood function. It has been supposed independent of α and ϕ because it will typically be reasonable to suppose that knowledge of the expectation and/or the variance of y (that is, of α and/or ϕ) will not alter one's knowledge of the dependence of y on x (that is, of β).

Known variance

Under these assumptions the joint posterior distribution of α, β and ϕ is given by (11), or more conveniently by (13). It is instructive to consider first the case where ϕ is known, equal to σ^2, say. Then, from (13) and (17),

$$\pi(\alpha, \beta \mid \mathbf{x}, \mathbf{y}, \sigma^2) \propto \exp[-\{n(\alpha-a)^2 + S_{xx}(\beta-b)^2\}/2\sigma^2]. \quad (18)$$

It is clear from (18) that α and β are now independent and have posterior distributions as follows:

$$\text{(i)} \quad \alpha \text{ is } N(a, \sigma^2/n), \quad (19)$$

$$\text{(ii)} \quad \beta \text{ is } N(b, \sigma^2/S_{xx}). \quad (20)$$

The posterior independence of α and β is the reason for writing the regression in the form (2). (See also §8.6(*c*) below.) The posterior expectations of α and β, a and b, are of importance. The former is simply $\bar{y}$, agreeing with our remark that α is the expectation of y averaged over the values of x. α has the usual variance σ^2/n: though notice that σ^2 here is $\mathscr{D}^2(y_i \mid \mathbf{x})$, not $\mathscr{D}^2(y_i)$. The expectation of β is $b = S_{xy}/S_{xx}$ and is usually called the *sample regression coefficient*; its variance is σ^2/S_{xx}. The variance increases with σ^2 but decreases with S_{xx}. In other words, the greater the spread of values of the independent variable the greater is the precision in the determination of β, which is in good agreement with intuitive ideas.

Least squares

The values a and b may be obtained in the following illuminating way. Suppose the pairs (x_i, y_i) plotted as points on a diagram with x as horizontal and y as vertical axes. Then the problem of estimating α and β can be thought of as finding a line in the diagram which passes as close as possible to all these points. Since it is the dependence of y on x expressed through the conditional variation of y for fixed x that interests us, it is natural to measure the closeness of fit of the line by the distances in a vertical direction of the points from the line. It is easier to work with the squares of the distances rather than their absolute values, so a possible way of estimating α and β is to find the values of them which minimize

$$\sum_{i=1}^{n} [y_i - \alpha - \beta(x_i - \bar{x})]^2. \tag{21}$$

This is the *principle of least squares*: the principle that says that the best estimates are those which make least the sum of squares of differences between observed values, y_i, and their expectations, $\alpha + \beta(x_i - \bar{x})$. Since (21) can be written in the form (12) it is obvious that a and b are the values which make (21) a minimum. They are called the *least squares estimates* of α and β. The regression line $y = a + b(x - \bar{x})$ is called the *line of best fit*. Notice, that from (10), a and b are also the maximum likelihood estimates of α and β. We shall have more to say about the principle of least squares in §8.3.

Unknown variance

Now consider what happens when σ^2 is unknown. In the case of β the $N(0, 1)$ variable, $(\beta - b)/(\sigma^2/S_{xx})^{\frac{1}{2}}$, used to make statements of posterior belief, such as confidence intervals or significance tests, is no longer available, and it is natural to expect, as in §5.4, that σ^2 would be replaced by an estimate of variance and the normal distribution by the t-distribution. This is exactly what happens here, and it is easy to see why if one compares equation (14) above for the joint density of β and ϕ, with equation 5.4.4, where the t-distribution was first derived. The form of the two equations is the same and the estimate of ϕ used here is $S^2/(n-2)$. Consequently the quantity having a t-distribution is $(\beta - b)/\{S^2/S_{xx}(n-2)\}^{\frac{1}{2}}$ agreeing with (4). Similar remarks apply to α. The form of S^2 and the degrees of freedom, $(n-2)$, associated with it, fit in naturally with the principle of least squares. The minimum value of (21) is easily seen from (12) to be S^2. Hence S^2 is the sum of squares of deviations from the line of best fit. Furthermore, since α and β have been estimated, only $(n-2)$ of these deviations need be given when the remaining two can be found. Hence S^2 has only $(n-2)$ degrees of freedom. (Notice that if n were equal to 2 the line would pass through all the points and S^2 would be zero with zero degrees of freedom.) According to the principle of least squares the estimate of variance is obtained by dividing the minimum of the sum of squares (called the *residual sum of squares*) by the number of observations less the number of parameters, apart from ϕ, estimated, here 2. Notice that the maximum likelihood estimate of ϕ is S^2/n.

In view of the close connexion between equations (14) and 5.4.4 it is clear that the posterior distribution of ϕ will be related to that of χ^2 in the usual way. This is expressed in part (iii) of theorem 1. As usual, the degrees of freedom will be the same as those of t, namely $(n-2)$.

Significance tests

Confidence intervals and significance tests for any of the parameters will follow in the same way as they did in chapter 5. For example, a 95 % confidence interval† for β is (cf. equation 5.4.9), with $\epsilon = 0{\cdot}025$,

$$b - t_\epsilon(n-2)\,S/\{S_{xx}(n-2)\}^{\frac{1}{2}} \leqslant \beta \leqslant b + t_\epsilon(n-2)\,S/\{S_{xx}(n-2)\}^{\frac{1}{2}}. \tag{22}$$

The significance test for $\beta = 0$; that is, for independence of x and y, can be put into an interesting form. The test criterion is (4), with $\beta = 0$, and since the t-distribution is symmetrical we can square it and refer

$$\frac{b^2}{S^2/S_{xx}(n-2)} = \frac{S_{xy}^2/S_{xx}}{[S_{yy} - S_{xy}^2/S_{xx}]/(n-2)} \tag{23}$$

to values of $t_\epsilon^2(n-2)$. (The alternative form follows from (8) and (9).) Now

$$S_{yy} = [S_{yy} - S_{xy}^2/S_{xx}] + [S_{xy}^2/S_{xx}]. \tag{24}$$

The left-hand side is the total sum of squares for the y's. The first bracket on the right is the residual sum of squares and also, when divided by $(n-2)$, the denominator of (23). The second bracket is the numerator of (23) and will be called the *sum of squares due to regression*. Furthermore, since (§6.2) t^2 is equal to F with 1 and $(n-2)$ degrees of freedom, we can write the calculations in the form of an analysis of variance table (§6.5)

† To avoid confusion between the average expectation of y_i and the significance level, ϵ, instead of α, has been used for the latter.

and use an F-test (table 8.1.1). The total variation in y has been split into two parts, one part of which, the residual, is unaffected by the regression. (24) should be compared with (3.2.23). These can be compared, by taking their ratio and using an F-test, to see whether the latter is large compared with the former. If it is, we can conclude that a significant regression effect exists. That is to say, we believe that y is influenced by x, or that $\beta \neq 0$. This is a test of the null hypothesis that the two variables are independent, granted the many assumptions of the theorem.

TABLE 8.1.1

	Sum of squares	Degrees of freedom	Mean square	F
Due to regression	$S_{xy}^2/S_{xx} = b^2 S_{xx}$	1	$b^2 S_{xx}$	$b^2/\{S^2/S_{xx}(n-2)\}$
Residual	$S_{yy} - S_{xy}^2/S_{xx} = S^2$	$n-2$	$S^2/(n-2)$	—
Total	S_{yy}	$n-1$	—	—

Sufficient statistics

The sufficient statistics in this problem are four in number; namely, S^2, a, b and S_{xx}. This is immediate from (10), (12) and the factorization theorem (theorem 5.5.2). Thus four statistics are needed to provide complete information about three parameters. The calculation of the sufficient statistics is most easily done by first finding the sums and sums of squares and products, Σx_i, Σy_i; Σx_i^2, $\Sigma x_i y_i$, Σy_i^2 and then

$$\bar{x}, \ \bar{y}, \ S_{xx} = \Sigma x_i^2 - (\Sigma x_i)^2/n,$$

S_{yy} similarly and

$$S_{xy} = \Sigma x_i y_i - (\Sigma x_i)(\Sigma y_i)/n.$$

From these S^2, a, b and S_{xx} are immediately obtained. The computation is easy using a desk machine but some people object to it and argue that as good a fit of the line can be obtained by eye using a transparent ruler. In skilled hands this is true, but what such people forget is that the method here described not only fits the line, that is, provides a and b, but also provides an idea of how much in error the line can be; for example, by enabling confidence limits to be obtained. This cannot be done readily when fitting by eye.

Posterior distribution of the conditional expectation

It is sometimes necessary to test hypotheses or make confidence statements about quantities other than α, β or ϕ separately. For example, we may require to give limits for the expected value of y for a given value x_0 of the independent variable, or to test the hypothesis that the regression line passes through a given point (x_0, y_0). These may easily be accomplished by using a device that was adopted in proving theorem 6.1.3: namely, to consider the required posterior distribution for fixed ϕ and then to average over the known posterior distribution of ϕ (equation (6)). In the examples quoted we require the posterior distribution of $\mathscr{E}(y \mid x_0) = \alpha + \beta(x_0 - \bar{x})$. Now, given $\phi = \sigma^2$, we saw in equations (18), (19) and (20) that α and β were independent normal variables with means and variances there specified. It follows (theorem 3.5.5) that $\mathscr{E}(y \mid x_0)$ is

$$N[a + b(x_0 - \bar{x}), \sigma^2\{n^{-1} + (x_0 - \bar{x})^2/S_{xx}\}].$$

Hence, by exactly the same argument as was used in proving theorem 6.1.3, the posterior distribution of $\mathscr{E}(y \mid x_0)$ is such that

$$\frac{\mathscr{E}(y \mid x_0) - a - b(x_0 - \bar{x})}{[S^2\{n^{-1} + (x_0 - \bar{x})^2/S_{xx}\}/(n-2)]^{\frac{1}{2}}} \tag{25}$$

has a t-distribution with $(n-2)$ degrees of freedom. Confidence limits for $\mathscr{E}(y \mid x_0)$ may be obtained in the usual way, and the null hypothesis that the line passes through (x_0, y_0), that is, that $\mathscr{E}(y \mid x_0) = y_0$, may be tested by putting $\mathscr{E}(y \mid x_0) = y_0$ in (25) and referring the resulting statistic to the t-distribution. Notice that the confidence limits derived from (25) will have width proportional to the denominator of (25) and will therefore be wider the more x_0 deviates from $\bar{x}$. It is more difficult to be precise about the line at points distant from $\bar{x}$ than it is near $\bar{x}$; which is intuitively reasonable.

Posterior distribution of the line

It is possible to make confidence statements about the line and not merely its value at a fixed value, x_0. To do this it is necessary to find the joint distribution of α and β. This can be done by integrating (13) with respect to ϕ. The result will clearly be a joint density which is constant for values of α and β for which $n(\alpha-a)^2+S_{xx}(\beta-b)^2$ is constant. To find joint confidence sets for α and β it is therefore only necessary to find the distribution of $n(\alpha-a)^2+S_{xx}(\beta-b)^2$. Now a comparison of equation (13) with equation 6.4.8, and the argument that led from the latter to the result concerning the F-distribution, will here show that

$$\frac{[n(\alpha-a)^2+S_{xx}(\beta-b)^2]/2}{S^2/(n-2)} \tag{26}$$

has an F-distribution with 2 and $(n-2)$ degrees of freedom. This is a special case of a general result to be established later (theorem 8.3.1). The hypothesis that the line has equation $y = \alpha_0+\beta_0(x-\bar{x})$ may be tested by putting $\alpha = \alpha_0$, $\beta = \beta_0$ in (26) and referring the statistic to the F-distribution.

Prediction

Another problem that often arises in regression theory is that of predicting the value of the dependent variable for a given value, x_0, of the independent variable: a scientist who has experimented at values $x_1, x_2, \ldots, x_n$ may wish to estimate what would happen if he were to experiment at x_0; or if the independent variable is time he may wish to predict what will happen at some point in the future. To answer this problem we need to have the density of y given a value x_0, the data $\mathbf{x}$, $\mathbf{y}$, and the prior knowledge: that is $\pi(y\mid\mathbf{x}, \mathbf{y}, x_0)$, where reference to the prior knowledge H is, as usual, omitted. To find this, consider first

$$\pi(y\mid\mathbf{x}, \mathbf{y}, x_0, \alpha, \beta, \phi).$$

This is $N(\alpha+\beta(x_0-\bar{x}), \phi)$. Also

$$\pi(\alpha\mid\mathbf{x}, \mathbf{y}, x_0, \beta, \phi) \quad \text{is} \quad N(a, \phi/n)$$

(equation (19)). Hence, for fixed $\mathbf{x}$, $\mathbf{y}$, x_0, β and ϕ; y and α have a joint normal density (§3.2, in particular equation 3.2.18) and

$$\pi(y\mid\mathbf{x}, \mathbf{y}, x_0, \beta, \phi) \quad \text{is} \quad N[a+\beta(x_0-\bar{x}), \phi+\phi/n].$$

Applying the same argument again to the joint density of y and β for fixed $\mathbf{x}$, $\mathbf{y}$, x_0 and ϕ we have, since

$$\pi(\beta\mid\mathbf{x}, \mathbf{y}, x_0, \phi) \quad \text{is} \quad N(b, \phi/S_{xx})$$

(equation (20)) that $\pi(y \mid \mathbf{x}, \mathbf{y}, x_0, \phi)$ is

$$N[a+b(x_0-\bar{x}),\ \phi+\phi/n+(x_0-\bar{x})^2\phi/S_{xx}].$$

Hence

$$\pi(y \mid \mathbf{x}, \mathbf{y}, x_0)$$
$$= \int \pi(y \mid \mathbf{x}, \mathbf{y}, x_0, \phi)\, \pi(\phi \mid \mathbf{x}, \mathbf{y}, x_0)\, d\phi$$
$$\propto \int \phi^{-\frac{1}{2}(n+1)} \exp\left[-\left(\frac{\{y-a-b(x_0-\bar{x})\}^2}{1+n^{-1}+(x_0-\bar{x})^2/S_{xx}}+S^2\right)\Big/2\phi\right] d\phi$$

from equation (17). The integration performed in the usual way (compare the passage from (14) to (15)) shows that

$$\frac{y-a-b(x_0-\bar{x})}{\{S^2[1+n^{-1}+(x_0-\bar{x})^2/S_{xx}]/(n-2)\}^{\frac{1}{2}}} \tag{27}$$

has a t-distribution with $(n-2)$ degrees of freedom. Hence confidence limits† for y may be found in the usual way. A comparison of (25) with (27) shows that the limits for y in (27) are wider than those for $\mathscr{E}(y \mid x_0)$ in (25) since there is an additional term, 1, in the multiplier of the estimate, $S^2/(n-2)$, of residual variance in the denominator of (27). This is reasonable since y equals $\mathscr{E}(y \mid x_0)$ plus an additional term which is $N(0, \phi)$ (cf. equation 3.2.22): indeed we could have derived (27) by using this remark for fixed ϕ, and then integrating as usual with respect to ϕ. Again these limits increase in width as x_0 departs from $\bar{x}$. Remember, too, that these methods are based on the assumption of linear regression which may not hold into the future even when it holds over the original range of x-values.

An alternative method

We conclude this section by giving an example of how the dependence of one random variable on an independent variable (random or not) can be investigated without making the assumptions of linear homoscedastic regression. The method described is by no means the only possible one. Suppose, for convenience

† Notice that y is not a parameter, so that the definition of a confidence interval given in §5.2 is being extended from the posterior distribution of a parameter to the posterior distribution of any random variable, here a future observation.

in explanation, that the x-values are chosen by the experimenter. Then the range of x-values may be divided into groups and the numbers in the groups, $r_{i.}$, say (the reason for the notation will be apparent in a moment), are fixed numbers chosen by the experimenter. Suppose the range of y similarly divided into groups and let r_{ij} be the numbers of observations in both the ith-group of x-values and the jth-group of y-values. Then

$$\sum_j r_{ij} = r_{i.}, \quad \text{and} \quad \sum_i r_{ij} = r_{.j}$$

is a random variable which is the number in the jth-group of y-values. Let ϕ_{ij} be the probabilty of an observation which belongs to the ith-group of x-values belonging to the jth-group of y-values. Then if y is independent of x, ϕ_{ij} does not depend on i and the null hypothesis of independence may be tested by the methods of theorem 7.6.2. If x were a random variable theorem 7.6.1 would be needed, but the resulting test will be the same. The group sizes will have to be chosen so that the expected numbers are sufficiently large for the limiting χ^2-distribution to be used as an approximation. In the case of the dependent variable this choice, which will have to be made after the results are available, will influence the randomness of the $r_{.j}$. The precise effect of this is not properly understood. Fisher and others claim that the results of § 7.6 are still valid even if the $r_{.j}$, as well as the $r_{i.}$, are fixed, but it is not clear from Bayesian arguments that this is so. Nevertheless, the effect of fixing the $r_{.j}$ is likely to be small and the test can probably be safely applied. Notice that this test makes no assumptions about linear homoscedastic regression of y on x.

8.2. Correlation coefficient

The topic of this section has nothing to do with least squares but is included here because of its close relationship to that of the previous section. The results are not used elsewhere in the book.

Theorem 1. *If* $(\mathbf{x}, \mathbf{y}) = (x_1, y_1; x_2, y_2; \ldots; x_n, y_n)$ *is a random sample from a bivariate normal density with* $\mathscr{E}(x_i) = \theta_1$, $\mathscr{E}(y_i) = \theta_2$, $\mathscr{D}^2(x_i) = \phi_1$, $\mathscr{D}^2(y_i) = \phi_2$ *and correlation coefficient* ρ; *and if the prior distributions of* θ_1, θ_2, $\ln \phi_1$ *and* $\ln \phi_2$ *are uniform* *and independent, and independent of the prior distribution of ρ which nowhere vanishes; then the posterior distribution of ρ is such that $\tilde{\omega} = \tanh^{-1}\rho$ is approximately ($n \to \infty$) normal with*

$$\textit{mean}: \tanh^{-1} r, \quad \textit{variance}: n^{-1}, \tag{1}$$

where $r = S_{xy}/\sqrt{(S_{xx}S_{yy})}$.

If $\pi(\rho)$ is the prior distribution of ρ then, in the usual way, the joint posterior density of θ_1, θ_2, ϕ_1, ϕ_2 and ρ is proportional to (cf. 3.2.17)

$$\frac{\pi(\rho)}{(\phi_1\phi_2)^{\frac{1}{2}n+1}(1-\rho^2)^{\frac{1}{2}n}} \exp\left[-\frac{1}{2(1-\rho^2)}\sum_{i=1}^{n}\left\{\frac{(x_i-\theta_1)^2}{\phi_1} - \frac{2\rho(x_i-\theta_1)(y_i-\theta_2)}{\phi_1^{\frac{1}{2}}\phi_2^{\frac{1}{2}}} + \frac{(y_i-\theta_2)^2}{\phi_2}\right\}\right]. \tag{2}$$

In the usual way we may write

$$\Sigma(x_i-\theta_1)^2 = \Sigma(x_i-\bar{x})^2 + n(\bar{x}-\theta_1)^2 = S_{xx} + n(\bar{x}-\theta_1)^2,$$

in the notation of equation 8.1.7, etc., and (2) becomes

$$\frac{\pi(\rho)}{(\phi_1\phi_2)^{\frac{1}{2}n+1}(1-\rho^2)^{\frac{1}{2}n}} \exp\left[-\frac{1}{2(1-\rho^2)}\left\{\frac{S_{xx}}{\phi_1} - \frac{2\rho S_{xy}}{\phi_1^{\frac{1}{2}}\phi_2^{\frac{1}{2}}} + \frac{S_{yy}}{\phi_2}\right\} - \frac{n}{2(1-\rho^2)}\left\{\frac{(\bar{x}-\theta_1)^2}{\phi_1} - \frac{2\rho(\bar{x}-\theta_1)(\bar{y}-\theta_2)}{\phi_1^{\frac{1}{2}}\phi_2^{\frac{1}{2}}} + \frac{(\bar{y}-\theta_2)^2}{\phi_2}\right\}\right]. \tag{3}$$

The integration with respect to θ_1 and θ_2 may now be carried out using the fact that the normal bivariate density necessarily has integral one. Consequently

$$\pi(\phi_1, \phi_2, \rho \mid \mathbf{x}, \mathbf{y}) \propto \frac{\pi(\rho)}{(\phi_1\phi_2)^{\frac{1}{2}(n+1)}(1-\rho^2)^{\frac{1}{2}(n-1)}} \exp\left[-\frac{1}{2(1-\rho^2)} \times \left\{\frac{S_{xx}}{\phi_1} - \frac{2\rho S_{xy}}{\phi_1^{\frac{1}{2}}\phi_2^{\frac{1}{2}}} + \frac{S_{yy}}{\phi_2}\right\}\right]. \tag{4}$$

If we substitute for ϕ_2 a multiple of ϕ_1, the integration with respect to ϕ_1 can be easily carried out. It is convenient, because of the form of the term in braces, to put $\phi_2 = \psi^2\phi_1(S_{yy}/S_{xx})$, when, remembering the Jacobian, we have

$$\pi(\phi_1, \psi, \rho \mid \mathbf{x}, \mathbf{y}) \propto \frac{\pi(\rho)}{\phi_1^n \psi^n (1-\rho^2)^{\frac{1}{2}(n-1)}} \exp\left[-\frac{1}{2(1-\rho^2)\phi_1} \times \left\{S_{xx} - \frac{2\rho}{\psi} S_{xy}\sqrt{\frac{S_{xx}}{S_{yy}}} + \frac{S_{xx}}{\psi^2}\right\}\right],$$

and the integration with respect to ϕ_1 gives (theorem 5.3.2)

$$\pi(\psi, \rho \mid \mathbf{x}, \mathbf{y}) \propto \frac{\pi(\rho)}{\psi^n(1-\rho^2)^{\frac{1}{2}(n-1)}} \frac{(1-\rho^2)^{n-1}}{(1-2\rho r/\psi+\psi^{-2})^{n-1}}$$

$$= \frac{\pi(\rho)(1-\rho^2)^{\frac{1}{2}(n-1)}}{\psi(\psi-2\rho r+\psi^{-1})^{n-1}}, \tag{5}$$

where $r = S_{xy}/\sqrt{(S_{xx}S_{yy})}$.

The integral with respect to ψ is not expressible in elementary functions so we attempt to find an approximation to it. The natural thing to do is to attempt a substitution for $\psi-2\rho r+\psi^{-1}$ and a convenient choice is to change from ψ to ξ defined by

$$\psi-2\rho r+\psi^{-1} = 2\frac{1-\rho r}{1-\xi}. \tag{6}$$

The factor $2(1-\rho r)$ is convenient because $\psi-2\rho r+\psi^{-1}$ ranges between $2(1-\rho r)$, when $\psi = 1$, and infinity: $(1-\xi)^{-1}$ leads to the power series expansion below. It is tedious but straightforward to verify that

$$\frac{1}{\psi}\frac{d\psi}{d\xi} = \left[\left(\rho r+\frac{1-\rho r}{1-\xi}\right)^2-1\right]^{-\frac{1}{2}}\frac{(1-\rho r)}{(1-\xi)^2},$$

and we finally obtain

$$\pi(\rho \mid \mathbf{x}, \mathbf{y}) \propto \frac{\pi(\rho)(1-\rho^2)^{\frac{1}{2}(n-1)}}{(1-\rho r)^{n-\frac{3}{2}}}\int_0^1 (1-\xi)^{n-2}\xi^{-\frac{1}{2}}[1-\tfrac{1}{2}(1+\rho r)\xi]^{-\frac{1}{2}}\,d\xi. \tag{7}$$

It is easy to show (see below) that $|\rho r| \leqslant 1$ so that the expression in square brackets may be expanded in a power series in ξ, when each term in the integrand may be integrated using the beta-integral (equation 5.4.7). The sth term in the integrated series is equal to the first term times a quantity of order $n^{-(s-1)}$, so that an approximation for large n may be obtained by retaining only the first term. Therefore, approximately,

$$\pi(\rho \mid \mathbf{x}, \mathbf{y}) \propto \frac{\pi(\rho)(1-\rho^2)^{\frac{1}{2}(n-1)}}{(1-\rho r)^{n-\frac{3}{2}}}. \tag{8}$$

Now make the substitutions

$$\rho = \tanh\varpi, \quad r = \tanh z, \tag{9}$$

so that†
$$\pi(\tilde{\omega} \mid \mathbf{x}, \mathbf{y}) \propto \frac{\pi(\tilde{\omega})}{\cosh^{\frac{1}{2}} \tilde{\omega} \cosh^{n-\frac{3}{2}}(\tilde{\omega}-z)}, \tag{10}$$

where $\pi(\tilde{\omega})$ is now the prior density of $\tilde{\omega}$. If n is large a further approximation may be made and the term

$$\pi(\tilde{\omega}) \cosh^{\frac{3}{2}}(\tilde{\omega}-z)/\cosh^{\frac{1}{2}} \tilde{\omega}$$

replaced by a constant, since it does not change with n. Then

$$\pi(\tilde{\omega} \mid \mathbf{x}, \mathbf{y}) \propto \cosh^{-n}(\tilde{\omega}-z).$$

Finally, put $\tilde{\omega}' = (\tilde{\omega}-z) n^{\frac{1}{2}}$ and expand the hyperbolic cosine, giving approximately

$$\pi(\tilde{\omega}' \mid \mathbf{x}, \mathbf{y}) \propto \left[1+\frac{1}{2} \frac{\tilde{\omega}'^2}{n}\right]^{-n} = e^{-\frac{1}{2}\tilde{\omega}'^2} \tag{11}$$

whence the result follows.

Improvements to the approximation which involve choice of the form of $\pi(\rho)$ will be given below.

Use of correlation methods

The regression methods discussed in the last section for investigating the association between two random variables are more useful than the results of this section, but the fact that they are not symmetrical in the two variables is sometimes an inconvenience. For example, in anthropometry, if two measurements such as nose length and arm length are being considered, there is no more reason to consider one regression than the other. Similarly, in education there is no reason to consider the influence of English marks on arithmetic marks rather than the other way around. The correlation coefficient (equation 3.1.9) is a measure of association which treats the two random variables symmetrically and is often used in these, and similar, situations.

Nevertheless, it is a coefficient that requires care in its use. For the bivariate normal distribution zero correlation means independence of the two variables (§3.2) and the dependence increases with the modulus of ρ, so that there ρ is satisfactory as a measure of dependence; but in other cases this may be far from the case. An example is given in §3.1. Also ρ is a much

† Readers may like to be reminded of the elementary result
$$1-\tanh x \tanh y = \cosh(x-y)/\cosh x \cosh y.$$

more difficult quantity to handle than the regression coefficients. Finally, notice that with the bivariate normal distribution the variance (equation 3.2.23) of one variable, say y, can be written as the sum of two parts, that due to x and the residual variance. The former is only a proportion ρ^2 of the total and, due to the square occurring here, ρ has to be very near one for this to amount to an appreciable proportion of the total variation. Consequently ρ tends to over-emphasize the association: with $\rho = 0{\cdot}70$, $\rho^2 = 0{\cdot}49$ and only about half the variation in one variable can be ascribed to the other. Unlike regression methods, correlation techniques only apply when both variables are random.

Derivation of the posterior distribution

The prior distributions of the means and variances in theorem 1 are as usual. The prior distribution of ρ has been chosen to be independent of these since it seems typically unlikely that, as it is dimensionless, knowledge of it would be influenced by knowledge of the separate variations of x and y. The first stage of the proof, leading to equation (4), serves to eliminate the means: a comparison of (4) with (2) shows that, in the usual way, lack of knowledge of the means effectively reduces the sample size by one. The quantity r appearing in (5) is called the *sample correlation coefficient* and its definition in terms of the sample is exactly the same as that of ρ in terms of the density (equation 3.1.9). It follows therefore since $|\rho| \leqslant 1$ that also $|r| \leqslant 1$. Equation (7) shows that the posterior distribution of ρ depends on the sample only through r. It is sometimes said that r is sufficient for ρ but this does not agree with the definition of sufficiency in §5.5. It is only for certain forms of prior distributions of the other parameters that the posterior distribution of ρ involves only r, and sufficiency is a concept which does not involve the prior distribution.

Approximations and prior distributions

In order to proceed beyond (7) it is simplest to introduce approximations. An alternative way of looking at the approximation given in the proof is to take logarithms of the posterior density, as we did in §7.1. It is then easy to see that the terms (apart from $\pi(\rho)$) outside the integral provide the dominant quantities, so that we obtain (8). From the logarithm of (8) we can evaluate its first and second derivatives and so obtain the normal approximation, again as in §7.1.

The form of (8) suggests prior densities of the former $(1-\rho^2)^c$, for some c, as being convenient, and the following considerations suggest the value of c. If the knowledge of ρ is very vague a sample of size 1 will not provide any information, indeed r is then undefined. A sample of size 2 has necessarily $r = \pm 1$, so only provides information about the sign of ρ. Consequently, if our prior knowledge of ρ is slight, we should not expect (8) to converge until $n = 3$. For this to happen calculation shows that we must have $-2 < c \leqslant -1$. If we confine ourselves to integers we must have $c = -1$ and

$$\pi(\rho) \propto (1-\rho^2)^{-1}. \tag{12}$$

This argument will be supported by a further consideration below.

It is possible to show that the posterior distribution in the form (8) tends to normality as $n \to \infty$ but the limit is approached very slowly owing to the extreme skewness of (8). Fisher suggested the transformation (9) to avoid this. The transformation has some interesting properties. First, the large-sample form of (8) is, with $\pi(\rho)$ given by (12), say,

$$\ln \pi(\rho \mid \mathbf{x}, \mathbf{y}) = C + n[\tfrac{1}{2}\ln(1-\rho^2) - \ln(1-\rho r)],$$

where C is a constant. This clearly has a maximum value at $\rho = r$ and the second derivative at the maximum is

$$\left\{\frac{d^2}{d\rho^2}\ln \pi(\rho \mid \mathbf{x}, \mathbf{y})\right\}_{\rho=r} = -\frac{n}{(1-r^2)^2}. \tag{13}$$

In the sense of maximum likelihood (13) is the negative of the information about ρ (equation 7.1.6) and depends heavily on r. Let us therefore find a transformation of ρ that has constant (that is, independent of r) information (compare the transformation for the binomial and Poisson distributions in §§7.2,

7.3). If $\tilde{\omega}(\rho)$ is such a function equation 7.1.24 says that it must satisfy, since the information is the inverse of the variance,

$$\left(\frac{d\tilde{\omega}}{d\rho}\right)^2 = a(1-\rho^2)^{-2}, \tag{14}$$

where a is a constant. A function satisfying this is

$$\tilde{\omega}(\rho) = \tfrac{1}{2}\ln\frac{1+\rho}{1-\rho} = \tanh^{-1}\rho, \tag{15}$$

the constant information being n. It is convenient to make a similar transformation of r, when the result (10) follows. A second property of the transformation is that if the prior distribution of ρ is (12) then $\tilde{\omega}$ is uniformly distributed over the whole real line. If inferences are to be expressed in terms of $\tilde{\omega}$ it is not unreasonable to assume $\tilde{\omega}$ so distributed before the results are available (compare the discussion in connexion with the variance in §5.3).

Further approximations

If one works in terms of $\tilde{\omega}$ with a uniform prior distribution it is often worthwhile to consider more accurate approximations to (10) than is provided by (11). We have

$$\ln\pi(\tilde{\omega}\,|\,\mathbf{x},\mathbf{y}) = C - \tfrac{1}{2}\ln\cosh\tilde{\omega} - (n-\tfrac{3}{2})\ln\cosh(\tilde{\omega}-z),$$

where C is a constant. The maximum of this density occurs where

$$\frac{\partial}{\partial\tilde{\omega}}\ln\pi(\tilde{\omega}\,|\,\mathbf{x},\mathbf{y}) = -\tfrac{1}{2}\tanh\tilde{\omega} - (n-\tfrac{3}{2})\tanh(\tilde{\omega}-z) = 0.$$

The root of this is approximately $\tilde{\omega} = z$ (from (11)) so write $\tilde{\omega} = z+\epsilon$ and retain only the terms of order ϵ: we have

$$-\tfrac{1}{2}\{\tanh z + \epsilon\,\mathrm{sech}^2 z\} - (n-\tfrac{3}{2})\epsilon = 0,$$

so that
$$\epsilon = -\tanh z/\{(2n-3)+\mathrm{sech}^2 z\}$$

or, to order n^{-1},
$$\epsilon = -r/2n. \tag{16}$$

The second derivative is

$$-\tfrac{1}{2}\mathrm{sech}^2\tilde{\omega} - (n-\tfrac{3}{2})\,\mathrm{sech}^2(\tilde{\omega}-z)$$

which, at the maximum $\tilde{\omega} = z - r/2n$, is approximately

$$-(n-\tfrac{3}{2})-\tfrac{1}{2}\operatorname{sech}^2 z. \tag{17}$$

Hence a slight improvement on the result of the theorem is the

Corollary. *The posterior distribution of* $\tilde{\omega} = \tanh^{-1}\rho$ *is approximately normal with*

$$\textit{mean}: \tanh^{-1} r - r/2n, \quad \textit{variance}: [n-\tfrac{3}{2}+\tfrac{1}{2}(1-r^2)]^{-1}. \tag{18}$$

The change in the variance between (1) and (18) is not appreciable but the correction to the mean of $-r/2n$ can be of importance if it is desired to combine estimates of ρ from different sources by means of theorem 6.6.1.

Several samples

The exact posterior distribution of ρ with $\pi(\rho) = \frac{1}{2}$ (equation (7)) has been tabulated (David, 1954), but the approximations are often more useful. The main merit of the approximate results is that they enable the usual normal methods developed in chapters 5 and 6 to be used when giving confidence limits or comparing several correlation coefficients. For example: if r_1 and r_2 are the sample correlation coefficients obtained from independent samples of sizes n_1 and n_2 respectively, the hypothesis that $\rho_1 = \rho_2$ (in an obvious notation) may be investigated by comparing the difference $(z_1 - r_1/2n_1) - (z_2 - r_2/2n_2)$ with its standard deviation $[n_1^{-1}+n_2^{-1}]^{\frac{1}{2}}$ in the usual way.

The approximation (18) differs a little from that usually given, namely $\tanh^{-1} r - r/2(n-1)$ for the mean and $(n-3)^{-1}$ for the variance, but the differences are trivial, or, more precisely, of smaller order than the terms retained.

8.3. Linear hypothesis

In this and the following sections of the present chapter we consider independent (except for §8.6(*d*)) normal random variables $(x_1, x_2, \ldots, x_n)$ with a common unknown variance ϕ. A statement that the expectations of these random variables are known linear functions of unknown parameters is called a *linear hypothesis*. The *column* vector of the x_i will be denoted by $\mathbf{x}$ and the linear hypothesis will be written

$$\mathscr{E}(\mathbf{x}) = \mathbf{A}\boldsymbol{\theta}, \tag{1}$$

where $\boldsymbol{\theta}$ is a column vector of s unknown parameters $(\theta_1, \theta_2, \ldots, \theta_s)$ and $\mathbf{A}$ is a known matrix of elements a_{ij} $(i = 1, 2, \ldots, n;$ $j = 1, 2, \ldots, s)$. $\mathbf{A}$ is often called the *design matrix*. We suppose $s < n$. For reasons which will appear below we consider the sum of squares of the differences between the random variables and their expectations (cf. 8.1.21); that is,

$$(\mathbf{x}-\mathbf{A}\boldsymbol{\theta})'(\mathbf{x}-\mathbf{A}\boldsymbol{\theta}) = \sum_{i=1}^{n}\left(x_i - \sum_{j=1}^{s} a_{ij}\theta_j\right)^2. \tag{2}$$

The least value of (2) for all real $\boldsymbol{\theta}$ but fixed $\mathbf{x}$ and $\mathbf{A}$ is called the *residual sum of squares* and will be denoted by S^2. We are interested in considering whether

$$\theta_{r+1} = \theta_{r+2} = \ldots = \theta_s = 0 \quad (0 \leqslant r < s),$$

and the least value of (2) when these θ's are put equal to zero will not be less than S^2. The difference between this least value and S^2 will therefore be non-negative and is called the *reduction in sum of squares due to* $\theta_{r+1}, \theta_{r+2}, \ldots, \theta_s$ (allowing for $\theta_1, \theta_2, \ldots, \theta_r$) and will be denoted by S_r^2. The words in brackets are often omitted when it is clear which parameters have been included in the restricted minimization. Throughout this chapter the square $(s \times s)$ matrix $\mathbf{A}'\mathbf{A}$ will be assumed to be non-singular.†

Theorem 1. *If the random variables* $\mathbf{x}$ *satisfy the linear hypothesis* (1), *with a non-singular matrix* $\mathbf{A}'\mathbf{A}$, *and if the parameters* $\theta_1, \theta_2, \ldots, \theta_s$ *and* $\ln\phi$ *have independent prior distributions which are uniform over the real line; then a significance test at level* α *of the hypothesis that*

$$\theta_{r+1} = \theta_{r+2} = \ldots = \theta_s = 0 \quad (0 \leqslant r < s) \tag{3}$$

is obtained by declaring the data significant if

$$[S_r^2/(s-r)]/[S^2/(n-s)] \tag{4}$$

exceeds $\bar{F}_\alpha(\nu_1, \nu_2)$ (*equation 6.2.6*); *with* $\nu_1 = s-r$, $\nu_2 = n-s$.

† The results can be extended to singular $\mathbf{A}'\mathbf{A}$, but the extensions will not be required in this book. The only situation with a singular matrix that will be studied is that of §8.5 with $K = 1$, which will be dealt with by a special argument.

The joint posterior distribution of $\boldsymbol{\theta}$ and ϕ is clearly

$$\pi(\boldsymbol{\theta}, \phi \mid \mathbf{x}) \propto \phi^{-\frac{1}{2}n-1} \exp\left[-\frac{1}{2\phi}\{(\mathbf{x}-\mathbf{A}\boldsymbol{\theta})'(\mathbf{x}-\mathbf{A}\boldsymbol{\theta})\}\right]. \quad (5)$$

As on other occasions (for example, in the proof of theorem 6.6.2) the expression in braces can be rearranged by collecting together the linear and quadratic terms in $\boldsymbol{\theta}$ and completing the square. In this way the expression can be rewritten

$$\boldsymbol{\theta}'\mathbf{A}'\mathbf{A}\boldsymbol{\theta} - 2\boldsymbol{\theta}'\mathbf{A}'\mathbf{x} + \mathbf{x}'\mathbf{x} = (\boldsymbol{\theta}-\hat{\boldsymbol{\theta}})'\mathbf{A}'\mathbf{A}(\boldsymbol{\theta}-\hat{\boldsymbol{\theta}}) + \mathbf{x}'\mathbf{x} - \hat{\boldsymbol{\theta}}'\mathbf{A}'\mathbf{A}\hat{\boldsymbol{\theta}}, \quad (6)$$

where $\boldsymbol{\theta}'\mathbf{A}'\mathbf{A}\hat{\boldsymbol{\theta}} = \boldsymbol{\theta}'\mathbf{A}'\mathbf{x}$: that is, since $\mathbf{A}'\mathbf{A}$ is non-singular,

$$\hat{\boldsymbol{\theta}} = (\mathbf{A}'\mathbf{A})^{-1}\mathbf{A}'\mathbf{x}, \quad (7)$$

a function of the data not involving the parameters. Furthermore, it is clear, by writing $\mathbf{A}(\boldsymbol{\theta}-\hat{\boldsymbol{\theta}}) = \mathbf{z}$, that the first term in (6) is the sum of squares of the z's and is therefore non-negative. Since it is zero when $\boldsymbol{\theta} = \hat{\boldsymbol{\theta}}$ the remaining terms in (6), which do not involve $\boldsymbol{\theta}$, must be the smallest value that (6) can take and are therefore equal to the residual sum of squares, S^2. Consequently

$$\pi(\boldsymbol{\theta}, \phi \mid \mathbf{x}) \propto \phi^{-\frac{1}{2}n-1} \exp\left[-\frac{1}{2\phi}\{(\boldsymbol{\theta}-\hat{\boldsymbol{\theta}})'\mathbf{A}'\mathbf{A}(\boldsymbol{\theta}-\hat{\boldsymbol{\theta}}) + S^2\}\right]. \quad (8)$$

A comparison of (8) with equation 3.5.17 shows that, for fixed ϕ, the θ's have a multivariate normal density. Now this density may alternatively be written in terms of the regressions, equation 3.5.10. Consider first the distribution of θ_s, then the distribution of θ_{s-1}, conditional on θ_s, and so on down to the distribution of θ_1 conditional on $\theta_2, \theta_3, \ldots, \theta_s$. (Notice we are dealing with the θ's in the order which is the reverse of that of the x's in equation 3.5.10: that is, θ_1 is equivalent to x_n, etc.) In this way we may write

$$\pi(\boldsymbol{\theta}, \phi \mid \mathbf{x}) \propto \phi^{-\frac{1}{2}n-1} \exp\left[-\frac{1}{2\phi}\{c_1(\theta_1-\alpha_1)^2 + c_2(\theta_2-\alpha_2)^2 + \ldots \right.$$

$$\left. + c_s(\theta_s-\alpha_s)^2 + S^2\}\right], \quad (9)$$

where the c's are constants and α_i is a linear function of $\theta_{i+1}, \theta_{i+2}, \ldots, \theta_s$ (in particular, α_s is a constant, the means in equation 3.5.10 being zero). Furthermore, $c_i^{-1}\phi$ is the variance of θ_i for fixed $\theta_{i+1}, \theta_{i+2}, \ldots, \theta_s$ and hence $c_i > 0$. Since θ_1 occurs only in the first term in braces, we may integrate with respect to it, and then with respect to θ_2, and so on up to θ_r finally obtaining

$$\pi(\theta_{r+1}, \theta_{r+2}, \ldots, \theta_s, \phi \mid \mathbf{x}) \propto \phi^{-\frac{1}{2}(n-r)-1} \exp\left[-\frac{1}{2\phi}\{c_{r+1}(\theta_{r+1}-\alpha_{r+1})^2 + \ldots + c_s(\theta_s-\alpha_s)^2 + S^2\}\right]. \quad (10)$$

A further integration with respect to ϕ, using theorem 5.3.2, gives

$$\pi(\theta_{r+1}, \theta_{r+2}, \ldots, \theta_s \mid \mathbf{x}) \propto \{c_{r+1}(\theta_{r+1}-\alpha_{r+1})^2 + \ldots + c_s(\theta_s-\alpha_s)^2 + S^2\}^{-\frac{1}{2}(n-r)}. \quad (11)$$

The result now follows in the same way that theorem 6.4.1 followed from equation 6.4.9. Denote the expression in braces in (11) by $S_r^2(\boldsymbol{\theta}) + S^2$. The posterior density, (11), is constant where $S_r^2(\boldsymbol{\theta})$ is constant. In the $(s-r)$-dimensional space of $\theta_{r+1}, \theta_{r+2}, \ldots, \theta_s$ the surfaces $S_r^2(\boldsymbol{\theta}) = c$ are ellipsoids, since the c_i are positive, and the density decreases as c increases; that is, as the distance from the common centre of the ellipsoids increases. Hence a confidence set is an ellipsoid and the same argument as used in proving theorem 6.4.1 shows that the posterior distribution of

$$\Phi = [S_r^2(\boldsymbol{\theta})/(s-r)]/[S^2/(n-s)]$$

is $F(\nu_1, \nu_2)$. The degrees of freedom, ν_1, is the dimension of the θ's, namely $(s-r)$, and a simple calculation following the lines of the derivation of equation 6.4.11 shows that $\nu_2 = n-s$.

The confidence set leads to a result which is significant if the null value $\theta_{r+1} = \theta_{r+2} = \ldots = \theta_s = 0$ does not belong to the confidence set. The confidence set is $\Phi \leqslant \bar{F}_\alpha(\nu_1, \nu_2)$ so that the result is significant if

$$\frac{S_r^2(\mathbf{0})/(s-r)}{S^2/(n-s)} > \bar{F}_\alpha(\nu_1, \nu_2).$$

To complete the proof of the theorem it remains only to prove that $S_r^2(\mathbf{0})$ is the reduction in sum of squares due to $\theta_{r+1}, \theta_{r+2}, \ldots, \theta_s$.

This is easily done by returning to the sum of squares, (2), in the form of the expression in braces in (9). When

$$\theta_{r+1} = \theta_{r+2} = \ldots = \theta_s = 0$$

this is $$c_1(\theta_1 - \alpha_1')^2 + \ldots + c_r(\theta_r - \alpha_r')^2 + S_r^2(\mathbf{0}) + S^2,$$

where α_i' is the value of α_i when $\theta_{r+1} = \theta_{r+2} = \ldots = \theta_s = 0$ $(i \leqslant r)$. Since the c's are positive this has a minimum, when $\theta_i = \alpha_i'$ $(i \leqslant r)$, of $S_r^2(\mathbf{0}) + S^2$. The equations $\theta_i = \alpha_i'$ $(i \leqslant r)$ are linear equations in $\theta_1, \theta_2, \ldots, \theta_r$ and will always have a solution. Hence $S_r^2(\mathbf{0})$ is the reduction, as required, and the theorem is proved.

Corollary 1. *The posterior distribution of* S^2/ϕ *is* χ^2 *with* $(n-s)$ *degrees of freedom.*

This follows from (9) on integrating with respect to all the θ's. We obtain

$$\pi(\phi \mid \mathbf{x}) \propto \phi^{-\frac{1}{2}(n-s)-1} \exp[-S^2/2\phi].$$

A comparison with equation 5.3.2 establishes the result.

Corollary 2. *The posterior distribution of a linear function,* $\sum_{i=1}^{s} g_i \theta_i$, *of the parameters is such that*

$$\mathbf{g}'(\boldsymbol{\theta} - \hat{\boldsymbol{\theta}})/[S^2\mathbf{g}'\mathbf{C}\mathbf{g}/(n-s)]^{\frac{1}{2}} \tag{12}$$

has a t-distribution with $\nu = n-s$ *degrees of freedom; where* $\mathbf{g}'$ *is the row vector* $(g_1, g_2, \ldots, g_s)$ *and* $\mathbf{C}$ *is the matrix* $(\mathbf{A}'\mathbf{A})^{-1}$.

For fixed ϕ, the θ's are multivariate normal, equation (8), with means $\hat{\boldsymbol{\theta}}$ and dispersion matrix $(\mathbf{A}'\mathbf{A})^{-1}\phi$, so that

$$\mathbf{g}'\boldsymbol{\theta} = \sum_{i=1}^{s} g_i \theta_i$$

is normal (a generalization to s variables of theorem 3.5.5) with mean $\mathbf{g}'\hat{\boldsymbol{\theta}}$ and variance $\phi\mathbf{g}'\mathbf{C}\mathbf{g}$ (theorem 3.3.2). Hence, by corollary 1,

$$\pi(\Sigma g_i \theta_i, \phi \mid x)$$
$$\propto \phi^{-\frac{1}{2}(n-s)-1} \exp[-S^2/2\phi]\, \phi^{-\frac{1}{2}} \exp[-\tfrac{1}{2}\{\mathbf{g}'(\boldsymbol{\theta} - \hat{\boldsymbol{\theta}})\}^2/\phi\mathbf{g}'\mathbf{C}\mathbf{g}].$$

The usual integration with respect to ϕ (cf. equation 5.4.5) establishes the result.

The quantities $\hat{\theta}_i$ (equation (7)) are called the *least squares estimates* of the θ_i. They are the means of the posterior distribution of the θ's. Similarly, $\mathbf{g}'\hat{\boldsymbol{\theta}}$ is the least squares estimate of $\mathbf{g}'\boldsymbol{\theta}$. The equations $(\mathbf{A}'\mathbf{A})\hat{\boldsymbol{\theta}} = \mathbf{A}'\mathbf{x}$, whose solution is (7), are called the *least squares equations*. It is often useful to find the variance of linear functions, $\mathbf{g}'\hat{\boldsymbol{\theta}}$, of the $\hat{\theta}_i$.

$$\begin{aligned}\mathscr{D}^2(\mathbf{g}'\hat{\boldsymbol{\theta}} \mid \boldsymbol{\theta}, \phi) &= \mathscr{D}^2(\mathbf{g}'\mathbf{CA}'\mathbf{x} \mid \boldsymbol{\theta}, \phi) \\ &= \mathbf{g}'\mathbf{CA}'\mathbf{ACg}\phi,\end{aligned}$$

since the x's are independent with common variance (corollary to theorem 3.3.2). Finally, since $\mathbf{C} = (\mathbf{A}'\mathbf{A})^{-1}$,

$$\mathscr{D}^2(\mathbf{g}'\hat{\boldsymbol{\theta}} \mid \boldsymbol{\theta}, \phi) = \mathbf{g}'\mathbf{Cg}\phi. \tag{13}$$

This provides a means of calculating the denominator in (12).

Corollary 3. *Under the same conditions as in theorem* 1 *a significance test at level* α *of the hypothesis that*

$$\sum_{j=1}^{s} b_{ij}\theta_j = 0 \quad (i = r+1, r+2, \ldots, s), \tag{14}$$

where $\mathbf{B}$, *a* $(s-r)\times s$ *matrix with elements* b_{ij}, *has rank* $(s-r)$, *is obtained by declaring the data significant if*

$$[T_r^2/(s-r)]/[S^2/(n-s)] \tag{15}$$

exceeds $\bar{F}_\alpha(\nu_1, \nu_2)$ *with* $\nu_1 = s-r$, $\nu_2 = n-s$. *Here* T_r^2 *is the reduction in sum of squares due to*

$$\sum_{j=1}^{s} b_{ij}\theta_j \quad (i = r+1, r+2, \ldots, s);$$

that is, $T_r^2 + S^2$ *is the minimum of* (2) *when* (14) *obtains.*

Since $\mathbf{B}$ has full rank we can find a non-singular $s \times s$ matrix $\mathbf{B}_0$ whose last $(s-r)$ rows agree with the rows of $\mathbf{B}$. Denote the elements of $\mathbf{B}_0$ by b_{ij}, i now running from 1 to s. Now change the parameters to $\boldsymbol{\psi} = \mathbf{B}_0\boldsymbol{\theta}$. Then $\mathscr{E}(\mathbf{x}) = \mathbf{A}\boldsymbol{\theta} = \mathbf{A}\mathbf{B}_0^{-1}\boldsymbol{\psi}$, which is a linear hypothesis in the ψ's, and the hypothesis to be tested, (14), is $\psi_{r+1} = \psi_{r+2} = \ldots = \psi_s = 0$. The result follows from the theorem.

Linear hypothesis

The theorem is one of the most important single results in statistics. We have met several special cases before and had we stated it earlier could have deduced them from it. However, we preferred to treat simple situations before passing to the general case. The linear hypothesis means that the random variables being observed are, apart from independent random variation of variance ϕ, known linear functions of unknown quantities. Since so many situations have a linear structure, or approximately so, we should expect to find the result widely applicable. In its general form, (14), the hypothesis to be tested is also linear. The test is carried out by minimizing the sum of squares, first allowing the parameters to range freely, obtaining the minimum S^2, and then restricting them according to the hypothesis, obtaining the minimum $S^2+S_r^2$ (or $S^2+T_r^2$ in the general case in corollary 3). It is important to observe, a point we will enlarge on below, that it is not necessary to consider the structure of $\mathbf{A}$ or $\mathbf{C} = (\mathbf{A}'\mathbf{A})^{-1}$: only the minima are required. Let us first see how cases already considered fit into the linear hypothesis concept.

Examples:

(i) *Normal means.* Take $s = 2$ and write $n = n_1+n_2$. Let $\mathbf{A}$ be a matrix whose first column is n_1 1's followed by n_2 0's and whose second column is n_1 0's followed by n_2 1's. That is,

$$\mathscr{E}(x_i) = \theta_1 \quad (1 \leqslant i \leqslant n_1), \qquad \mathscr{E}(x_i) = \theta_2 \quad (n_1 < i \leqslant n), \qquad (16)$$

and we have two independent samples of sizes n_1 and n_2 from $N(\theta_1, \phi)$ and $N(\theta_2, \phi)$ respectively. Consider a test of the hypothesis that $\theta_1 = \theta_2$, that the means are equal (theorem 6.1.3). The sum of squares, (2), is

$$\sum_{i=1}^{n_1}(x_i-\theta_1)^2+\sum_{i=n_1+1}^{n}(x_i-\theta_2)^2,$$

with unrestricted minimum, equal to the residual, of

$$S^2 = \sum_{i=1}^{n_1}(x_i-\bar{x}_1)^2+\sum_{i=n_1+1}^{n}(x_i-\bar{x}_2)^2,$$

where $\bar{x}_1, \bar{x}_2$ are the means of the two samples. If $\theta_1 = \theta_2$ we have to minimize with respect to the common value θ

$$\sum_{i=1}^{n}(x_i-\theta)^2,$$

with the result $$\sum_{i=1}^{n} (x_i - \bar{x})^2,$$

where $\bar{x}$ is the mean of the combined sample of n observations. It is easy to verify that the difference between these two minima is

$$(\bar{x}_1 - \bar{x}_2)^2 (n_1^{-1} + n_2^{-1})^{-1}. \tag{17}$$

Hence (with $s = 2$, $r = 1$), the F-statistic, (15), is

$$(\bar{x}_1 - \bar{x}_2)^2 (n_1^{-1} + n_2^{-2})^{-1} / [S^2/(n-2)], \tag{18}$$

with 1 and $(n-2)$ degrees of freedom. This is the same as the significance test derived from equation 6.1.6: the relationship between the two equations is easily established by putting $\delta = 0$, $s^2 = S^2/(n-2)$ in equation 6.1.6 and remembering that $F = t^2$ when the value of $\nu_1 = 1$ (§6.2). Furthermore, the present corollary 1 is equivalent in this special case to theorem 6.1.2 and corollary 2, with $\mathbf{g}'\boldsymbol{\theta} = \theta_1 - \theta_2$, to theorem 6.1.3. We leave it to the reader to verify that theorem 6.4.1, for testing the difference between several normal means, is also a special case of the present theorem 1: indeed, we have used the method of proof of the earlier theorem to prove the new one.

(ii) *Weighing.* Another special case is the weighing example of §3.3. In §6.6, equations (12) and (13), the situation was expressed as a linear hypothesis with design matrix given in the latter equation. However, there ϕ was supposed known, equal to σ^2. If ϕ is unknown, since $n = s = 4$, there are no degrees of freedom for error $(n - s = 0)$ and no inferences are possible. But if the whole experiment were repeated, so that $n = 8$, still with $s = 4$, then the results of this section could be applied. Corollary 1 gives the posterior distribution of the precision (§5.1) and corollary 2 gives the posterior distribution of any weight or of any difference of weights. Corollary 3 could be used, for example, to test the hypothesis that all the weights were equal $(\theta_1 = \theta_2 = \theta_3 = \theta_4)$.

(iii) *Linear regression.* As a final special case consider linear homoscedastic normal regression and the results obtained in §8.1. Replace the x's of the present section by y's and (θ_1, θ_2) by (α, β). Then if the ith row of $\mathbf{A}$ is

$$(1, x_i - \bar{x}), \tag{19}$$

the linear hypothesis is that given by equation 8.1.2. The matrix $\mathbf{A'A}$ is

$$\mathbf{A'A} = \mathbf{C}^{-1} = \begin{pmatrix} n & 0 \\ 0 & \Sigma(x_i - \bar{x})^2 \end{pmatrix} = \begin{pmatrix} n & 0 \\ 0 & S_{xx} \end{pmatrix}. \tag{20}$$

The least squares estimates are $\hat{\boldsymbol{\theta}} = \mathbf{CA'x}$, which here give

$$\begin{pmatrix} \hat{\alpha} \\ \hat{\beta} \end{pmatrix} = \begin{pmatrix} n^{-1} & 0 \\ 0 & S_{xx}^{-1} \end{pmatrix} \begin{pmatrix} \Sigma y_i \\ \Sigma y_i(x_i - \bar{x}) \end{pmatrix} = \begin{pmatrix} \bar{y} \\ S_{xy}/S_{xx} \end{pmatrix} = \begin{pmatrix} a \\ b \end{pmatrix} \tag{21}$$

agreeing with the least squares estimates of §8.1. Alternatively, we can consider the sums of squares. Either by minimization, or by inserting the least squares estimates, the residual is $\Sigma(y_i - a - b(x_i - \bar{x}))^2 = S_{yy} - S_{xy}^2/S_{xx}$, as before. If $\beta = 0$, the restricted minimum is S_{yy} and the difference between these two minima is $S_{xy}^2/S_{xx} = b^2 S_{xx}$. The F-test of our present theorem is therefore the same as that of table 1 in §8.1. The present corollaries similarly provide results equivalent to those of theorem 8.1.1.

Prior distribution

New applications of the theorem are postponed until later sections: here we make some comments on the assumptions, the proof and the conclusions. The form of the prior distribution is important. It is equivalent to saying that our knowledge of each parameter is vague in comparison with that to be obtained from the observations, which may often be reasonable, but also that these parameters are independent; an assumption which is often unreasonable. We may know, for example, that all the θ's have about the same value, a case that was discussed in §6.6 in connexion with between and within analyses. (It will be shown below that this is a special case of the results of this section.) Alternatively, we may know that a few (most likely one or two) of the θ's may differ from the rest, which are approximately equal. No simple methods are known which are available for treating this situation when ϕ is unknown. The methods of theorem 6.6.3, based on the multivariate normal distribution, are available if ϕ is known to be equal to σ^2. The form which has here been assumed for the prior distribution should be remembered in all applications. With this form of prior distribution the results obtained in this book agree with standard methods based on the considerations of the sample space only (§5.6) when the θ's are assumed constants—the so-called 'fixed effects' model (§8.5).

Design matrix

The design matrix is so-called because it describes the experimental method without reference either to the unknown values (the θ's) that the experiment is designed to investigate, or to the observations (the x's) that will result from any performance of it. For example, we saw in §6.6 that the design matrix for the weighing example was given by equation (13) there: had each object been weighed separately the design matrix would have been the unit matrix, $\mathbf{I}$, a diagonal matrix with 1's in the diagonal. Hence these two matrices distinguish the two possible ways of carrying out the experiment. $\mathbf{A}$ is typically under our control and there is a large literature on the design of experiments which, mathematically, reduces to the choice of $\mathbf{A}$.

Discussion of the proof

The first stage of the proof consists in rewriting the posterior density in the form of equation (8). Notice that to do this it is necessary that $\mathbf{A}'\mathbf{A}$ be non-singular: otherwise $\hat{\boldsymbol{\theta}}$ is undefined. When this is so the distribution of the θ's, *for fixed* ϕ, equal to σ^2, say, is multivariate normal with means equal to the least squares estimates, $\hat{\boldsymbol{\theta}}$, and non-singular dispersion matrix $(\mathbf{A}'\mathbf{A})^{-1}\sigma^2$. This is a special case of theorem 6.6.3. That theorem was concerned with a linear hypothesis and it is only necessary to take the matrix, there denoted by $\mathbf{C}$, to be $\mathbf{I}\sigma^2$ and the matrix $\mathbf{C}_0$ to be a multiple of the unit matrix (the multiple being supposed to tend to infinity, so as to obtain the uniform prior distribution) to obtain the present situation, but with ϕ known equal to σ^2.

If $\phi = \sigma^2$, then we obtain from (10),

$$\pi(\theta_{r+1}, \theta_{r+2}, \ldots, \theta_s \mid \mathbf{x}) \propto \exp\left[-\frac{1}{2\sigma^2}\left\{\sum_{i=r+1}^{s} c_i(\theta_i - \alpha_i)^2\right\}\right],$$

and this result replaces (11). In the proof just given we extended the proof of theorem 6.4.1 to obtain an F-distribution. If $\phi = \sigma^2$

we can similarly extend the proof of theorem 6.4.2 to obtain a χ^2-distribution. As a consequence of this, a significance test of the hypothesis that $\theta_{r+1} = \theta_{r+2} = \ldots = \theta_s$ is provided by referring S_r^2/σ^2 to the χ^2-distribution with $\nu = s-r$ degrees of freedom. This differs from the test of the present section in that $S^2/(n-s)$ is used as an estimate of σ^2 and the F-distribution replaces χ^2.

The second stage in the proof consists of rewriting the multivariate normal distribution in its regression form. This was the way the distribution was introduced in §3.5. The reason for writing it this way is that the variables, here the θ's, are introduced successively and may be integrated in the *reverse* order. Here we begin with θ_s, then introduce θ_{s-1}, depending on θ_s, and so on: whence θ_1 only appears in the final stage where we may integrate with respect to it. This expression in regression form will form the basis of the method of numerical calculation to be described in the next section. With the integration of the nuisance parameters $\theta_1, \theta_2, \ldots, \theta_r, \phi$ carried out we are left with (11) and from there we proceed in the way discussed in detail in §6.4.

Distribution of a linear form

Corollary 2 provides the distribution of any linear function of the θ's, in particular of any θ_i. (The joint distribution is given by (11) but is usually too complicated to be used.) As in previous situations, the relevant distribution is a t-distribution. The variance is estimated by the residual sum of squares divided by $(n-s)$, and the mean by the least squares estimate. The denominator in (12) can be troublesome because of the term $\mathbf{g}'\mathbf{C}\mathbf{g}$. A simple way of calculating this term is provided by equation (13), which says that $\mathbf{g}'\mathbf{C}\mathbf{g}\phi$ is the variance of the least squares estimate, $\mathbf{g}'\hat{\boldsymbol{\theta}}$, of $\mathbf{g}'\boldsymbol{\theta}$. Since $\hat{\boldsymbol{\theta}}$ is a linear function of the x's this variance may be found by the methods of §3.3. An example is given in §8.5. The result being used here is an extension of that discussed in §5.1 where two distinct statements (a) and (b) are easily confused: the posterior variance of $\mathbf{g}'\boldsymbol{\theta}$ is the same as the sampling variance of its estimate $\mathbf{g}'\hat{\boldsymbol{\theta}}$.

In many situations the corollary is more useful than the theorem. The latter gives a test of a rather complicated hypothesis and not a complete statement of a posterior distribution. The former provides a posterior distribution for a linear function of the parameters. Complete specification for one function is often more relevant than a partial specification for many. It is true that the theorem can easily be adapted to yield a posterior distribution but the result is too complicated for appreciation. It is commonly sensible to use the theorem in order to eliminate unnecessary parameters, and then to use the corollary to investigate those that contribute most to the variation of the dependent variable.

General linear constraints

Corollary 3 extends the test of the theorem to general linear constraints amongst the parameters. Notice that **B** must have full rank, or alternatively that the $(s-r)$ constraints in (14) must be linearly independent. If this were not so then some of the constraints would be implied by the others and the effective number would be less than $(s-r)$, thereby influencing the degrees of freedom in the F-test.

Analysis of variance

The F-test can be written as an analysis of variance test. Table 8.3.1 is a generalization of table 8.1.1. The total sum of squares refers to the minimum of the sum of squares of differences between the x's and their expectations when these latter are expressed in terms of the nuisance parameters $\theta_1, \theta_2, \ldots, \theta_r$ only. The total can be broken down into two parts: the reduction due to introducing the additional parameters $\theta_{r+1}, \theta_{r+2}, \ldots, \theta_s$, and a residual which is the corresponding minimum when all parameters are used. If the null hypothesis that the extra parameters are all zero is correct the two corresponding mean squares should be of comparable magnitude since one is breaking down quite arbitrarily a total which is entirely due to random error: on the other hand, if the null hypothesis is false the former should be substantially greater than the residual. The F-statistic is, apart from a multiplier involving the degrees of freedom, the ratio of these two parts and the data are significant if F is some amount in excess of 1; the exact amount depending on the F-distribution. The degrees of freedom, $n-s$, for the residual are obvious from corollary 1 and equally the degrees of freedom for the total are $n-r$. The remaining degrees of freedom for the reduction are defined to have the same additivity as the sum of squares and the F-statistic is, as before, a ratio of mean squares.

TABLE 8.3.1

	Sums of squares	Degrees of freedom	Mean squares	F
Reduction due to $\theta_{r+1}, \theta_{r+2}, \ldots, \theta_s$ (allowing for $\theta_1, \theta_2, \ldots, \theta_r$)	S_r^2	$s-r$	$S_r^2/(s-r)$	$\dfrac{[S\ /(s-r)]}{[S^2/(n-s)]}$
Residual (fitting $\theta_1, \theta_2, \ldots, \theta_s$)	S^2	$n-s$	$S^2/(n-s)$	—
Total (for $\theta_1, \theta_2, \ldots, \theta_r$)	$S^2+S_r^2$	$n-r$	—	—

Breakdown of sum of squares

The sum of squares can be broken down still further, but the interpretation requires considerable care. Let $r < t < s$ and let the three groups of parameters

$$(\theta_1, \theta_2, \ldots, \theta_r), \quad (\theta_{r+1}, \theta_{r+2}, \ldots, \theta_t), \quad (\theta_{t+1}, \theta_{t+2}, \ldots, \theta_s)$$

be referred to as the first, second and third groups, respectively. Let $S_{\bar{1}\bar{3}}^2$ be the minimum of the sum of squares when the expectations are assumed to contain only the first and third groups of parameters, the second group being zero. Define other sums of squares similarly: thus $S_{\bar{1}}^2$ is now the total sum of squares for $\theta_1, \theta_2, \ldots, \theta_r$ and $S_{\bar{1}\bar{2}\bar{3}}^2$ is the residual. In table 8.3.1 we have written

$$S_{\bar{1}}^2 = S_{\bar{1}\bar{2}\bar{3}}^2 + (S_{\bar{1}}^2 - S_{\bar{1}\bar{2}\bar{3}}^2).$$

We may further write

$$S_{\bar{1}}^2 = S_{\bar{1}\bar{2}\bar{3}}^2 + (S_{\bar{1}\bar{2}}^2 - S_{\bar{1}\bar{2}\bar{3}}^2) + (S_{\bar{1}}^2 - S_{\bar{1}\bar{2}}^2). \tag{22}$$

The three terms on the right-hand side are, respectively, the residual, allowing for all three groups, the reduction due to the third group allowing for the first two, and the reduction due to the second group allowing for the first. Equally we can write

$$S_{\bar{1}}^2 = S_{\bar{1}\bar{2}\bar{3}}^2 + (S_{\bar{1}\bar{3}}^2 - S_{\bar{1}\bar{2}\bar{3}}^2) + (S_{\bar{1}}^2 - S_{\bar{1}\bar{3}}^2), \tag{23}$$

where the roles of the second and third groups have been interchanged. In both (22) and (23) the ratio of the mean squares of the second term on the right-hand side to that of the first term gives a valid F-test: in (22) of the null hypothesis that the parameters in the third group all vanish, in (23) of the null hypothesis that those of the second group all vanish. This follows from the theorem. The remaining terms on the right-hand sides do not provide tests, under the assumptions of the theorem, since, for example in (22), in general $S_1^2 - S_{12}^2$ will not be equal to $S_{13}^2 - S_{123}^2$. These are both reductions due to the second group but the former does not take account of the possible existence of the third group. However, it can happen that

$$S_1^2 - S_{12}^2 = S_{13}^2 - S_{123}^2, \tag{24}$$

and consequently also

$$S_1^2 - S_{13}^2 = S_{12}^2 - S_{123}^2, \tag{25}$$

and the two decompositions, (22) and (23), are equivalent. Equation (25) says that the reduction in sum of squares due to the third group is the same irrespective of whether only the first group, or both the first and second groups, have been allowed for. Equation (24) is the same with the second and third groups interchanged. In this situation we often refer to the reduction due to the second or third groups without saying what has been allowed for. In these special circumstances table 8.3.1 can be extended to the form shown in table 8.3.2. From this table, in which the sums of squares and degrees of freedom are additive, it is possible to derive valid F-tests for the hypotheses that the second and third

TABLE 8.3.2

	Sums of squares	Degrees of freedom	Mean squares	F
Reduction due to $\theta_{r+1}, \ldots, \theta_t$	$S_{13}^2 - S_{123}^2$	$t-r$	$(S_{13}^2 - S_{123}^2)/(t-r)$	Mean squares divided by s^2
Reduction due to $\theta_{t+1}, \ldots, \theta_s$	$S_{12}^2 - S_{123}^2$	$s-t$	$(S_{12}^2 - S_{123}^2)/(s-t)$	
Residual	S_{123}^2	$n-s$	$S_{123}^2/(n-s) = s^2$	—
Total (for $\theta_1, \ldots, \theta_r$)	S_1^2	$n-r$	—	—

groups of parameters separately vanish. But it must be remembered that this is only sensible if (24), and therefore (25), obtain.

A sufficient, but not necessary, condition for (24) is easily obtained from the expression in (8) for the sum of squares. Suppose that the matrix† $\mathbf{B} = \mathbf{A}'\mathbf{A}$ can be partitioned into submatrices corresponding to the three groups of parameters in the following way:

$$\begin{array}{r} r \\ t-r \\ s-t \end{array}\left(\begin{array}{c|c|c} \mathbf{B}_{11} & 0 & 0 \\ \hline 0 & \mathbf{B}_{22} & 0 \\ \hline 0 & 0 & \mathbf{B}_{33} \end{array}\right) = \mathbf{B}. \qquad (26)$$
$$\quad r \quad t-r \quad s-t$$

(The letters at the side and bottom indicate the sizes of the submatrices: thus $\mathbf{B}_{11}$ has r rows and r columns.) In terms of the submatrices $\mathbf{B}$ is diagonal and therefore in (8) the three groups of parameters appear separately and, for given ϕ, they are independent. Consequently, minimization with respect to the parameters of one group does not affect those of another and (24) holds. An example of this is the weighing example of §§3.3, 6.6, where we saw (equation 6.6.13) that $\mathbf{B} = \mathbf{A}'\mathbf{A}$ was diagonal so that each parameter may be investigated separately, irrespective of the others. A design matrix $\mathbf{A}$ which is such that $\mathbf{A}'\mathbf{A}$ has the form (26) is said to be *orthogonal* with respect to the three groups of parameters (and generally for any number of groups). Other things being equal, orthogonal experiments are to be preferred.

Although, in general, $S_1^2 - S_{12}^2$ does not provide a test under the assumptions of the theorem, because it ignores the third group of parameters, it can under different assumptions. If $S_{12}^2 - S_{123}^2$ does not yield a significant result, so that the parameters of the third group are possibly all zero, one might wish to test whether those of the second group are also zero under the assumption that those of the third group are. Clearly $S_1^2 - S_{12}^2$ then does provide such a test in comparison with S_{12}^2 (not S_{123}^2). This test is particularly useful where there is a natural order in which the groups should be introduced—first, second, third—

† $\mathbf{B}$ should not be confused with the matrix occurring in corollary 3.

and examples will be found in §§8.6(*a*) and (*c*). Compare, also, this argument with that suggested in §6.3 for deciding whether to use Behrens's test or a *t*-test.

Least squares estimates

It follows from (8) that the statistics $\hat{\boldsymbol{\theta}}$ and S^2 are jointly sufficient for $\boldsymbol{\theta}$ and ϕ. The least square estimates and the residual are therefore important quantities to calculate in any analysis of a linear hypothesis. The calculation can be done in two ways: by finding $(\mathbf{A}'\mathbf{A})^{-1} = \mathbf{C}$ and hence $\hat{\boldsymbol{\theta}}$ from (7), and then

$$S^2 = \mathbf{x}'\mathbf{x} - \hat{\boldsymbol{\theta}}'\mathbf{A}'\mathbf{A}\hat{\boldsymbol{\theta}}, \quad \text{from (6)}, \tag{27}$$

$$= \mathbf{x}'\mathbf{x} - \mathbf{x}'\mathbf{A}\mathbf{C}\mathbf{A}'\mathbf{x}, \quad \text{from (7)} \tag{28}$$

(from which it appears that $\mathbf{x}'\mathbf{x}$ and $\mathbf{A}'\mathbf{x}$ are also jointly sufficient); or alternatively by actual minimization of the sum of squares, the minimum being S^2, the values of $\boldsymbol{\theta}$ at the minimum being $\hat{\boldsymbol{\theta}}$. The former method is to be preferred in the case of a general $\mathbf{A}$ and we discuss it in the next section: the calculation depending on the inversion of $\mathbf{A}'\mathbf{A}$. The latter method is more convenient when the structure of $\mathbf{A}$ is particularly simple, as it often is in well-designed (for example, orthogonal) experiments. Examples are given in §§8.5, 8.6.

8.4. Computational methods

In this section we consider how the computations necessary to perform the tests of the previous section may be arranged when the matrix $\mathbf{A}'\mathbf{A}$ is general and it is therefore not possible to take advantage of any special structure that it might have. The methods are designed for use with desk calculating machines.

If $\mathbf{f}'$ and $\mathbf{g}'$ are two row-vectors, or simply rows,

$$\mathbf{f}' = (f_1, f_2, \ldots, f_n), \quad \mathbf{g}' = (g_1, g_2, \ldots, g_n),$$

each of n elements; then by the *product* of these two rows we mean the scalar product $\sum_{i=1}^{n} f_i g_i$. The product of a row and column or two columns is defined similarly. If all the elements of $\mathbf{f}$ and $\mathbf{g}$ are known, except for g_n, and we require the product to equal c, we speak of '*making-up*' $\mathbf{g}$ so that $\sum_{i=1}^{n} f_i g_i = c$.

The stages in the computation are as follows. They should be studied in connexion with the numerical example given below:

(1) The quantities $\mathbf{x}'\mathbf{x}$, $\mathbf{u} = \mathbf{A}'\mathbf{x}$ and $\mathbf{B} = \mathbf{A}'\mathbf{A}$ are calculated.

(2) The theory was based on rewriting (8.3.8) in the form (8.3.9) and the same device is used in the computations. Let $\mathbf{A}'\mathbf{A}$ be written $\mathbf{\Gamma}'\mathbf{\Gamma}$, where $\mathbf{\Gamma}$ is an $s \times s$ upper triangular matrix: that is, a matrix with all elements below the leading diagonal zero; $\gamma_{ij} = 0$ for $i > j$. Then

$$(\boldsymbol{\theta}-\hat{\boldsymbol{\theta}})'\,\mathbf{A}'\mathbf{A}(\boldsymbol{\theta}-\hat{\boldsymbol{\theta}}) = (\boldsymbol{\theta}-\hat{\boldsymbol{\theta}})'\,\mathbf{\Gamma}'\mathbf{\Gamma}(\boldsymbol{\theta}-\hat{\boldsymbol{\theta}}) = \boldsymbol{\xi}'\boldsymbol{\xi},$$

where $\boldsymbol{\xi} = \mathbf{\Gamma}(\boldsymbol{\theta}-\hat{\boldsymbol{\theta}})$ and, because of the triangular form of $\mathbf{\Gamma}$, ξ_j involves only $\theta_j, \theta_{j+1}, \ldots, \theta_s$. It easily follows that, in the notation of equation 8.3.9, $\xi_j^2 = c_j(\theta_j - \alpha_j)^2$. The fact that we know such a transformation is possible and unique, establishes the existence and uniqueness of $\mathbf{\Gamma}$.

The equations for $\mathbf{\Gamma}$, $\mathbf{\Gamma}'\mathbf{\Gamma} = \mathbf{B}$, may be written:

$$\left.\begin{aligned} &(i\text{th row of } \mathbf{\Gamma}') \times (j\text{th column of } \mathbf{\Gamma}) = b_{ij} \\ \text{or}\quad &(i\text{th column of } \mathbf{\Gamma}) \times (j\text{th column of } \mathbf{\Gamma}) = b_{ij}. \end{aligned}\right\} \tag{1}$$

If the calculations are carried out in the order $i = 1, j = 1, 2, \ldots, s$; $i = 2, j = 2, 3, \ldots, s$ and so on, on each occasion one element in the product of the two columns will be unknown and may be found by 'making-up': this follows because of the zero elements in $\mathbf{\Gamma}$. (If $i = j$ two equal elements are unknown and the 'making-up' involves a square root.)

It is advisable, as always, to have a check on the calculations. This is here conveniently provided by using the relation, which easily follows from (1), that

$$\begin{aligned} &(i\text{th column of } \mathbf{\Gamma}) \times (\text{column of row sums of } \mathbf{\Gamma}) \\ &\qquad = (i\text{th element of row sums of } \mathbf{B}) \end{aligned}$$

after each row of $\mathbf{\Gamma}$ has been found.

(3) The vector $\boldsymbol{\omega}$, of elements ω_i, satisfying

$$\boldsymbol{\Gamma}'\boldsymbol{\omega} = \mathbf{u} \tag{2}$$

is computed. This is done by thinking of (2) as

$$(i\text{th column of } \boldsymbol{\Gamma}) \times (\text{the column } \boldsymbol{\omega}) = (i\text{th element of } \mathbf{u}). \tag{3}$$

Since the first column of $\boldsymbol{\Gamma}$ contains only one non-zero element, ω_1 may be found; and then $\omega_2, \omega_3, \ldots$. A check is provided by

$$(\text{column of row sums of } \boldsymbol{\Gamma}) \times (\text{the column } \boldsymbol{\omega})$$
$$= (\text{sum of elements of } \mathbf{u}).$$

(4) The reduction in sum of squares due to $\theta_{r+1}, \theta_{r+2}, \ldots, \theta_s$ is $\sum_{i=r+1}^{s} \omega_i^2$: the residual sum of squares is $S^2 = \mathbf{x}'\mathbf{x} - \sum_{i=1}^{s} \omega_i^2$. The latter result may be proved by remarking that since

$$\boldsymbol{\Gamma}'\boldsymbol{\Gamma}\hat{\boldsymbol{\theta}} = \mathbf{u} \quad (\text{from equation 8.3.7}),$$

$$\boldsymbol{\Gamma}\hat{\boldsymbol{\theta}} = \boldsymbol{\omega}, \tag{4}$$

from (2). Hence, from equation 8.3.6,

$$S^2 = \mathbf{x}'\mathbf{x} - \hat{\boldsymbol{\theta}}'\mathbf{A}'\mathbf{A}\hat{\boldsymbol{\theta}} = \mathbf{x}'\mathbf{x} - \hat{\boldsymbol{\theta}}'\boldsymbol{\Gamma}'\boldsymbol{\Gamma}\hat{\boldsymbol{\theta}}$$
$$= \mathbf{x}'\mathbf{x} - \boldsymbol{\omega}'\boldsymbol{\omega}, \tag{5}$$

as required. The former result follows by remarking that just as

$$\mathbf{x}'\mathbf{x} - \sum_{i=1}^{s} \omega_i^2$$

is the minimum after including all the θ's, so

$$\mathbf{x}'\mathbf{x} - \sum_{i=1}^{r} \omega_i^2$$

is the minimum if only $\theta_1, \theta_2, \ldots, \theta_r$ are included and the remainder put equal to zero. This is obvious if the computing scheme with this reduced number of θ's is considered: it will mean merely considering only the first r rows and columns of the matrices. Hence the minimum is reduced still further by $\sum_{i=r+1}^{s} \omega_i^2$ if $\theta_{r+1}, \theta_{r+2}, \ldots, \theta_s$ are included.

(5) The statistic (8.3.4) is then

$$\left[\sum_{i=r+1}^{s} \omega_i^2/(s-r)\right] \Big/ [S^2/(n-s)].$$

(6) The least squares estimates, $\hat{\boldsymbol{\theta}}$, are found from (4), which may be thought of as

$$(i\text{th row of } \boldsymbol{\Gamma}) \times (\text{the column } \hat{\boldsymbol{\theta}}) = \omega_i. \tag{6}$$

Since the last row of $\boldsymbol{\Gamma}$ contains only one non-zero element, $\hat{\theta}_s$ may be found first, then $\hat{\theta}_{s-1}, \hat{\theta}_{s-2}, \ldots$. A check is provided by

$$(\text{row of column sums of } \boldsymbol{\Gamma}) \times (\text{the column } \hat{\boldsymbol{\theta}}) = \sum_{i=1}^{s} \omega_i.$$

(7) Since

$$\mathbf{A}'\mathbf{A} = \boldsymbol{\Gamma}'\boldsymbol{\Gamma}$$

$$(\mathbf{A}'\mathbf{A})^{-1} = \boldsymbol{\Gamma}^{-1}(\boldsymbol{\Gamma}')^{-1}$$

and

$$(\mathbf{A}'\mathbf{A})^{-1}\boldsymbol{\Gamma}' = \boldsymbol{\Gamma}^{-1}.$$

This may be thought of as

$$(i\text{th row of } \mathbf{B}^{-1}) \times (j\text{th row of } \boldsymbol{\Gamma}) = \gamma^{ij},$$

where γ^{ij} is the typical element of $\boldsymbol{\Gamma}^{-1}$. Now we know that since $\boldsymbol{\Gamma}$ is upper triangular so also is $\boldsymbol{\Gamma}^{-1}$. Hence $\gamma^{ij} = 0$ for $i > j$. Furthermore, $\gamma^{ii} = \gamma_{ii}^{-1}$. Thus all elements of $\boldsymbol{\Gamma}^{-1}$ on the leading diagonal and below it are known and consequently with

$$i = s, \quad j = s, s-1, \ldots, 1; \qquad i = s-1, \quad j = s-1, s-2, \ldots, 1$$

and so on, we can find the elements of $\mathbf{B}^{-1}$, known to be symmetric, successively. A check is provided by

$$(i\text{th row of } \mathbf{B}^{-1}) \times (\text{column of row sums of } \mathbf{B}) = 1.$$

All the relevant quantities are now available.

Arrangement of the calculations

The following example illustrates the procedure and gives the most convenient arrangement of the calculations on the paper.

We consider first the purely numerical part, and leave the statistical considerations until later.

$$\mathbf{B} = \begin{pmatrix} +5{\cdot}5140 & +5{\cdot}0958 & -1{\cdot}5150 & -1{\cdot}4446 \\ & +6{\cdot}6257 & -0{\cdot}8510 & -1{\cdot}1755 \\ & & +5{\cdot}9980 & +1{\cdot}7614 \\ & & & +7{\cdot}3071 \end{pmatrix}$$

$$\begin{array}{ll} \boldsymbol{\Gamma} = \begin{pmatrix} +2{\cdot}3482 & +2{\cdot}1701 & -0{\cdot}6452 & -0{\cdot}6152 \\ & +1{\cdot}3843 & +0{\cdot}3967 & +0{\cdot}1153 \\ & & +2{\cdot}3290 & +0{\cdot}5662 \\ & & & +2{\cdot}5680 \end{pmatrix} & \begin{array}{r} \text{row sums} \\ +3{\cdot}2579 \\ +1{\cdot}8963 \\ +2{\cdot}8952 \\ +2{\cdot}5680 \end{array} \end{array}$$

$$\begin{array}{lrrrr} \text{column sums} & +2{\cdot}3482 & +3{\cdot}5544 & +2{\cdot}0805 & +2{\cdot}6343 \\ \gamma_{ii}^{-1} & +0{\cdot}4259 & +0{\cdot}7224 & +0{\cdot}4294 & +0{\cdot}3894 \end{array}$$

$$\mathbf{u} = \begin{pmatrix} 6979{\cdot}32 \\ 7583{\cdot}66 \\ 3488{\cdot}51 \\ 6262{\cdot}42 \end{pmatrix} \quad \boldsymbol{\omega} = \begin{pmatrix} 2972{\cdot}20 \\ 818{\cdot}96 \\ 2181{\cdot}75 \\ 2632{\cdot}86 \end{pmatrix} \quad \hat{\boldsymbol{\theta}} = \begin{pmatrix} 1437{\cdot}52 \\ 309{\cdot}18 \\ 687{\cdot}53 \\ 1025{\cdot}26 \end{pmatrix}$$

sum 24313·91 sum 8605·77

$$\begin{array}{ll} \mathbf{B}^{-1} = \begin{pmatrix} +0{\cdot}6873 & -0{\cdot}5112 & +0{\cdot}0918 & +0{\cdot}0315 \\ & +0{\cdot}5370 & -0{\cdot}0523 & -0{\cdot}0021 \\ & & +0{\cdot}1933 & -0{\cdot}0369 \\ & & & +0{\cdot}1516 \end{pmatrix} & \begin{array}{r} \text{row sums} \\ \text{of } \mathbf{B} \\ +7{\cdot}6502 \\ +9{\cdot}6950 \\ +5{\cdot}3934 \\ +6{\cdot}4484 \end{array} \end{array}$$

Notes. (1) **B** is symmetric, so only the leading diagonal and terms above it need be written out. Nevertheless, in forming the row sums, for checking, the omitted elements should not be forgotten. These row sums are most conveniently placed at the end of the computations next to $\mathbf{B}^{-1}$, since that is where they are needed in checking stage (7) of the calculation.

(2) The omitted elements of $\boldsymbol{\Gamma}$ are, of course, zero. Both row sums (in stage (3)) and column sums (in stage (6)) are needed for checking.

(3) Notice that aside from the elements required for checking and the final quantities required (the ω_i^2's, $\hat{\boldsymbol{\theta}}$ and $\mathbf{B}^{-1}$) only the

matrix $\mathbf{\Gamma}$, the vector $\mathbf{u}$ and the inverses $\gamma_{ii}^{-1} = \gamma^{ii}$ have to be recorded. Since the main source of error with desk machines lies in the transfer from paper to machine and vice versa the opportunities for error are reduced by this procedure.

(4) All operations consist of row or column multiplications with 'making-up'. The use of only one basic operation makes for simplicity.

(5) Experience shows that this method is of high accuracy. It is usually enough to take one more significant figure in the calculations than will be needed in the final answer to allow for rounding-off errors.

Multiple regression

The case of general $\mathbf{A}$ most usually arises in a generalization of the situation of §8.1. Consider s variables $(y, x_1, x_2, \ldots, x_{s-1})$: y is a random variable whose distribution depends on the x's in a way to be described. The x's may be random or not; if random, then they are ancillary (see the discussion in §8.1). If the x's take values $x_{i1}, x_{i2}, \ldots, x_{i,s-1}$ $(i = 1, 2, \ldots, n)$ the random variable y, with value y_i, is normally distributed with mean

$$\mathscr{E}(y \mid x_{i1}, x_{i2}, \ldots, x_{i,s-1}) = \alpha + \sum_{j=1}^{s-1} \beta_j (x_{ij} - x_{.j}) \tag{7}$$

and

$$\mathscr{D}^2(y \mid x_{i1}, x_{i2}, \ldots, x_{i,s-1}) = \phi, \tag{8}$$

where, as usual, $x_{.j} = \sum_{i=1}^{n} x_{ij}/n$.

In these circumstances y is said to have a (*linear homoscedastic*) *multiple regression* on the x's. The words in brackets are usually omitted. If the y_i are, for these values of the x's, independent, we have a random sample from the multiple regression. β_j is called the *multiple regression coefficient* of y on x_j. The notation for it is an abbreviation since it also depends on the other variables included in the regression and, in general, would change if any of them were excluded or others introduced. The coefficient measures the change in the dependent variable caused by a unit change in x_j, the others being held constant.

The situation is clearly a linear hypothesis: in the notation of §8.3:

$$\mathbf{x}' = (y_1, y_2, \ldots, y_n), \tag{9}$$

$$\boldsymbol{\theta}' = (\alpha, \beta_1, \ldots, \beta_{s-1}) \tag{10}$$

and

$$\mathbf{A} = \begin{pmatrix} 1 & x_{11}-x_{.1} & x_{12}-x_{.2} & \ldots & x_{1,s-1}-x_{.s-1} \\ 1 & x_{21}-x_{.1} & x_{22}-x_{.2} & \ldots & x_{2,s-1}-x_{.s-1} \\ \ldots & \ldots & \ldots & \ldots & \ldots \\ 1 & x_{n1}-x_{.1} & x_{n2}-x_{.2} & \ldots & x_{n,s-1}-x_{.s-1} \end{pmatrix} \tag{11}$$

(compare the linear regression example in §8.3 especially equation 8.3.19). Typically the x_{ij} cannot be conveniently chosen, because the x's are random or are otherwise not under control, and the design matrix has a general form, apart from the first column. The test of theorem 8.3.1 is a test that

$$\beta_r = \beta_{r+1} = \ldots = \beta_{s-1} = 0:$$

that is, a test of the hypothesis that the random variable y does not depend on the variables $x_r, x_{r+1}, \ldots, x_{s-1}$.

The fact that the first column of $\mathbf{A}$ consists entirely of 1's can be exploited. We easily see that

$$\mathbf{A}'\mathbf{A} = \begin{pmatrix} n & 0 & 0 & \ldots & 0 \\ 0 & b_{11} & b_{12} & \ldots & b_{1,s-1} \\ \ldots & \ldots & \ldots & \ldots & \ldots \\ 0 & b_{1,s-1} & b_{2,s-1} & \ldots & b_{s-1,s-1} \end{pmatrix}, \tag{12}$$

where
$$b_{ij} = \sum_{k=1}^{n} (x_{ki}-x_{.i})(x_{kj}-x_{.j}), \tag{13}$$

the sum of squares ($i = j$) and products ($i \neq j$) of the x's about their means. Because of the zeros in $\mathbf{A}'\mathbf{A}$; in equation 8.3.8 the term in θ_1 (here $\theta_1 = \alpha$) is separate from the rest and we may perform an integration with respect to it to obtain (in the multiple regression notation)

$$\pi(\beta_1, \beta_2, \ldots, \beta_{s-1}, \phi \mid \mathbf{x}) \propto \phi^{-\frac{1}{2}(n-1)-1}\exp\left[-\{(\boldsymbol{\beta}-\hat{\boldsymbol{\beta}})'\mathbf{B}(\boldsymbol{\beta}-\hat{\boldsymbol{\beta}})+S^2\}/2\phi\right], \tag{14}$$

which is the same form as before with $\mathbf{B}$ the matrix whose typical element is b_{ij}, equation (13). It is also clear from equation 8.3.7 that $\hat{\alpha} = \bar{y} = \sum_{i=1}^{n} y_i/n$, and hence, from equation 8.3.6, that the residual sum of squares is

$$S^2 = \mathbf{x}'\mathbf{x} - \hat{\boldsymbol{\theta}}'\mathbf{A}'\mathbf{A}\boldsymbol{\theta} = \Sigma y_i^2 - n\bar{y}^2 - \hat{\boldsymbol{\beta}}'\mathbf{B}\hat{\boldsymbol{\beta}}$$
$$= \Sigma(y_i - \bar{y})^2 - \hat{\boldsymbol{\beta}}'\mathbf{B}\hat{\boldsymbol{\beta}}. \qquad (15)$$

Hence the form of (14) is the same as that of (8.3.8) with n reduced by one and the observed variables replaced by deviations from their means. We have the usual phenomenon that the presence of an unknown mean reduces the degrees of freedom by 1. The calculations can be carried through with $\beta_1, \beta_2, \ldots, \beta_{s-1}$ only, and deviations from the means.

Numerical example

The numerical example concerned a situation of the multiple regression form, where a sample of thirty small farms of similar character was studied over a period of three years. The variable y was income of a farm, and it was desired to see how this was affected by x_1, the size of the farm; x_2, the standardized production and by two indices, x_3 and x_4, of yield and feeding.

The matrix **B** of the numerical example gives the sums of squares and products, equation (13), of the x variables usually called the independent variables. These have been given here in suitable units to make the four diagonal entries of **B** of the same order. Such changes of scale are desirable in order to simplify the calculations. The vector **u** is the vector of sums of products of y with the x's: namely the ith element is

$$u_i = \sum_{k=1}^{n} y_k(x_{ki} - x_{.i}) = \sum_{k=1}^{n} (y_k - \bar{y})(x_{ki} - x_{.i}). \qquad (16)$$

The total sum of squares of the y's is $\Sigma(y_i - \bar{y})^2 = 40, 572, 526$. Hence the residual sum of squares is this less

$$\sum_{i=1}^{4} \omega_i^2 = 21, 196, 653,$$

giving $S^2 = 19, 375, 873$ on $(29-4) = 25$ degrees of freedom. As an example of the test of theorem 8.3.1 consider a test of whether $\beta_3 = \beta_4 = 0$; that is, of whether the income depends on the two indices. This is provided by referring

$$[(\omega_3^2+\omega_4^2)/2]/[S^2/25] \tag{17}$$

to the F-distribution with 2 and 25 degrees of freedom. The numerical value of (17) is 7·54 and the upper 1 % value of F is 5·57. The result is therefore significant at 1 %: or, in Bayesian terms, we are at least 99 % confident that the yield does depend on the indices. The significance level is greater than 0·1 %, however, since at that level the F value is 9·22.

A test of dependence on any one of the x's is provided by corollary 2 of the last section. For example, a test of the dependence on x_2, the standardized production, is obtained by referring

$$309{\cdot}18/[S^2 \times 0{\cdot}5370/25]^{\frac{1}{2}} = 0{\cdot}48$$

to the t-distribution with 25 degrees of freedom. ($\hat{\beta}_2 = 309{\cdot}18$ and 0·5370 is the appropriate element in $\mathbf{B}^{-1} = \mathbf{C}$.) The 5 % value for t is 2·06 so that the result is not significant. There is, therefore, fair evidence to suggest that the income is not affected by the standardized production, after allowance for the other factors.

Consequences of non-orthogonality

The only hypotheses that it is possible to test using the computational lay-out suggested above are that

$$\beta_{r+1} = \beta_{r+2} = \ldots = \beta_4 = 0 \quad \text{for} \quad r = 0, 1, 2, 3$$

(by the theorem) or

$$\beta_i = 0 \quad \text{for} \quad i = 1, 2, 3, 4$$

(by corollary 2). It is not immediately possible, for example, to test the hypothesis that $\beta_1 = \beta_2 = 0$. To do this it would be necessary to rearrange the matrix $\mathbf{B}$ and the vector $\mathbf{u}$ so that the last two rows and columns referred to x_1 and x_2; when a test would be obtained by using what would then be $\omega_3^2+\omega_4^2$, corresponding to the last two parameters introduced. Before beginning the computations it is therefore important to consider which hypotheses are the most important to test and to order the rows and columns accordingly. Notice that it is not permissible to arrange the calculations in the form of an analysis of variance table (table 8.3.2) in order to obtain a test, for example, of $\beta_1 = \beta_2 = 0$. If the calculations are so arranged all one could obtain would be a test that $\beta_1 = \beta_2 = 0$ *assuming* $\beta_3 = \beta_4 = 0$, since the reduction due to β_1 and β_2 in the computations, as arranged above, has not allowed for the presence of β_3 and β_4. This test might have been useful had the test of $\beta_3 = \beta_4 = 0$ not been significant.

The argument used above in describing stage (2) of the calculations shows when the analysis of variance table is useful in providing significance tests. Consider the notation as in table 8.3.2. Suppose that $\mathbf{\Gamma}$ has the form

$$\begin{array}{c} r \\ t-r \\ s-t \end{array} \left(\begin{array}{c|c|c} \mathbf{\Gamma}_{11} & \mathbf{\Gamma}_{12} & \mathbf{\Gamma}_{13} \\ \hline 0 & \mathbf{\Gamma}_{22} & 0 \\ \hline 0 & 0 & \mathbf{\Gamma}_{33} \end{array}\right) \qquad (18)$$
$$\qquad r \qquad t-r \qquad s-t$$

(cf. 8.3.26): that is $\mathbf{\Gamma}_{23} = 0$. If the sum of squares is written in the form $\Sigma\xi_i^2 + S^2$, where S^2 is the residual and $\boldsymbol{\xi} = \mathbf{\Gamma}(\boldsymbol{\theta} - \hat{\boldsymbol{\theta}})$, we see from (18) that ξ_j^2, for $r < j \leqslant t$, has no terms in $\theta_{t+1}, \ldots, \theta_s$. That is, it involves only terms in the second group, just as ξ_j^2 for $j > t$ has only terms in the third group. This equally applies to $c_j(\theta_j - \alpha_j) = \xi_j$ (equation 8.3.9). Consequently the reduction in the sum of squares due to the second group is $\sum_{j=r+1}^{t} c_j(0 - \alpha_j')^2$ irrespective of the values in the third group. (Here α_j' is the value of α_j when the θ's in the second group are zero.) Consequently this is the genuine reduction due to the second group, $S_1^2 - S_{12}^2 = S_{13}^2 - S_{123}^2$, and the tests for the two groups are both valid.

Joint distribution of regression coefficients

If ϕ is known, equal to σ^2, the matrix $\mathbf{B}^{-1}\sigma^2$ is the dispersion matrix of the posterior distribution of the regression coefficients (from 8.3.8). The diagonal elements are the variances needed in the t-tests for the individual coefficients. It is also useful to calculate the correlations from the off-diagonal terms, the covariances, so that one can see how far the distribution of one coefficient is affected by another. The correlations here are

$$\begin{pmatrix} - & -0{\cdot}8415 & +0{\cdot}2519 & +0{\cdot}0976 \\ - & - & -0{\cdot}1623 & -0{\cdot}0074 \\ - & - & - & -0{\cdot}2155 \\ - & - & - & - \end{pmatrix}. \tag{19}$$

The only noteworthy one is the negative correlation between x_1 and x_2, the larger farms having the smaller standardized production. The result above, that x_2 probably had no effect, may have arisen because of this correlation. The effect investigated by the test is the effect on y of varying x_2, keeping x_1 constant, and it is difficult to estimate this accurately with such a high correlation between x_1 and x_2.

It is possible to proceed in much the same way that we did in §8.1 and obtain confidence limits for quantities like the expected value of y when $x_i = x_i^{(0)}$ $(i = 1, 2, \ldots, s-1)$. Confidence limits for the individual β's may similarly be obtained. Joint confidence sets for the four β's (or the three significant ones, β_2 not being significant) are complicated in form and it is perhaps better to look at the form (19) which gives a picture of these sets had ϕ been known.

8.5. Two-way classification

In this section we discuss an application of the general theory of §8.3 when the design matrix is particularly simple and the full computational technique of §8.4 is not needed. Observations, x_{ijk}, normal and independently distributed with common unknown variance ϕ, are said to form a *two-way classification* if

$$\mathscr{E}(x_{ijk}) = \theta_{ij}, \tag{1}$$

for $i = 1, 2, \ldots, I; j = 1, 2, \ldots, J$ and $k = 1, 2, \ldots, K$, where I, J and K all exceed one. All the observations for a fixed value of i are said to be at the ith *level* of the *first factor*: all those for a fixed value of j are at the jth *level* of the *second factor*. The K observations at both the ith level of the first, and the jth level of the second factor, are *replications*, identically distributed. Using the usual 'dot' notation for an average we may write

$$\begin{aligned}\theta_{ij} &= (\theta_{ij}-\theta_{i.}-\theta_{.j}+\theta_{..})+(\theta_{i.}-\theta_{..})+(\theta_{.j}-\theta_{..})+\theta_{..}\\ &=\theta'_{ij}+\theta'_{i.}+\theta'_{.j}+\theta_{..},\end{aligned} \tag{2}$$

where the terms with primes correspond to the three terms in brackets in the previous line. Necessarily we have

$$\sum_i \theta'_{i.} = \sum_j \theta'_{.j} = \sum_i \theta'_{ij} = \sum_j \theta'_{ij} = 0, \tag{3}$$

and, provided these relations obtain, the correspondence between the θ's and the θ''s plus $\theta_{..}$ is one to one. $\theta'_{i.}$ is the *main effect* of the first factor at the ith level: if $\theta'_{i.} = 0$, all i, then the first factor has no main effect. Similarly, for the second factor. θ'_{ij} is the *interaction* of the two factors at the ith and jth levels: if $\theta'_{ij} = 0$, all i and j, then there is no interaction of the two factors.

Theorem 1. *In a two-way classification, as just defined, the analysis of variance table (table 8.5.1) provides, in an obvious way, significance tests of the null hypotheses* (i) *no main effect of the first factor,* (ii) *no main effect of the second factor, and* (iii) *no interaction of the two factors.*

TABLE 8.5.1

Sums of squares	Degrees of freedom	Mean squares	F
Main effect of first factor			
$S_I^2 = JK\sum_i (x_{i..}-x_{...})^2$	$I-1$	$s_I^2 = S_I^2/(I-1)$	s_I^2/s^2
Main effect of second factor			
$S_J^2 = IK\sum_j (x_{.j.}-x_{...})^2$	$J-1$	$s_J^2 = S_J^2/(J-1)$	s_J^2/s^2
Interaction			
$S_{IJ}^2 = K\times \sum_{i,j}(x_{ij.}-x_{i..}-x_{.j.}+x_{...})^2$	$(I-1)(J-1)$	$s_{IJ}^2 = S_{IJ}^2/(I-1)(J-1)$	s_{IJ}^2/s^2
Residual			
$S^2 = \sum_{i,j,k}(x_{ijk}-x_{ij.})$	$IJ(K-1)$	$s^2 = S^2/IJ(K-1)$	—
Total $\sum_{i,j,k}(x_{ijk}-x_{...})^2$	$IJK-1$	—	—

The proof is based on the following breakdown of the total sum of squares, suggested by equation (3):

$$\begin{aligned}\sum_{i,j,k}(x_{ijk}-\theta_{ij})^2 &= \sum_{i,j,k}[(x_{ijk}-x_{ij.})+(x_{ij.}-x_{i..}-x_{.j.}+x_{...}-\theta'_{ij})\\ &\quad +(x_{i..}-x_{...}-\theta'_{i.})+(x_{.j.}-x_{...}-\theta'_{.j})\\ &\qquad +(x_{...}-\theta_{..})]^2\\ &= \sum_{i,j,k}(x_{ijk}-x_{ij.})^2\\ &\quad +K\sum_{i,j}(x_{ij.}-x_{i..}-x_{.j.}+x_{...}-\theta'_{ij})^2\\ &\quad +JK\sum_i(x_{i..}-x_{...}-\theta'_{i.})^2\\ &\quad +IK\sum_j(x_{.j.}-x_{...}-\theta'_{.j})^2\\ &\quad +IJK(x_{...}-\theta_{..})^2,\end{aligned} \tag{4}$$

in which all the product terms vanish.

Consider first a test of the null hypothesis that the first factor has no main effect: that is, $\theta'_{i.}=0$ for all i. This is a linear hypothesis and the result of corollary 3 to theorem 8.3.1 may be applied. To do this the total sum of squares has first to be unrestrictedly minimized with respect to the θ's; this may be replaced by a minimization with respect to the θ''s and $\theta_{..}$ provided the relations (3) obtain. The values

$$\left.\begin{aligned}\hat{\theta}'_{ij} &= x_{ij.}-x_{i..}-x_{.j.}+x_{...},\\ \hat{\theta}'_{i.} &= x_{i..}-x_{...},\\ \hat{\theta}'_{.j} &= x_{.j.}-x_{...},\\ \hat{\theta}_{..} &= x_{...},\end{aligned}\right\} \tag{5}$$

satisfy (3) and with these values (4) reduces simply to its first term. Hence the residual sum of squares is $S^2=\sum_{i,j,k}(x_{ijk}-x_{ij.})^2$. The degrees of freedom for S^2 are $IJK-IJ=IJ(K-1)$. If $\theta'_{i.}=0$ for all i, and we minimize with this restriction the values $\hat{\theta}'_{ij}$, $\hat{\theta}'_{.j}$ and $\hat{\theta}_{..}$ still obviously provide the least values and the minimum, with this restriction, is $S^2+JK\sum_i(x_{i..}-x_{...})^2$. Hence the reduction in sum of squares due to $\theta'_{1.}, \theta'_{2.}, \ldots, \theta'_{I.}$ (allowing for the other θ''s and $\theta_{..}$) is

$$S_I^2 = JK\sum_i(x_{i..}-x_{...})^2. \tag{6}$$

The null hypothesis is $\theta_{1.} = \theta_{2.} = \ldots = \theta_{I.}$ which imposes $(I-1)$ linear relations amongst the θ's, and hence

$$F = [S_I^2/(I-1)]/[S^2/IJ(K-1)] \tag{7}$$

with $(I-1)$ and $IJ(K-1)$ degrees of freedom provides the required significance test. Similar remarks apply to testing that there is no main effect of the second factor.

Consider next a test of the null hypothesis that the interaction is zero: $\theta'_{ij} = 0$ for all i and j. Minimizing with this restriction the values $\hat{\theta}'_{i.}$, $\hat{\theta}'_{.j}$ and $\hat{\theta}_{..}$ still provide the least values and the reduction in sum of squares due to all the θ'_{ij} is

$$S_{IJ}^2 = K \sum_{i,j} (x_{ij.} - x_{i..} - x_{.j.} + x_{...})^2. \tag{8}$$

The degrees of freedom need a little care. In terms of the θ's the null hypothesis is that

$$\theta_{ij} - \theta_{i.} - \theta_{.j} + \theta_{..} = 0 \tag{9}$$

for all i and j. For any fixed j this means that

$$\theta_{1j} - \theta_{1.} = \theta_{2j} - \theta_{2.} = \ldots = \theta_{Ij} - \theta_{I.}: \tag{10}$$

that is, $(I-1)$ independent relations amongst the θ's. Consider these for all j except $j = J$: that is, $(I-1)(J-1)$ relations in all. Then the same relation, (10), for $j = J$ can be deduced from them, since from (3),

$$\sum_{j=1}^{J-1} (\theta_{ij} - \theta_{i.}) + \theta_{iJ} - \theta_{i.} = 0.$$

Hence (9) implies that the matrix **B** of corollary 3 to theorem 8.3.1 has rank $(I-1)(J-1)$ and these are therefore the degrees of freedom to be associated with S_{IJ}^2. The F-test proceeds in the usual way and the three tests described in the table have all been obtained. Because of the breakdown of the sum of squares, (4), the design is orthogonal with respect to the four sets of parameters, $\{\theta'_{ij}\}$, $\{\theta'_{i.}\}$, $\{\theta'_{.j}\}$ and $\theta_{..}$ as explained in §8.3, and the sums of squares do not need to specify the other parameters allowed for. The total in table 8.5.1 refers to that allowing for $\theta_{..}$, which is not usually of interest. The sums of squares are additive, again by (4).

Relation with one-way classifications

The situation considered in this section is a generalization of that of §§6.4, 6.5. The between and within analysis there discussed can be put into a linear hypothesis form. In the notation of those sections $\mathscr{E}(x_{ij}) = \theta_i$, the notation only differing from that of this section in the use of double suffixes for the observations: i and j here, for the single suffix i there. The null hypothesis to be tested is that all the θ's are equal. The observations may be thought of as being classified in one way according to the sample to which they belong. The observations in this section are classified in two ways according to the two factors. A practical example of such a situation is provided by the preparation of a drug, in which two factors may be of importance in determining the quality; the amount of glucose used and the temperature of the reaction. If K preparations of the drug are made at each of I levels of the amount of glucose and J levels of temperature, the IJ different combinations produce in all IJK observations on the quality of the drug which might satisfy the linear hypothesis, equation (1). If we wished to test the null hypothesis that all the θ's were equal, the methods of §§6.4, 6.5 could be used. There are IJ samples each of size K. But we are usually more interested in testing hypotheses specific to one of the factors; for example, that the temperature does not affect the drug. We consider how this can be done.

Main effects and interactions

The meanings to be attached to the terms 'main effect' and 'interaction' require careful consideration. To take the main effect first: from the definition $\theta_{i.} = \sum_j \theta_{ij}/J$, $\theta_{i.}$ is the average expectation at the ith level of the first factor, the average being taken over the J levels of the other factor. $\theta'_{i.}$ is the deviation of this average from the average over all levels of both factors, $\theta_{..} = \sum_{i,j} \theta_{ij}/IJ$. Consequently the main effect of the first factor is only defined with reference to the second. For example, if one of the levels of the second factor were to be omitted, as might happen if a calamity overtook the readings at that level, then the definition of the main effect of the first factor might change. Thus, if there is no main effect, $\theta'_{i.} = 0$ for all i, or $\theta_{i.}$ does not depend on i, and, we ought, in full, to say 'the first factor does not influence the results (when averaged over the particular levels of the other factor used in the experiment)'. The words in brackets are often omitted.

There are circumstances in which the averaging effect may be irrelevant and to discuss this consider the interaction. The meaning to be attached to saying that the interaction is zero (one null hypothesis tested in the analysis of variance) can most easily be understood from equation (10). $\theta_{1j} - \theta_{1.}$ is the difference between the expectation at the jth level of the second factor and the average over the second factor *when the first factor is at level* 1: briefly, the effect of the second factor at level j when $i = 1$. Equation (10) says that this is the same for all i and is consequently equal to the main effect $\theta_{.j} - \theta_{..} = \theta'_{.j}$. Thus the interaction being zero means that this is true for all i and j. The two factors can be interchanged, and consequently the significance test investigates whether the effect of one factor depends upon the other or not. If it does the two factors are said to *interact*. If there is no interaction then the main effect of one factor is the same as the effect at all levels of the other factor included in the experiment, and may therefore be quoted without reference to the averaging over it. Consequently it is only when the interaction is zero that it is satisfactory to talk of a main effect. Even then it only refers to the levels used in the experiment. If there is an interaction then it is probably better not to use a main effect but to think in terms of the effects $\theta_{ij} - \theta_{i.}$ at the different levels of the other factor. Thus, if in our example, the temperature only influences the drug at high glucose concentrations, so that an interaction is present; then rather than quote a main effect of temperature (that is, averaged over the glucose levels) it would be better to quote the temperature effects separately at high and low levels of glucose (the latter being zero). Notice that to say that a main effect is zero does not mean that the factor has no effect, for it may interact with the other factor in opposite directions at different levels of the other factor, whilst keeping the average zero. To say that the first factor had no effect would be to say that $\theta_{ij} = \theta_{.j}$ for all i, j: that is, $\theta'_{i.} = \theta'_{ij} = 0$, or that both main effect and interaction are zero.

Additivity

An interaction being zero also means (equation (2)) that

$$\mathscr{E}(x_{ij}) = \theta'_{i.} + \theta'_{.j} + \theta_{..}, \tag{11}$$

so that the effects are often described as being *additive*: the expectation is the total of the overall mean plus the two main effects. Notice that whether or not the interaction is zero depends upon the random variable being considered. For example, suppose

$$\theta_{ij} = \alpha_i \beta_j \tag{12}$$

for suitable α's and β's. Then (11) is not true and there is, in general, an interaction. But

$$\ln \theta_{ij} = \ln \alpha_i + \ln \beta_j, \tag{13}$$

so that if the logarithms are used the effects are additive and the interaction vanishes.

Orthogonality

Another reason for discussing main effects and interactions is the breakdown of the sum of squares into the five separate sums in equation (4). The first of these refers only to the random variation and is unaffected by the θ's; the second to the interaction, the third and fourth to the main effects of the two factors separately, and the last to the deviation from the mean, $\theta_{..}$. The last is not usually of interest because experiments of this form are usually comparative ones: that is, one wishes to compare one level of a factor with another level, rather than to consider the value of any θ_{ij}. Because of this breakdown the reduction due to one set of θ''s is unaffected by the values of the other θ''s and the design is orthogonal with respect to the three groups of parameters (cf. §8.3). Consequently the analysis of variance table is useful and the only difficulty in proving theorem 1 lies in deriving the appropriate F-tests. In table 8.5.1 the reduction in sum of squares due to $\theta'_{1.}, \theta'_{2.}, \ldots, \theta'_{I.}$ has simply been referred to as that of the main effect of the first factor, and the others similarly.

Numerical calculations

The arithmetical calculation of table 8.5.1 follows the same method as used for table 6.1 in §6.5. The calculation of the sum of squares for the main effect of the first factor is performed by thinking of the factor as dividing the observations into I groups with JK readings in each, when the quantity required is simply the between groups sum of squares, found as in (6) of §6.5. Similarly, for the other factor. Equally the observations may be thought of as being made up of IJ groups with K readings in each. The between sum of squares with this grouping will combine the two main effects and the interaction, so, by the additivity of the sums, the interaction sum of squares is obtained by subtracting from this 'between' sum the two previous 'between' sums for the main effects. The residual is most easily obtained by subtraction from the total, although it could be found by adding the IJ separate sums of squares $\sum_k (x_{ijk} - x_{ij.})^2$ from each of the IJ samples.

Breakdowns of the sum of squares

The resulting table is capable of a simple and informative interpretation: like all analysis of variance tables it provides a breakdown of the total variation present in the readings, as measured by the total sum of squares, into components with separate meanings. Here it has four components, one of which, the residual, is unaffected by the θ's and provides, in effect, our knowledge of ϕ (corollary 1 to theorem 8.3.1). The other three refer to the main effects and interaction as already explained. The separation of these components can be appreciated by remarking, for example, that a change in one main effect will not alter the sum of squares for the other, nor for the interaction. Thus if all the readings with $i = 1$ were increased by the same amount the only sum of the four in table 8.5.1 to be affected would be the main effect of the first factor.

One should not stick slavishly to the orthogonal breakdown of table 8.5.1 and its associated tests. There are other possibilities: for example, one could test the null hypothesis that the first factor had no effect by adding the first main effect and the interaction sums of squares, similarly adding the degrees of freedom together, and using the F-test derived from these. Alternatively, since the method is an extension of that of § 6.5, it is possible, as in that section, to break up the sum of squares into further orthogonal parts, if this seems physically appropriate. For example, if one of the levels of temperature is the standard temperature, then the main effect for temperature may have its sum broken into two parts; one comparing the standard against the average for the remaining levels, and one amongst those levels. The interaction may be similarly subdivided.

Confidence limits

Corollary 2 to theorem 8.3.1 enables confidence limits to be assigned to any linear function of the θ's. Suppose, in line with the suggestion in the last paragraph, we wished to consider the value of $\alpha = \theta_{1.} - \sum_{i>1} \theta_{i.}/(I-1)$, the difference between the first level and the average for the other levels of the first main effect. Since θ_{ij} has posterior mean $\hat{\theta}_{ij} = x_{ij.}$, α will have mean

$$\hat{\alpha} = x_{1..} - \sum_{i>1} x_{i..}/(I-1).$$

The posterior variance of α (the quantity $\mathbf{g}'\mathbf{C}\mathbf{g}\phi$ of the corollary) is most easily found from equation 8.3.13. Since the x_{ijk} have variance ϕ and are independent

$$\mathscr{D}^2(\hat{\alpha} \mid \{\theta_{ij}\}, \phi) = \phi/JK + \phi/JK(I-1) = \phi I/JK(I-1). \quad (14)$$

Consequently
$$(\alpha - \hat{\alpha})/[s^2 I/JK(I-1)]^{\frac{1}{2}} \quad (15)$$
has a t-distribution with $\nu = IJ(K-1)$ degrees of freedom.

Random effects

There are situations which superficially look like a two-way classification of the type here considered but require a different treatment. Suppose that the levels of one or both of the factors are obtained by random sampling from a population of such levels. For example, we may be interested in the effect of different human operators on the quality of the drug, so that one factor would correspond to operators, and the levels of that factor to the different people, chosen at random from the population. In that situation we would wish to make inferences about the population of operators, and not merely about those operators who took part in the experiment, so that a different analysis is required. The model we are studying, under the name linear hypothesis, is a *fixed effects model* (the θ_{ij} have no random variation): the other is called a *random effects model.* If one factor is fixed and the other random we have a *mixed model.* Only the fixed effects model will be discussed in this book. The others are considered, from a practical viewpoint by O. L. Davies (1957, chapter 6) and the theory is given by H. Scheffé (1959, part II): though neither practice nor theory are in the same definitive form as is that of the fixed effects model based on the linear hypothesis.

No replication

The case $K = 1$, so far excluded, is interesting and of common occurrence. The observations may be denoted by x_{ij} instead of x_{ij1}. Since each is about a possibly different mean θ_{ij} there is no information about the random error ϕ. This is reflected in table 8.5.1 where the degrees of freedom for the residual are zero. However, if some prior knowledge of the θ's is available inferences are possible. A common form for this prior knowledge to take is for it to be known that the interaction is zero. Then (11) obtains and the two main effects are additive. This equation is a linear hypothesis in certain parameters $\{\theta'_{i.}\}$, $\{\theta'_{.j}\}$ and $\theta_{..}$ which themselves satisfy two linear constraints, from equation (3),

$$\sum_i \theta'_{i.} = \sum_j \theta'_{.j} = 0. \tag{16}$$

Because of these constraints it is not of the usual linear hypothesis form but may be put into the form in two fewer parameters by eliminating, say $\theta'_{1.}$ and $\theta'_{.1}$, using (16). The tests could then be obtained in this modified system by the appropriate minimizations. But it is not necessary to do this because it is clear that the procedure just mentioned is equivalent to minimizing the sum of squares in the original system of $\{\theta'_{i.}\}$, $\{\theta'_{.j}\}$ and $\theta_{..}$ subject to the constraints (16) and any additional ones needed in connexion with the hypothesis being tested.

Corresponding to (4) we have

$$\sum_{i,j} (x_{ij}-\theta'_{i.}-\theta'_{.j}-\theta_{..})^2 = \sum_{i,j} (x_{ij}-x_{i.}-x_{.j}+x_{..})^2 + J\sum_i (x_{i.}-x_{..}-\theta'_{i.})^2 + I\sum_j (x_{.j}-x_{..}-\theta'_{.j})^2 + IJ(x_{..}-\theta_{..})^2, \tag{17}$$

and the same type of argument that justified table 8.5.1 produces the analysis of variance table 8.5.2. The first two rows are exactly as in table 8.5.1 with $K = 1$. The third row only differs from the corresponding row of table 8.5.1 in that, instead of referring to an interaction, now known to be zero, it is the residual. The F-tests now compare the main effects mean squares with this new residual. Essentially then the case $K = 1$ is only distinguished from $K > 1$ by the fact that the interaction is used as a residual in default of the usual residual: which procedure is only valid if the true interaction, $\{\theta'_{ij}\}$, is known to be zero.

TABLE 8.5.2

Sums of squares	Degrees of freedom	Mean squares	F
Main effect of first factor $S_I^2 = J\sum_i (x_{i.}-x_{..})^2$	$I-1$	$s_I^2 = S_I^2/(I-1)$	s_I^2/s^2
Main effect of second factor $S_J^2 = I\sum_j (x_{.j}-x_{..})^2$	$J-1$	$s_J^2 = S_J^2/(J-1)$	s_J^2/s^2
Residual $S^2 = \sum_{i,j} (x_{ij}-x_{i.}-x_{.j}+x_{..})^2$	$(I-1)(J-1)$	$s^2 = S^2/(I-1)(J-1)$	—
Total $\sum_{i,j} (x_{ij}-x_{..})^2$	$IJ-1$	—	—

Another related situation that often arises is a two-way classification in which the numbers of observations at each combination of factor levels are not all equal: that is, k runs from 1 to

K_{ij} in (1). The simple orthogonal breakdown given in equation (4) is no longer possible and it is necessary to resort to a fairly complicated minimization of the sums of squares using theorem 8.3.1. We shall not enter into details: these will be found in Scheffé (1959).

8.6. Further applications of linear hypothesis theory

(a) *Comparison of regression lines*

Theorem 1. *If* $\mathbf{x} = (x_{ij};\, i = 1, 2, \ldots, n_j;\, j = 1, 2, \ldots, m)$ *is a set of real numbers, and if, for fixed* $\mathbf{x}$, *the random variables* y_{ij} *(with the same ranges of suffixes) are independent normal random variables with*

$$\mathscr{E}(y_{ij}\,|\,\mathbf{x}) = \alpha_j + \beta_j(x_{ij} - x_{..}) \tag{1}$$

and

$$\mathscr{D}^2(y_{ij}\,|\,\mathbf{x}) = \phi; \tag{2}$$

then, if the prior distributions of $\{\alpha_j\}$, $\{\beta_j\}$ *and* $\ln\phi$ *are independent and uniform, a significance test at level* α *of the hypothesis that*

$$\beta_1 = \beta_2 = \ldots = \beta_m \tag{3}$$

is obtained by declaring the data significant if

$$\frac{\left[\sum_j \{S^2_{xyj}/S_{xxj}\} - \{\sum_j S_{xyj}\}^2/\sum_j S_{xxj}\right]/(m-1)}{\sum_j S_j^2/\sum_j (n_j-2)} \tag{4}$$

exceeds $\bar{F}_\alpha(\nu_1, \nu_2)$ *with* $\nu_1 = m-1$, $\nu_2 = \sum_{j=1}^{m} (n_j - 2)$.

The notation used in (4) *is* (*cf. theorem* 8.1.1)

$$\left.\begin{aligned} S_{xxj} &= \sum_i (x_{ij} - x_{.j})^2, \quad S_{xyj} = \sum_i (x_{ij} - x_{.j})(y_{ij} - y_{.j}), \\ S_{yyj} &= \sum_i (y_{ij} - y_{.j})^2, \end{aligned}\right\} \tag{5}$$

$$S_j^2 = S_{yyj} - S^2_{xyj}/S_{xxj}. \tag{6}$$

The conditions of the theorem ensure that we are dealing with a linear hypothesis and the general theorem 8.3.1 may be applied.† The θ's of that theorem are here the α_j and β_j, $2m$ in all. The null hypothesis (equation (3)), places $(m-1)$ restrictions

† Notice that the variables denoted by y here correspond to those denoted by x in the general result.

on these. Hence, in the notation of theorem 8.3.1, corollary 3, $n = \Sigma n_j$, $s = 2m$, $r = m+1$, and the degrees of freedom for F are as stated. It remains to calculate the minima of the sums of squares.

First, notice that the regression lines (1) have been written in a slightly different form from that used in §8.1 (equation 8.1.2). The overall mean $x_{..}$ has been used and not $x_{.j}$ ($\bar{x}$ in (8.1.2)). The reason for this is that we wish (see theorem 2 below) to test a hypothesis concerning the α_j in the form (1). To recover the form of §8.1, and hence to be able to use the results of that section, we have only to write (1) as

$$\begin{aligned}\mathscr{E}(y_{ij}\,|\,\mathbf{x}) &= \alpha_j + \beta_j(x_{.j} - x_{..}) + \beta_j(x_{ij} - x_{.j}) \\ &= \alpha'_j + \beta_j(x_{ij} - x_{.j}), \quad \text{say.}\end{aligned}$$

To calculate the residual we have to minimize

$$\sum_{j=1}^{m} \sum_{i=1}^{n_j} [y_{ij} - \alpha_j - \beta_j(x_{ij} - x_{..})]^2.$$

We may minimize for each j separately and the result, from equation 8.1.8, is

$$S^2 = \sum_{j=1}^{m} S_j^2 = \sum_j S_{yyj} - \sum_j \{S^2_{xyj}/S_{xxj}\}. \tag{7}$$

In order to find the minimum when the hypothesis (3) obtains we have to minimize

$$\sum_{j=1}^{m} \sum_{i=1}^{n_j} [y_{ij} - \alpha_j - \beta(x_{ij} - x_{..})]^2, \tag{8}$$

where β is the common value of the β_j. Differentiating (8) with respect to α_j and β, and denoting the values at which the minimum is attained by a_j and b, we have

$$\sum_{i=1}^{n_j} [y_{ij} - a_j - b(x_{ij} - x_{..})] = 0 \tag{9}$$

and
$$\sum_{j=1}^{m} \sum_{i=1}^{n_j} [y_{ij} - a_j - b(x_{ij} - x_{..})]\,(x_{ij} - x_{..}) = 0. \tag{10}$$

From (9)
$$a_j = y_{.j} - b(x_{.j} - x_{..}), \tag{11}$$

so that, inserting these values into (10),

$$\sum_{j=1}^{m} \sum_{i=1}^{n_j} [(y_{ij} - y_{.j}) - b(x_{ij} - x_{.j})]\,(x_{ij} - x_{.j} + x_{.j} - x_{..}) = 0,$$

whence
$$b = \sum_{j=1}^{m} S_{xyj} \Big/ \sum_{j=1}^{m} S_{xxj} \tag{12}$$

since
$$\sum_{i=1}^{n_j} (x_{ij} - x_{.j}) = \sum_{i=1}^{n_j} (y_{ij} - y_{.j}) = 0.$$

Replacing α_j in (8) by a_j given by (11), the minimum of the sum of squares, constrained by (3), is

$$\sum_{=1}^{m} \sum_{i=1}^{n_j} [(y_{ij} - y_{.j}) - b(x_{ij} - x_{.j})]^2$$
$$= \sum_j S_{yyj} - 2b \sum_j S_{xyj} + b^2 \sum_j S_{xxj}$$
$$= \sum_j S_{yyj} - \{\sum_j S_{xyj}\}^2 / \sum_j S_{xxj}, \quad \text{from (12).} \tag{13}$$

The reduction in the sum of squares due to different β's (allowing for the α_j and a common β) is the difference between (13) and (7), namely
$$S_\beta^2 = \sum_j \{S_{xyj}^2 / S_{xxj}\} - \{\sum_j S_{xyj}\}^2 / \sum_j S_{xxj}, \tag{14}$$

and the F-ratio is as given in (4), so proving the theorem.

Theorem 2. *Under the same conditions as theorem 1, except that the β's are known to be all equal, and their common value β has a uniform prior distribution independent of $\{\alpha_j\}$ and* ln ϕ, *a significance test at level α of the hypothesis that*

$$\alpha_1 = \alpha_2 = \dots = \alpha_m \tag{15}$$

is obtained by declaring the data significant if

$$\frac{[\{S_{yy} - S_{xy}^2/S_{xx}\} - [\sum_j S_{yyj} - \{\sum_j S_{xyj}\}^2/\sum_j S_{xxj}]]/(m-1)}{[\sum_j S_{yyj} - \{\sum_j S_{xyj}\}^2/\sum_j S_{xxj}]/(\sum_j n_j - m - 1)} \tag{16}$$

exceeds $\bar{F}_\alpha(\nu_1, \nu_2)$ with $\nu_1 = m-1$, $\nu_2 = \sum_j n_j - m - 1$.

The notation used in (16) *is*

$$\left. \begin{aligned} S_{xx} &= \sum_{i,j} (x_{ij} - x_{..})^2, \\ S_{xy} &= \sum_{i,j} (x_{ij} - x_{..})(y_{ij} - y_{..}), \\ S_{yy} &= \sum_{i,j} (y_{ij} - y_{..})^2. \end{aligned} \right\} \tag{17}$$

This is another linear hypothesis. In the notation of theorem 8.3.1, corollary 3, $n = \Sigma n_j$, $s = m+1$, $r = 2$ and the degrees of freedom for F are as stated.

The residual, the unrestricted minimum, is the minimum of (8) given in (13). The residual, with the constraint (15), is the residual after fitting a common regression line to all the data, which, from equation 8.1.8, is

$$S_{yy} - S_{xy}^2/S_{xx}. \tag{18}$$

The reduction in sum of squares due to different α's (allowing for a common value for β) is the difference between (18) and 13), namely

$$S_\alpha^2 = [S_{yy} - S_{xy}^2/S_{xx}] - [\sum_j S_{yyj} - \{\sum_j S_{xyj}\}^2/\sum_j S_{xxj}], \tag{19}$$

and the F-ratio is as given in (16)

Test for slopes

The results of this subsection are a straightforward generalization of those of §8.1 from one regression line to several. It was pointed out in §8.3 that the topic of §8.1 was a particular case of the linear hypothesis: here the general theory is applied to give the required results. In the formulation of theorem 1 there are m sets of data, to each of which the usual linear homoscedastic regression situation applies. Furthermore, the variance, ϕ, about the regression line is the same for each set: this important assumption is similar to that used in the between and within analysis of §6.5. In theorem 1 a test is developed of the null hypothesis that the m regression lines have the same slope. This is often useful when it is desired to test whether the relationship between two variables is the same under m different sets of conditions, without making assumptions about the equality of the means of the variables under the different conditions: in particular, without assuming the ordinates, α_j, to be equal. Notice, however, that a significant result when this test is applied does not necessarily mean that the relationship between x and y changes with the conditions. For example, suppose that the regression of y on x is not linear but that it is almost linear over each of two non-overlapping ranges of x-values. If two sets of data have x-values in the different ranges then the regression coefficients will differ. To avoid such troubles the x-values should be about the same in each set: but the test is still satisfactory even if they are not; it is the interpretation of it that requires attention.

Test for ordinates

If the slopes of the regression lines are unequal, the difference between the ordinates changes with the value of x. But in the special case of equal slopes, the differences in ordinates are the same for all x-values. It is then of interest to know if the ordinates differ. Theorem 2 enables this equality to be tested. Notice that both tests involve the sums of squares and products *within* each set S_{xxj}, S_{xyj} and S_{yyj} (equations (5)), and the latter also involves the *total* sums of squares and products S_{xx}, S_{xy} and S_{yy} (equations (17)) obtained by pooling all the data. From the fundamental relationship expressed in equation 6.5.1, here generalized to the case of unequal sample sizes, the total may be expressed as the sum of the within and the *between*: thus,

$$S_{xx} = \sum_j S_{xxj} + \sum_j n_j(x_{.j} - x_{..})^2 \tag{20}$$

(with a similar expression with y in place of x) and

$$S_{xy} = \sum_j S_{xyj} + \sum_j n_j(x_{.j} - x_{..})(y_{.j} - y_{..}). \tag{21}$$

Analysis of variance

Both tests may be incorporated into an analysis of variance table, but since the two reductions, S^2_α and S^2_β, are not orthogonal some care must be exercised, as explained in §8.3. Table 8.6.1 shows the arrangement. The total sum of squares is that about the common line—the values of which we do not wish to test. This may be reduced, first by fitting different ordinates and then by fitting different slopes. The first breaks the total up into S^2_α for the ordinates and $S^2 + S^2_\beta$. The second breaks the latter up into S^2 (equation (7)) and S^2_β. The title 'between slopes sum of squares', S^2_β, is an abbreviation for the reduction due to different slopes (allowing for different α's and β). The title 'between ordinates sum of squares', S^2_α, is an abbreviation for the reduction due to different ordinates (allowing for a common line). This is not equal to the reduction due to different ordinates (allowing for different β's and α); as explained above this is not usually of interest. What one usually does is to test for the slopes (theorem 1) first. If this is not significant one can test for the ordinates, acting as if the slopes were the same. The point being that there is a natural ordering of the parameters here: the α's naturally occurring first in the reduction, before the β's.

TABLE 8.6.1

	Sums of squares	Degrees of freedom	Mean squares	F
Between ordinates	S_α^2	$m-1$	$s_\alpha^2 = S_\alpha^2/(m-1)$	$s_\alpha^2/\bar{s}^2$
Between slopes	S_β^2	$m-1$	$s_\beta^2 = S_\beta^2/(m-1)$	s_β^2/s^2
Residual	S^2	$\Sigma(n_j-2)$	$s^2 = S^2/\Sigma(n_j-2)$	—
Ordinates residual	$S^2+S_\beta^2$	Σn_j-m-1	$\bar{s}^2 = (S^2+S_\beta^2)/(\Sigma n_j-m-1)$	—
Total	$S_{yy}-S_{xy}^2/S_{xx}$	Σn_j-2	—	—

(b) *Analysis of covariance*

Theorem 3. *Let $R(\mathbf{x}, \mathbf{x})$ denote the residual sum of squares from the linear hypothesis (in the notation of §8.3)*

$$\mathscr{E}(\mathbf{x}) = \mathbf{A}\boldsymbol{\theta} \quad or \quad \mathscr{E}(x_i) = \sum_j a_{ij}\theta_j. \tag{22}$$

From equation 8.3.28

$$R(\mathbf{x}, \mathbf{x}) = \mathbf{x}'\mathbf{x} - \mathbf{x}'\mathbf{A}(\mathbf{A}'\mathbf{A})^{-1}\mathbf{A}'\mathbf{x}. \tag{23}$$

Then if the linear hypothesis

$$\mathscr{E}(\mathbf{x}) = \mathbf{A}\boldsymbol{\theta} + \beta\mathbf{z} \quad or \quad \mathscr{E}(x_i) = \sum_j a_{ij}\theta_j + \beta z_i \tag{24}$$

is considered, the residual sum of squares is

$$R(\mathbf{x}, \mathbf{x}) - R(\mathbf{x}, \mathbf{z})^2/R(\mathbf{z}, \mathbf{z}). \tag{25}$$

(Here β is an additional parameter and **z** *is a known vector linearly independent of the columns of* **A**.)

The expression to be minimized is

$$(\mathbf{x} - \mathbf{A}\boldsymbol{\theta} - \beta\mathbf{z})'(\mathbf{x} - \mathbf{A}\boldsymbol{\theta} - \beta\mathbf{z}).$$

Let us minimize it in two stages, first with respect to $\boldsymbol{\theta}$ for fixed β, and then with respect to β. If $\mathbf{y} = \mathbf{x}-\beta\mathbf{z}$ the first stage gives

$$R(\mathbf{y}, \mathbf{y}) = \mathbf{y}'\mathbf{y}-\mathbf{y}'\mathbf{A}(\mathbf{A}'\mathbf{A})^{-1}\mathbf{A}'\mathbf{y}$$
$$= (\mathbf{x}-\beta\mathbf{z})'(\mathbf{x}-\beta\mathbf{z})-(\mathbf{x}-\beta\mathbf{z})'\mathbf{A}(\mathbf{A}'\mathbf{A})^{-1}\mathbf{A}'(\mathbf{x}-\beta\mathbf{z}),$$

which is a quadratic in β,

$$R(\mathbf{x}, \mathbf{x})-2\beta R(\mathbf{x}, \mathbf{z})+\beta^2 R(\mathbf{z}, \mathbf{z}),$$

with minimum at $b = R(\mathbf{x}, \mathbf{z})/R(\mathbf{z}, \mathbf{z})$ and minimum value given by (25).

Use of analysis of covariance

This theorem describes how the residual sum of squares is reduced by the introduction of a single additional parameter, β, and, as such, gives an alternative test for this single parameter β which is identical with that obtained in corollary 2 to theorem 8.3.1. The corollary is the most convenient form when this test is required: the present form is more suitable in a connexion now to be described.

The term 'analysis of covariance' is a rather loose one. 'Analysis of variance' most commonly refers to the case of a design matrix $\mathbf{A}$ having specially simple properties (usually orthogonality with respect to sets of parameters) such as was discussed in §8.5. The matrix $\mathbf{A}$ is truly designed, that is the elements are known before experimentation, unlike the elements in the design matrix $\mathbf{A}$ of the multiple regression situation of §8.4 which, depending on the x's (in the notation of that section), are not systematic. Analysis of covariance usually refers to the model given in (24), where $\mathbf{A}$ is truly designed and $\mathbf{z}$ is (irrelevantly) random. Some, or all, of the θ's are of interest, β is not. The situation arises when one is investigating the dependence of $\mathbf{x}$ on certain factors, expressed in the θ's, but $\mathbf{x}$ is also known to be influenced by another quantity $\mathbf{z}$. Thus, in an experiment to investigate the effect of type of wool and type of machine on the breakage rate in spinning textile yarn, a two-way classification as in §8.5 might be used with these two factors. However, the breakage rate is obviously affected by the weight per unit length of yarn, a variable which it is difficult to control. An analysis of covariance might therefore be used, based on the model in (24), with the weight as the 'z' variable.

Regression reduction

The calculations necessary for an analysis of covariance follow directly from those for the corresponding analysis of variance, in which the z-variable is omitted, by using theorem 3. Suppose the θ's are divided into two groups, called the first and second, and it is desired to test the hypothesis that those in the second group are all zero. Then in the analysis of variance $S_1^2 - S_{12}^2$ must be compared with S_{12}^2 (the notation is as in §8.3). In the analysis of covariance $S_{1\beta}^2 - S_{12\beta}^2$ must be compared with $S_{12\beta}^2$, where $S_{12\beta}^2$ is the residual fitting both groups and β, and $S_{1\beta}^2$ similarly. But both $S_{12\beta}^2$ and $S_{1\beta}^2$ are easily obtained from S_{12}^2 and S_1^2 respectively by use of theorem 3.

The quantity $R(\mathbf{x}, \mathbf{x})$ is the residual for the $\mathbf{x}$'s with $\beta = 0$. $R(\mathbf{z}, \mathbf{z})$ is the same expression with $\mathbf{z}$ replacing $\mathbf{x}$. $R(\mathbf{x}, \mathbf{z})$ is the same expression with every squared $\mathbf{x}$ replaced by a product of an $\mathbf{x}$ with its corresponding $\mathbf{z}$. Consequently, if we calculate the analysis of variance for $\mathbf{x}$ and add the similar analysis for $\mathbf{z}$ and for the product of $\mathbf{x}$ and $\mathbf{z}$, we shall have all the terms necessary for (25). The operation of subtracting $R(\mathbf{x}, \mathbf{z})^2/R(\mathbf{z}, \mathbf{z})$ from the original residual to obtain the new one will be called the *regression reduction.* Both $S_{12\beta}^2$ and $S_{1\beta}^2$ can be obtained from S_{12}^2 and S_1^2 respectively by regression reduction. Table 8.6.2 shows how the calculations may be laid out for the two-way classification in order to test the interaction. The notation corresponds to that of table 8.5.1 with additions to make clear which variable is being used: only the sums of squares and products are included. Arrows show the order in which the terms are calculated. Thus $S_{IJ}^2(\mathbf{x})$ denotes $K \sum_{i,j} (x_{ij.} - x_{i..} - x_{.j.} + x_{...})^2$ and so $S_{IJ}^2(\mathbf{x}, \mathbf{z})$ denotes

$$K \sum_{i,j} (x_{ij.} - x_{i..} - x_{.j.} + x_{...})\,(z_{ij.} - z_{i..} - z_{.j.} + z_{...}).$$

The interaction effects and the residuals are added together to give the minima of the sums of squares with zero interaction,

$S_{IJ}^2(\mathbf{x})+S^2(\mathbf{x})$, etc. Both minima suffer a regression reduction: thus

$$S_\beta^2 = S^2(\mathbf{x})-\{S^2(\mathbf{x},\mathbf{z})\}^2/S^2(\mathbf{z})$$

and

$$S_{IJ\beta}^2+S_\beta^2 = S_{IJ}^2(\mathbf{x})+S^2(\mathbf{x})-\{S_{IJ}^2(\mathbf{x},\mathbf{z})+S^2(\mathbf{x},\mathbf{z})\}^2/\{S_{IJ}^2(\mathbf{z})+S^2(\mathbf{z})\}.$$

Then $S_{IJ\beta}^2$ is obtained by subtraction. The test for the interaction proceeds as before except that the residual S_β^2 has one less degree of freedom than $S^2(\mathbf{x})$ since an additional parameter, β, has been incorporated. The F-statistic is therefore

$$\{S_{IJ\beta}^2/(I-1)(J-1)\} \quad \text{divided by} \quad S_\beta^2/[IJ(K-1)-1].$$

Exactly similar methods enable the two main effects to be tested. The sums of squares and products will be added to the residual, will undergo a regression reduction, and have S_β^2 subtracted from them.

TABLE 8.6.2

	Sums of squares and products				
	$\mathbf{x}$	$\mathbf{x}$ and $\mathbf{z}$	$\mathbf{z}$		Covariance
Interaction	$S_{IJ}^2(\mathbf{x})$	$S_{IJ}^2(\mathbf{x},\mathbf{z})$	$S_{IJ}^2(\mathbf{z})$		$S_{IJ\beta}^2$ ↑
Residual	$S^2(\mathbf{x})$ ↓	$S^2(\mathbf{x},\mathbf{z})$ ↓	$S^2(\mathbf{z})$ ↓	→	S_β^2
Interaction + residual	$S_{IJ}^2(\mathbf{x}) + S^2(\mathbf{x})$	$S_{IJ}^2(\mathbf{x},\mathbf{z}) + S^2(\mathbf{x},\mathbf{z})$	$S_{IJ}^2(\mathbf{z}) + S^2(\mathbf{z})$	→	$S_{IJ\beta}^2+S_\beta^2$

Notice that the test of theorem 2 may be regarded as an analysis of covariance, where the corresponding analysis of variance is the simple between and within analysis of §6.5. To see this we note that the linear hypothesis is

$$\begin{aligned}\mathscr{E}(y_{ij}\,|\,\mathbf{x}) &= \alpha_j+\beta(x_{ij}-x_{..})\\ &= \alpha_j'+\beta x_{ij}, \quad \text{say},\end{aligned}$$

and without β it is $\qquad \mathscr{E}(y_{ij}\,|\,\mathbf{x}) = \alpha_j'.$

There are m samples of sizes $n_1, n_2, \ldots, n_m$, with means α_j'; and the null hypothesis is that all the α_j' are equal. We leave the reader to verify that the F-test based on (16) results from the analysis of covariance.

(c) *Polynomial regression*

There are many situations, where the relationship between two variables is being considered, in which the regression $\mathscr{E}(y \mid x)$ is not linear and the results of §8.1 are inadequate because of curvature in the regression. Indeed, one of the oldest of statistical problems is that of 'fitting a curve to data'; that is, to a set of (x, y)-points. In modern language we consider a set $\mathbf{x} = (x_1, x_2, \ldots, x_n)$ and suppose that for fixed $\mathbf{x}$ the random variables $y_1, y_2, \ldots, y_n$ are independent normal variables with

$$\mathscr{E}(y_i \mid \mathbf{x}) = \alpha_0 + \alpha_1 x_i + \alpha_2 x_i^2 + \ldots + \alpha_k x_i^k \tag{26}$$

and

$$\mathscr{D}^2(y_i \mid \mathbf{x}) = \phi. \tag{27}$$

The assumptions are those of homoscedastic (equation (27)) normal regression of polynomial form. If the α_i and $\ln \phi$ have independent uniform prior distributions, then the model is in the standard linear hypothesis form with $\boldsymbol{\theta}' = (\alpha_0, \alpha_1, \ldots, \alpha_k)$ and a design matrix with ith row $(1, x_i, x_i^2, \ldots, x_i^k)$. In view of the general form of the design matrix the general computational methods of §8.4 have to be used; and we perform a multiple *linear* regression on the quantities $x, x^2, \ldots, x^k$. (In the notation of equation 8.4.7, $x_{ij} = x_i^j$.) It was pointed out in that section that the order in which the variables were introduced was important because it affected the significance tests that could easily be performed. In polynomial regression there is a natural order, namely that of increasing powers of x. The parameters will then be introduced in the order $\alpha_0, \alpha_1, \ldots, \alpha_k$ and it is possible to test that $\alpha_{s+1} = \alpha_{s+2} = \ldots = \alpha_k = 0$ for any s. This provides a test that, given the polynomial is at most of degree k, it is of degree at most s. Usual practice is to take $s = k-1$; that is, to test whether $\alpha_k = 0$. If this is not significant then k is reduced by one and one tests whether $\alpha_{k-1} = 0$, finishing up with the polynomial of least degree not exceeding k. The tests are easily carried out since the reduction in sum of squares due to including α_s, allowing for $\alpha_0, \alpha_1, \ldots, \alpha_{s-1}$ is, in the notation of the computational scheme of §8.4, simply ω_s^2 (stage (4) of the computation). When a test is significant then the polynomial of that degree must be used, and the estimates of the coefficients α_i have, for fixed ϕ, a dispersion matrix given by the inverse matrix $\mathbf{B}^{-1}$ (in the notation again of §8.4) of the order corresponding to the degree of the polynomial.

A difficulty here is knowing what value of k to use to start. Even if the tests of $\alpha_s, \alpha_{s+1}, \ldots, \alpha_k$ are all insignificant there is always the possibility that the inclusion of a term of higher degree than k would result in an even better 'fit': or, in modern language, would reduce the residual sum of squares appreciably. The key to this problem lies in the form of this residual. In any of these tests ω_s^2 is compared with the residual, which is the sum of squares of deviations about the curve of best 'fit' of degree s. Therefore, rather loosely, the tests compare whether the introduction of x^s results in an appreciable reduction compared with what remains, but there is no way of judging whether what remains is adequate. The situation is most satisfactorily resolved if the value of ϕ is known, equal to σ^2. In this case corollary 1 to theorem 8.3.1 may be used, in conjunction with the method of §5.3, to test the adequacy of the 'fit'. For, if the correct degree of polynomial has been used a test of the null hypothesis that $\phi = \sigma^2$ should be nonsignificant. Consequently, when $\phi = \sigma^2$, the powers of x may be introduced in ascending order until the residual sum of squares is compatible with the known amount of variation, σ^2. It would be pointless to attempt to improve the 'fit' so that the residual was less than the known random error. This method may be modified if an independent estimate of ϕ is available such that c/ϕ, for appropriate c, has a χ^2-distribution. (Such an estimate results from most normal distribution situations as we have repeatedly seen.) The χ^2-test from §5.3, used above, may then be replaced by the F-test of §6.2, to compare the two estimates of variance: one from the residual and the other independent one.

Orthogonal polynomials

One unpleasant feature of the polynomial regression, in the form of equation (26), is that if the degree of the polynomial is changed, say from s to $s-1$, then all the estimates of the parameters will, in general, be changed. This is because the design is not orthogonal and just as a test for α_i will be influenced by how many other α's have been allowed for, so will the posterior distribution of α_i be similarly changed. A way out of this difficulty is to rewrite equation (26) in the form

$$\mathscr{E}(y_i \mid \mathbf{x}) = \beta_0 P_0(x_i) + \beta_1 P_1(x_i) + \beta_2 P_2(x_i) + \ldots + \beta_k P_k(x_i), \quad (28)$$

where $P_s(x)$ is a known polynomial of degree s in x, to be determined. Equation (26) is the special case $P_s(x) = x^s$. The design matrix, $\mathbf{A}$, now has typical row $(P_0(x_i), P_1(x_i), \ldots, P_k(x_i))$ and the elements of $\mathbf{A}'\mathbf{A}$ are $\sum_{u=1}^{n} P_i(x_u)\, P_j(x_u)$. Suppose that the polynomials are chosen so that

$$\sum_{u=1}^{n} P_i(x_u)\, P_j(x_u) = 0 \quad (i \neq j). \quad (29)$$

Then $\mathbf{A}'\mathbf{A}$ is a diagonal matrix and (see equation 8.3.26) the design is orthogonal with respect to each of the parameters separately. Consequently if the degree of the polynomial (28) is increased from k to $k+1$ none of the β's in (28) is altered. Such polynomials are called *orthogonal polynomials*.

It remains to show that orthogonal polynomials exist. $P_0(x)$ may be put equal to 1. Then $P_1(x)$ must satisfy (29) with $i = 0$, $j = 1$. If $P_1(x) = ax+b$, for some a and b, this gives

$$\sum_{u=1}^{n} (ax_u + b) = 0$$

which is satisfied, for example, by

$$a = 1, \quad b = -\sum_{u=1}^{n} x_u/n = \bar{x}.$$

Hence, if $k = 1$, (28) becomes

$$\mathscr{E}(y_i \mid \mathbf{x}) = \beta_0 + \beta_1(x_i - \bar{x}),$$

exactly the form used in §8.1. Indeed the term $-\beta\bar{x}$ was introduced into equation 8.1.2 so that the estimates of α and β should be orthogonal and the posterior distributions of α and β be independent, given ϕ. Suppose $P_0(x), \ldots, P_{s-1}(x)$ have been found: we show how $P_s(x)$ can be found. Since $P_s(x)$ is a polynomial of degree x^s it may be written

$$P_s(x) = a_s x^s + a_{s-1} P_{s-1}(x) + a_{s-2} P_{s-2}(x) + \ldots + a_0 P_0(x), \quad (30)$$

and it has to satisfy the equations $\sum_{u=1}^{n} P_s(x_u)\, P_j(x_u) = 0$ for all $j < s$. Let $a_s = 1$. Since orthogonal polynomials exist up to degree $s-1$, substituting the expression (30) into these equations, we have

$$\sum_{u=1}^{n} P_s(x_u)\, P_j(x_u) = \sum_{u=1}^{n} x_u^s P_j(x_u) + a_j \sum_{u=1}^{n} P_j(x_u)^2 = 0.$$

The coefficient of a_j is necessarily non-zero and

$$a_j = \sum_{u=1}^{n} x_u^s P_j(x_u) \Big/ \sum_{u=1}^{n} P_j(x_u)^2.$$

Hence a_j is determined and $P_s(x)$ is found from (30). Hence, by the usual inductive argument, the existence of orthogonal polynomials is established. Notice that the polynomials depend on the values $x_1, x_2, \ldots, x_n$. For the special case where

$$x_s = x_1 + (s-1)h$$

for some h, so that the values are equally spaced at distance h apart, the polynomials have been tabulated by Fisher and Yates (1963) and Pearson and Hartley (1958).

(d) *Weighted least squares*

We have already mentioned the close connexion between the linear hypothesis theory and the results on combination of observations in §6.6. It is possible to extend significantly all the linear hypothesis theory to the case of correlated observations, as in §6.6. Suppose all the conditions stated at the beginning of §8.3 for a linear hypothesis are retained except that instead of the x's being independent and normal they have a multivariate normal distribution with dispersion matrix $\mathbf{V}\phi$, where $\mathbf{V}$ is known but ϕ is not. The case already studied is $\mathbf{V} = \mathbf{I}$, the unit matrix. Then we know (§3.5) that it is possible to find a linear transformation of the x's, say $\mathbf{y} = \mathbf{Tx}$ such that the y's are independent and normal with a common variance. Then

$$\mathscr{E}(\mathbf{y}) = \mathscr{E}(\mathbf{Tx}) = \mathbf{T}\mathscr{E}(\mathbf{x}) = \mathbf{TA\theta}$$

and the y's obey the original conditions on a linear hypothesis with design matrix $\mathbf{TA}$ instead of $\mathbf{A}$. Consequently the whole theory can be carried through with $\mathbf{A}$ replaced by $\mathbf{TA}$. Notice, that unfortunately the orthogonality of any design based on $\mathbf{A}$ will be thereby destroyed. Since $\mathbf{x} = \mathbf{T}^{-1}\mathbf{y}$ the dispersion matrix of the x's will be $\mathbf{T}^{-1}(\mathbf{T}^{-1})'\phi = \mathbf{V}\phi$, so that this establishes the relationship between the new and old design matrices.

The sum of squares to be minimized (with or without constraints) is

$$\begin{aligned}(\mathbf{y}-\mathbf{TA\theta})'(\mathbf{y}-\mathbf{TA\theta}) &= (\mathbf{x}-\mathbf{A\theta})'\mathbf{T}'\mathbf{T}(\mathbf{x}-\mathbf{A\theta}) \\ &= (\mathbf{x}-\mathbf{A\theta})'\mathbf{V}^{-1}(\mathbf{x}-\mathbf{A\theta}). \qquad (31)\end{aligned}$$

Hence the term weighted least squares, since the terms are weighted with the elements of $\mathbf{V}^{-1}$ (compare the weighting in theorem 6.6.1).

Suggestions for further reading

A valuable text on least squares is that of Plackett (1960) and a related one on the analysis of variance is by Scheffé (1959).

The important and interesting topic of the design of experiments was founded in its modern form by Fisher (1960). More modern works in this field are those of Cochran and Cox (1957) and Kempthorne (1952); the latter being the more mathematical. An excellent introductory, non-mathematical account is given by Cox (1958).

Sampling methods are discussed in the book by Cochran (1963).

Exercises

1. The *Macaroni* penguin lays clutches of two eggs which are markedly different in size. The following are the weights, in grammes, of the eggs in eleven clutches. Fit a regression line of weight of the larger egg on the smaller egg and test whether the slope differs significantly from unity.

Smaller egg, x	Larger egg, y	Smaller egg, x	Larger egg, y
79	133	96	162
93	143	109	170
100	164	70	127
105	171	71	133
101	165	87	148
96	159		

(The following values of the sums, and sums of squares and products, of the above readings, after 100 has been subtracted from the values of y, may be used in your answer: $\Sigma x = 1007$, $\Sigma y = 575$, $\Sigma x^2 = 93{,}939$, $\Sigma y^2 = 32{,}647$, $\Sigma xy = 54{,}681$.)

A further clutch gives weights 75 and 115 g. Test the suspicion that this clutch was not that of a *Macaroni* penguin. (Camb. N.S.)

2. Obtain a 95 % confidence interval for the regression coefficient of y on x from the following data:

x	-2	-1	0	1	2
y	$-2\cdot1$	$-0\cdot9$	0	$1\cdot1$	$1\cdot9$

Obtain a similar interval for the value of y to be expected when $x = 4$. (Camb. N.S.)

3. Single wool fibres are measured for fibre diameter x; and their breaking load y is then determined. The following table shows pairs of measurements x, y made on a sample of 29 fibres. Use these data to examine how breaking load depends on diameter, giving posterior standard deviations for the parameters of any equation you fit to the observations:

Fibre diameter, x	Breaking load, y	Fibre diameter, x	Breaking load, y	Fibre diameter, x	Breaking load, y
24	3·2	28	9·5	38	12·0
41	18·3	38	7·8	35	15·0
20	1·2	40	19·0	14	0·6
38	10·5	22	2·0	19	3·1
12	0·6	42	9·0	31	12·0
13	1·1	10	0·8	29	4·0
33	7·8	11	0·7	28	7·1
18	1·1	32	5·8	21	5·0
15	2·1	41	17·0	24	5·2
30	5·8	42	11·8		

(Lond. B.Sc.)

4. The sample correlation between height of plant and total leaf area was obtained for each of six sites. The results were:

No. of plants	22	61	7	35	49	21
Correlation	+0·612	+0·703	+0·421	+0·688	+0·650	+0·592

Find an approximate 95 % confidence interval for the correlation.

5. Three groups of children were each given two psychological tests. The numbers of children and the sample correlations between the two test scores in each group were as follows:

No. of children	51	42	67
Correlation	+0·532	+0·477	+0·581

Is there any evidence that the association between the two tests differs in the three groups?

6. The mean yield $\mathscr{E}(\bar{y})$ of a chemical process is known to be a quadratic function of the temperature T. Observations of the yield are made with an error having a normal distribution of zero mean and constant variance σ^2. An experiment is performed consisting of m observations at $T = T_0$ and n observations at each of $T = T_0 - 1$ and $T_0 + 1$. Writing the quadratic relation in the form

$$\mathscr{E}(\bar{y}) = \alpha + \beta(T - T_0) + \gamma(T - T_0)^2,$$

show that the least squares estimates of β, γ are given by

$$\hat{\beta} = \tfrac{1}{2}(\bar{y}_1 - \bar{y}_{-1}),$$

$$\hat{\gamma} = \tfrac{1}{2}(\bar{y}_1 - 2\bar{y}_0 + \bar{y}_{-1}),$$

where $\bar{y}_{-1}$, $\bar{y}_0$, $\bar{y}_1$ are the averages of the observations at $T_0 - 1$, T_0, $T_0 + 1$, respectively.

Show that, if $\gamma < 0$, $\mathscr{E}(\bar{y})$ is maximum when $T = T_M = T_0 - \tfrac{1}{2}\beta/\gamma$ and that when σ^2 is small, the variance of the estimate $T_0 - \tfrac{1}{2}\hat{\beta}/\hat{\gamma}$ of T_M is approximately

$$\frac{\sigma^2}{8\gamma^2}\left[\frac{1}{n} + \frac{\beta^2}{\gamma^2}\left(\frac{1}{n} + \frac{2}{m}\right)\right].$$

For a fixed large number of observations $N = 2n + m$, show that if $T_M - T_0$ is known to be large the choice of m which minimizes the latter variance is approximately $\tfrac{1}{2}N$. (Wales Dip.)

7. In an agricultural experiment, k treatments are arranged in a k by k Latin square and y_{ij} is the yield from the plot in the ith row and jth column, having treatment $T_{(ij)}$; x_{ij} is a measure of the fertility of this plot obtained in the previous year. It is assumed that

$$y_{ij} = \mu + r_i + c_j + t_{(ij)} + \beta x_{ij} + \gamma x_{ij}^2 + \epsilon_{ij},$$

where the $\{\epsilon_{ij}\}$ are normally and independently distributed variables with mean zero and variance σ^2. Show how you would test whether $\gamma = 0$.

[A Latin square is an arrangement such that each treatment occurs once in each row and once in each column.] (Camb. Dip.)

8. In an experiment there are a control C and t treatments $T_1, \ldots, T_t$. The treatments are characterized by parameters $\theta_1, \ldots, \theta_t$ to be thought of as measuring the difference between the treatments and the control. For each observation an individual is tested under two conditions and the *difference* in response taken as the observation for analysis. Observations are subject to uncorrelated normal random errors of zero mean and constant variance σ^2.

There are k observations comparing T_1 with C, i.e. having expectation θ_1; k observations comparing T_2 with C, i.e. having expectation θ_2, and so on for $T_3, \ldots, T_t$. Then there are l observations comparing T_2 with T_1 and having expectation $\theta_2 - \theta_1$, and so on for every possible comparison of treatments, i.e. there are for every $i > j$, l observations comparing T_i with T_j and having expectation $\theta_i - \theta_j$.

Obtain the least squares estimate $\hat{\theta}_i$ and prove that

$$\mathscr{D}^2(\hat{\theta}_i - \hat{\theta}_j) = \frac{2\sigma^2}{(k+lt)} \quad (i \neq j).$$

(Lond. M.Sc.)

9. The random variables $Y_{11}, \ldots, Y_{nn}$ are independently normally distributed with constant unknown variance σ^2 and with

$$\mathscr{E}(Y_{ij}) = \{i - \tfrac{1}{2}(n+1)\}\alpha + \{j - \tfrac{1}{2}(n+1)\}\beta + x_{ij}\gamma,$$

where α, β, γ are unknown parameters and x_{ij} are given constants. Set out the calculations for (i) obtaining confidence intervals for γ, (ii) testing the hypothesis $\alpha = \beta = 0$. (Lond. M.Sc.)

10. In order to estimate two parameters θ and ϕ it is possible to make observations of three types: (i) the first type have expectation $\theta + \phi$; (ii) the second type have expectation $2\theta + \phi$; (iii) the third type have expectation $\theta + 2\phi$. All observations are subject to uncorrelated normal errors of mean zero and constant variance σ^2.

If n observations of type 1, m observations of type 2 and m observations of type 3 are made, obtain the least squares estimates of θ and ϕ and prove that, given the data,

$$\mathscr{D}^2(\theta) = \mathscr{D}^2(\phi) = \frac{n+5m}{m(2n+9m)}\sigma^2.$$

(Lond. B.Sc.)

11. The set of normal variables $(y_1, y_2, \ldots, y_n)$ has

$$\mathscr{E}(y_j) = \theta x_j, \quad \mathscr{C}(y_j, y_k) = v_{jk}, \quad \mathscr{D}^2(y_j) = v_{jj},$$

the quantities x_j, v_{jk} being known. Obtain the posterior distribution of θ under the usual assumptions.

Suppose

$$y_i = \theta + \eta_j - \eta_{j-1},$$

where the η_j are independent normal random variables with mean zero and variance σ^2. By solving the system of equations

$$\sum_{k=1}^{n} v_{jk}\xi_k = 1$$

in ξ_k, for this case, obtain the posterior distribution of θ. (Camb. Dip.)

12. The following experiment, due to Ogilvie *et al.*, was done to determine if there is a relation between a person's ability to perceive detail at low levels of brightness, and his absolute sensitivity to light. For a given size of target, which is measured on the visual acuity scale (1/angle subtended at eye) and denoted x_1, a threshold brightness (denoted y) for seeing this target was determined [x_1 was restricted to be either 0·033 or 0·057]. In addition, the absolute light threshold (denoted x_2) was determined for each subject.

It is known that y depends on x_1, and this dependence can be taken to be linear. Use multiple regression to test whether y is significantly dependent on x_2 (for the purposes of the test, assume that any such dependence would be linear), i.e. use the model in which the expected value of y is

$$\alpha+\beta_1(x_1-\bar{x}_1)+\beta_2(x_2-\bar{x}_2),$$

where $\bar{x}_1$ and $\bar{x}_2$ denote the averages of x_1 and x_2, respectively.

Examine graphically (without carrying out a test of significance), whether the dependence of y on x_2 is in fact linear.

Subject	(*A*)	(*B*)	(*C*)	Subject	(*A*)	(*B*)	(*C*)
1	3·73	4·72	5·17	14	4·09	5·21	6·74
2	3·75	4·60	4·80	15	4·10	4·97	5·41
3	3·86	4·86	5·63	16	4·17	4·82	5·22
4	3·88	4·74	5·35	17	4·20	5·03	5·49
5	3·89	4·42	4·73	18	4·22	5·44	6·71
6	3·90	4·46	4·92	19	4·23	4·64	5·14
7	3·92	4·93	5·15	20	4·24	4·81	5·37
8	3·93	4·96	5·42	21	4·29	4·58	5·11
9	3·98	4·73	5·14	22	4·29	4·98	5·40
10	3·99	4·63	5·92	23	4·31	4·94	5·71
11	4·04	5·20	5·51	24	4·32	4·90	5·41
12	4·07	5·31	5·89	25	4·42	5·10	5·51
13	4·07	5·06	5·44				

Column (*A*) gives values of x_2. Columns (*B*) and (*C*) give values of y for $x_1 = 0{\cdot}033$ and $x_1 = 0{\cdot}057$, respectively. (Lond. Psychol.)

13. In an experiment into the laboratory germination of *Hypericum perforatum* (by E. W. Tisdale *et al.*, *Ecology*, **40**, 54, 1959), 9 batches of seeds were buried in siltloam soil at depths of ½, 1 and 3 in. (3 batches at each depth). One batch at each depth was left buried for 1 year, one for 2 years and one for 3 years. At the end of these times the seeds were recovered and tested for germination. The experiment was repeated three times.

The percentages germinating were as follows (each figure being the average of 3 replications):

No. of years buried	Depth (in.) ½	1	3
1	20·6	27·3	25·2
2	30·6	42·0	62·0
3	9·6	45·0	52·0

Test for differences in percentages germinating between the different depths and the different lengths of time for which the seeds were buried. Test also for quadratic dependence of percentage germinating on the length of time buried.

State what additional information could have been obtained had the individual percentages for the replicates been available, and comment on the statistical advantage, if any, which might have been gained from burying seeds at a depth of 1¾ in. rather than of 1 in. (Leic. Stat.)

14. In an experiment on the effect of radiation on the synthesis of deoxyribonucleic acid (DNA) four rats were partially hepatectomized and irradiated. The DNA contents of samples of cells were determined by three different methods; the table gives the mean contents per cell for the four rats by each of the three methods.

	Rat				
Method	1	2	3	4	Total
1	217	283	239	262	1001
2	206	269	226	274	975
3	231	298	256	252	1037
Total	654	850	721	788	

Investigate whether the rats differed in DNA content, and whether values obtained by the three methods could be treated as comparable.

(Camb. N.S.)

15. In a poison assay doses $x_1, \ldots, x_n$ of poison are given to groups of k mice and the numbers dying in each group are $r_1, \ldots, r_n$; r_i may be assumed to have a binomial distribution with parameter p_i, where

$$f(p_i) = \alpha + \beta(x_i - \bar{x}),$$

where $f(u)$ is a known function, $\bar{x}$ is the mean of $x_1, \ldots, x_n$, and α and β are constants. Set up the maximum likelihood equations for the estimation of α and β, and show that if α_0, β_0 are approximate solutions of these equations the process of obtaining better solutions is equivalent to a certain least squares estimation problem. (Camb. N.S.)

16. In an experiment on the effect of cultivation on the absorption by bean plants of radioactive contamination deposited on the soil surface, three systems of cultivation were used. Twelve plots were arranged in four blocks, each block containing one plot of each system. The values of the radioactivity per unit weight of pod for the twelve plots, and the totals for blocks and systems were:

System of cultivation	Block 1	2	3	4	Total
A	14	10	19	17	60
B	8	7	12	12	39
C	9	5	9	10	33
Total	31	22	40	39	

The sums of squares of the plot values, the block totals and the system totals are respectively 1634, 4566 and 6210. Determine what evidences there are that the system used affects the amount of radioactivity taken up by the plant, and the arrangement of the plots in blocks increases the precision of the experiment. (Camb. N.S.)

17. An agricultural experiment to compare five different strains of cereal, *A*, *B*, *C*, *D* and *E*, was carried out using five randomized blocks, each containing five plots. The resulting yields per plot (lb) are shown below.

Analyse these data and discuss what conclusions regarding differences in yields can be legitimately drawn from them.

Block	Strain A	B	C	D	E
1	36·5	47·1	53·5	37·1	46·5
2	48·4	43·0	58·0	40·9	41·3
3	50·9	52·4	66·0	47·9	48·8
4	60·9	65·5	67·1	56·1	55·7
5	46·3	50·0	58·0	44·5	45·3

(Lond. B.Sc.)

18. In an agricultural trial a variety of wheat was grown for 3 years in succession on 4 plots in each of 3 areas, and the yield per plot was measured each year. Different randomly selected plots in each area were used each year. The yields per plot (kg) are given in the table below:

	Area 1	Area 2	Area 3
Year 1	14·20	13·60	16·96
	14·44	16·28	16·10
	17·46	17·22	16·62
	16·80	15·40	14·30
Year 2	14·98	17·02	12·16
	15·90	14·36	14·34
	15·80	13·06	16·84
	17·58	13·10	13·46
Year 3	14·14	12·00	17·88
	12·14	14·74	18·98
	11·86	14·50	16·22
	15·24	12·86	15·12

Analyse these data and report on your conclusions. (Lond. B.Sc.)

19. The table shows three sets of observed values y_{ij} of a variable y corresponding to fixed values x_j of a related variable x ($j = 1, 2, \ldots, 6$; $i = 1, 2, 3$). The standard linear regression model may be assumed, viz. that the y_{ij} belong independently to normal distributions with means $a_i + \beta_i x_j$ and common variance σ^2. Show that a_1, a_2, a_3 may reasonably be taken to have a common value, a, and calculate 98 % confidence limits for a.

x_j	y_{1j}	y_{2j}	y_{3j}
2·6	16·5	12·0	28·2
3·6	21·5	15·2	33·7
4·4	25·3	12·4	40·6
5·3	27·7	14·4	43·2
5·7	25·8	17·1	46·9
6·0	29·0	19·1	47·8

(Manch. Dip.)

20. A random variable $\tilde{y}$ is normally distributed with mean $a+bx$ and known variance σ^2. At each of n values of x a value of $\tilde{y}$ is observed giving n pairs of observations, (x_1, y_1), (x_2, y_2), ..., (x_n, y_n). Obtain the least squares estimates of a and b and find the posterior standard deviation of b.

Suppose that observations are made on two different days, n_1 on the first day and n_2 on the second. There are practical reasons for supposing that conditions are not constant from day to day but vary in such a way as to affect a but not b. Show that the least squares estimate of the common b derived from the two samples is

$$\hat{b}_{12} = \frac{w_1\hat{b}_1 + w_2\hat{b}_2}{w_1 + w_2},$$

where $\hat{b}_1$ and $\hat{b}_2$ are the least squares estimates of b from the first and second samples, respectively, and w_1 and w_2 are the corresponding sums of squares of the x's about their means. Obtain the posterior standard deviation of b.

Hence show that this posterior standard deviation is not less than that which would have been obtained if the combined samples had been treated as if a had remained constant. Comment briefly on the practical implications of this result giving special consideration to the case when the standard deviations are equal. (Wales Dip.)

21. The $(n \times 1)$ vector random variable $\mathbf{x}$ is such that $\mathscr{E}(\mathbf{x}) = \mathbf{A\theta}$ and $\mathscr{D}^2(\mathbf{x}) = \sigma^2\mathbf{V}$, where $\mathbf{A}$ is a known $(n \times p)$ matrix of rank $p < n$, $\boldsymbol{\theta}$ is a $(p \times 1)$ vector of parameters and $\mathbf{V}$ is a non-singular matrix.

Show that the method of least squares gives

$$\hat{\boldsymbol{\theta}} = (\mathbf{A'V^{-1}A})^{-1}\mathbf{A'V^{-1}x},$$

as an estimate of $\boldsymbol{\theta}$.

Let $\mathbf{u}$ be a $(n \times 1)$ vector each of whose elements is unity and let $\mathbf{I}$ be the unit matrix. If $\mathbf{V} = (1-\rho)\mathbf{I} + \rho\mathbf{uu'}$ and $\mathbf{A} = \mathbf{u}$, so that $\boldsymbol{\theta}$ is a scalar, show that

$$\mathbf{V}^{-1} = \mathbf{I}/(1-\rho) - \rho\mathbf{uu'}/(1-\rho)\,[1+(n-1)\rho].$$

Hence show that, in this case,

$$\hat{\boldsymbol{\theta}} = \mathbf{u'x}/n,$$

is the least squares estimate of $\boldsymbol{\theta}$ and determine the variance of this estimate. (Leic. Gen.)

22. The distance between the shoulders of the larger left valve and the lengths of specimens of *Bairdia oklahomaensis* from two different geological levels (from R. H. Shaver, *J. Paleontology*, **34**, 656, 1950) are given in the following table:

Level 1		Level 2	
Distance between shoulders (μ)	Length (μ)	Distance between shoulders (μ)	Length (μ)
631	1167	682	1257
606	1222	631	1227
682	1278	631	1237
480	1045	707	1368
606	1151	631	1227
556	1172	682	1262
429	970	707	1313
454	1166	656	1283
		682	1298
		656	1283
		672	1278

Test for any differences between the two levels in respect of the regression of distance on length, and obtain 95 % confidence intervals for each of the regression coefficients. (Leic. Stat.)

23. F. C. Steward and J. A. Harrison (*Ann. Bot.*, N.S., 3, 1939) considered an experiment on the absorption and accumulation of salts by living plant cells. Their data pertain to the rate of uptake of rubidium (Rb) and bromide (Br) ions by potato slices after immersion in a solution of rubidium bromide for various numbers of hours. The uptake was measured in the number of milligramme equivalents per 1000 g of water in the potato tissue.

	Time of immersion (hours)	Mg. equivalents per 1000 g of water in the tissue	
		Rb	Br
	21·7	7·2	0·7
	46·0	11·4	6·4
	67·0	14·2	9·9
	90·2	19·1	12·8
	95·5	20·0	15·8
Total	320·4	71·9	45·6

On the assumption that the rates of uptake of both kinds of ions are linear with respect to time, determine the two regression equations giving the rates of change, and test the hypothesis that the two rates are, in fact, equal. Also, determine the mean uptake of the Rb and Br ions and test their equality.

Give a diagrammatic representation of the data and of the two regression lines. (Leic. Gen.)

24. An experimenter takes measurements, y, of a property of a liquid while it is being heated. He takes the measurements at minute intervals for 15 min beginning 1 min after starting the heating apparatus. He repeats the experiment with a different liquid and obtains 15 measurements at minute intervals as before. The 30 measurements of y are given in the table below:

Time (min)	1st experiment, y	2nd experiment, y	Time (min)	1st experiment, y	2nd experiment, y
1	1·51	1·35	9	4·53	3·84
2	3·80	1·86	10	3·00	4·99
3	4·39	3·16	11	3·83	4·62
4	1·97	0·63	12	4·80	4·35
5	3·34	0·69	13	2·54	5·93
6	3·39	3·00	14	5·24	7·16
7	4·86	4·53	15	6·11	5·82
8	3·81	2·38			

Before carrying out the experiments he had anticipated from theoretical considerations that:

(1) the slope of the regression lines of y on time would be greater than zero for both liquids;

(2) the slopes of the two regression lines would differ from one another;
(3) the difference between the slopes would be 0·25.

Examine whether the data are consistent with these hypotheses.

The experimenter wishes to estimate the values of y at the start of the experiments (time 0) for both liquids and also the time at which the values of y are the same for both liquids.

Derive these estimates attaching standard deviations to the first two.

(Lond. B.Sc.)

25. The $n \times 1$ vector of observations $\mathbf{Y}$ has expectation $\mathbf{a}\boldsymbol{\theta}$, where $\mathbf{a}$ is an $n \times p$ matrix of known constants of rank $p < n$, and $\boldsymbol{\theta}$ is a $p \times 1$ vector of unknown parameters. The components of $\mathbf{Y}$ are uncorrelated and have equal variance σ^2. The residual sum of squares is denoted $\mathbf{Y}'\mathbf{r}\mathbf{Y}$. Prove that

$$\mathbf{r}^2 = \mathbf{r}, \quad \mathbf{r}\mathbf{a} = \mathbf{0}.$$

Suppose now that $\mathbf{Y}$ has expectation

$$\mathbf{a}\boldsymbol{\theta} + \mathbf{b}\phi,$$

where $\mathbf{b}$ is an $n \times 1$ column linearly independent of the columns of $\mathbf{a}$ and ϕ is an unknown scalar. Prove that the least squares estimate of ϕ is

$$\frac{\mathbf{b}'\mathbf{r}\mathbf{Y}}{\mathbf{b}'\mathbf{r}\mathbf{b}}$$

and has variance $\sigma^2/(\mathbf{b}'\mathbf{r}\mathbf{b})$. Show how to estimate σ^2. (Lond. M.Sc.)

26. In each of two laboratories the quadratic regression of a variable y on a variable x is estimated from eleven observations of y, one at each of the values 0, 1, 2, ..., 10 for x. The fitted regressions are

$$y = 2{\cdot}11 + 0{\cdot}71x + 0{\cdot}12x^2 \quad \text{in the first laboratory,}$$
$$y = 2{\cdot}52 + 0{\cdot}69x + 0{\cdot}13x^2 \quad \text{in the second laboratory.}$$

The residual mean square in the first laboratory is 0·24 and in the second laboratory, 0·29. Assuming that the regression of y on x is, in fact, quadratic, and that variation of y about the regression curve is normal, estimate the probability that, if one further measurement is carried out in each laboratory with $x = 5$, the value of y observed in the first laboratory will be less than the value observed in the second laboratory.

(Lond. Dip.)

27. To investigate the effect of two quantitative factors A, B on the yield of an industrial process an experiment consisting of 13 independent runs was performed using the following pairs of levels (x_1, x_2) the last being used five times:

x_1	-1	-1	1	1	$-\sqrt{2}$	$\sqrt{2}$	0	0	0
x_2	-1	1	-1	1	0	0	$-\sqrt{2}$	$\sqrt{2}$	0

It is assumed that the effect of level x_1 of A and level x_2 of B can be represented by a second-degree polynomial

$$\phi(x_1, x_2) = \beta_0 + \beta_1 x_1 + \beta_2 x_2 + \beta_{11} x_1^2 + \beta_{12} x_1 x_2 + \beta_{22} x_2^2,$$

and that variations in yield from run to run are distributed independently with variance σ^2. Show that the least squares estimates of the coefficients in ϕ are given by

$$10\hat{\beta}_0 = 2\Sigma y_j - (\Sigma y_j x_{1j}^2 + \Sigma y_j x_{2j}^2),$$

$$8\hat{\beta}_i = \Sigma y_j x_{ij} \quad (i = 1, 2),$$

$$160\hat{\beta}_{11} = 23\Sigma y_j x_{1j}^2 + 3\Sigma y_j x_{2j}^2 - 16\Sigma y_j,$$

$$160\hat{\beta}_{22} = 3\Sigma y_j x_{1j}^2 + 23\Sigma y_j x_{2j}^2 - 16\Sigma y_j,$$

$$4\hat{\beta}_{12} = \Sigma y_j x_{1j} x_{2j},$$

where y_j, x_{1j}, x_{2j} are respectively the yield and the levels of A, B for run j. Show also that the least squares estimate of $\phi(x_1, x_2)$ for any given combination of levels (x_1, x_2) has variance

$$\sigma^2(\tfrac{1}{5} - \tfrac{3}{40}\rho^2 + \tfrac{23}{160}\rho^4), \quad \text{where} \quad \rho^2 = x_1^2 + x_2^2.$$

Indicate briefly how you would use the results of this experiment in recommending values of x_1 and x_2 for a future experiment. (Camb. Dip.)

28. A surveyor measures angles whose true values are λ and μ and then makes a determination of the angle $\lambda + \mu$. His measurements are x, y and z, and have no bias, but are subject to independent random errors of zero mean and variance σ^2. Apply the method of least squares to derive estimates $\hat{\lambda}$, $\hat{\mu}$, and show that the variance of $\hat{\lambda}$ is $\frac{2}{3}\sigma^2$.

Suppose now that all measurements are possibly subject to an unknown constant bias β. Show how to estimate β from x, y and z, and, assuming σ^2 known and the errors normally distributed, give a significance test of the hypothesis $\beta = 0$. (Lond. B.Sc.)

29. To determine a law of cooling, observations of temperature T are made at seven equally spaced instants t, which are taken to be

$$t = -3, -2, -1, 0, 1, 2, 3.$$

It is supposed that the temperature readings T are equal to a quadratic function $f(t)$ of t, plus a random observation error which is normally distributed with zero mean and known variance σ^2, the errors of all readings being independent. Taking the form

$$f(t) = a + bt + c(t^2 - 4),$$

find the estimates of a, b and c which minimize the sum of squares of the differences $T - f(t)$. Prove that the posterior variances of a, b and c are

$$\frac{\sigma^2}{7}, \quad \frac{\sigma^2}{28} \quad \text{and} \quad \frac{\sigma^2}{84},$$

respectively.

Indicate the corresponding results when $f(t)$ is expressed in the alternative form

$$f(t) = a' + b't + c't^2.$$

(Camb. N.S.)

30. The random variables $Y_1, \ldots, Y_n$ have covariance matrix $\sigma^2\mathbf{v}$, where $\mathbf{v}$ is a known positive definite $n \times n$ matrix,

$$\mathscr{E}(Y_i) = \theta x_i,$$

where $x_1, \ldots, x_n$ are known constants and θ is an unknown (scalar) parameter. Derive from first principles the least squares estimate, $\hat{\theta}$, of θ and obtain the variance of $\hat{\theta}$. Compare this variance with that of the 'ordinary' least squares estimate $\Sigma x_i Y_i / \Sigma x_i^2$, when $\mathbf{v}$ is the matrix with 1 in the diagonal elements and ρ in all off-diagonal elements.

(Lond. M.Sc).

APPENDIX

Two-sided tests for the χ^2-distribution

ν	5 %		1 %		0·1 %	
	$\underline{\chi}^2$	$\bar{\chi}^2$	$\underline{\chi}^2$	$\bar{\chi}^2$	$\underline{\chi}^2$	$\bar{\chi}^2$
1	$0{\cdot}0^{2}31593$	7·8168	$0{\cdot}0^{3}13422$	11·345	$0{\cdot}0^{5}14026$	16·266
2	0·084727	9·5303	0·017469	13·285	$0{\cdot}0^{2}18055$	18·468
3	0·29624	11·191	0·101048	15·127	0·022097	20·524
4	0·60700	12·802	0·26396	16·901	0·083097	22·486
5	0·98923	14·369	0·49623	18·621	0·19336	24·378
6	1·4250	15·897	0·78565	20·296	0·35203	26·214
7	1·9026	17·392	1·1221	21·931	0·55491	28·004
8	2·4139	18·860	1·4978	23·533	0·79722	29·754
9	2·9532	20·305	1·9069	25·106	1·0745	31·469
10	3·5162	21·729	2·3444	26·653	1·3827	33·154
11	4·0995	23·135	2·8069	28·178	1·7185	34·812
12	4·7005	24·525	3·2912	29·683	2·0791	36·446
13	5·3171	25·900	3·7949	31·170	2·4620	38·058
14	5·9477	27·263	4·3161	32·641	2·8651	39·650
15	6·5908	28·614	4·8530	34·097	3·2865	41·225
16	7·2453	29·955	5·4041	35·540	3·7248	42·783
17	7·9100	31·285	5·9683	36·971	4·1786	44·325
18	8·5842	32·607	6·5444	38·390	4·6468	45·854
19	9·2670	33·921	7·1316	39·798	5·1281	47·370
20	9·9579	35·227	7·7289	41·197	5·6218	48·874
21	10·656	36·525	8·3358	42·586	6·1269	50·366
22	11·361	37·818	8·9515	43·967	6·6428	51·848
23	12·073	39·103	9·5755	45·340	7·1688	53·320
24	12·791	40·383	10·2073	46·706	7·7043	54·782
25	13·514	41·658	10·846	48·064	8·2487	56·236
26	14·243	42·927	11·492	49·416	8·8016	57·682
27	14·977	44·192	12·145	50·761	9·3625	59·119
28	15·716	45·451	12·803	52·100	9·9310	60·549
29	16·459	46·707	13·468	53·434	10·507	61·972
30	17·206	47·958	14·138	54·762	11·089	63·388
31	17·958	49·205	14·813	56·085	11·678	64·798
32	18·713	50·448	15·494	57·403	12·274	66·202
33	19·472	51·688	16·179	58·716	12·875	67·599
34	20·235	52·924	16·869	60·025	13·482	68·991
35	21·001	54·157	17·563	61·330	14·094	70·378
36	21·771	55·386	18·261	62·630	14·712	71·759
37	22·543	56·613	18·964	63·927	15·335	73·136
38	23·319	57·836	19·670	65·219	15·963	74·507
39	24·097	59·057	20·380	66·508	16·595	75·874
40	24·879	60·275	21·094	67·793	17·232	77·236

Two-sided tests for the χ^2-distribution (*cont.*)

ν	5 %		1 %		0·1 %	
	$\underline{\chi}^2$	$\overline{\chi}^2$	$\underline{\chi}^2$	$\overline{\chi}^2$	$\underline{\chi}^2$	$\overline{\chi}^2$
41	25·663	61·490	21·811	69·075	17·873	78·595
42	26·449	62·703	22·531	70·354	18·518	79·948
43	27·238	63·913	23·255	71·629	19·168	81·298
44	28·029	65·121	23·982	72·901	19·821	82·645
45	28·823	66·327	24·712	74·170	20·478	83·987
46	29·619	67·530	25·445	75·437	21·139	85·326
47	30·417	68·731	26·181	76·700	21·803	86·661
48	31·218	69·931	26·919	77·961	22·471	87·992
49	32·020	71·128	27·660	79·220	23·142	89·321
50	32·824	72·323	28·404	80·475	23·816	90·646
51	33·630	73·516	29·150	81·729	24·494	91·968
52	34·439	74·708	29·898	82·979	25·174	93·287
53	35·248	75·897	30·649	84·228	25·858	94·603
54	36·060	77·085	31·403	85·474	26·544	95·916
55	36·873	78·271	32·158	86·718	27·233	97·227
56	37·689	79·456	32·916	87·960	27·925	98·535
57	38·505	80·639	33·675	89·200	28·620	99·840
58	39·323	81·820	34·437	90·437	29·317	101·142
59	40·143	83·000	35·201	91·673	30·016	102·442
60	40·965	84·178	35·967	92·907	30·719	103·74
61	41·787	85·355	36·735	94·139	31·423	105·03
62	42·612	86·531	37·504	95·369	32·130	106·33
63	43·437	87·705	38·276	96·597	32·839	107·62
64	44·264	88·878	39·049	97·823	33·551	108·91
65	45·092	90·049	39·824	99·048	34·264	110·19
66	45·922	91·219	40·600	100·271	34·980	111·48
67	46·753	92·388	41·379	101·492	35·698	112·76
68	47·585	93·555	42·159	102·71	36·418	114·04
69	48·418	94·722	42·940	103·93	37·140	115·32
70	49·253	95·887	43·723	105·15	37·864	116·59
71	50·089	97·051	44·508	106·36	38·590	117·87
72	50·926	98·214	45·294	107·58	39·317	119·14
73	51·764	99·376	46·081	108·79	40·047	120·41
74	52·603	100·536	46·870	110·00	40·778	121·68
75	53·443	101·696	47·661	111·21	41·511	122·94
76	54·284	102·85	48·452	112·42	42·246	124·21
77	55·126	104·01	49·245	113·62	42·983	125·47
78	55·969	105·17	50·040	114·83	43·721	126·73
79	56·814	106·32	50·836	116·03	44·461	127·99
80	57·659	107·48	51·633	117·23	45·203	129·25
81	58·505	108·63	52·431	118·44	45·946	130·51
82	59·352	109·79	53·230	119·64	46·690	131·76
83	60·200	110·94	54·031	120·84	47·436	133·02
84	61·049	112·09	54·833	122·03	48·184	134·27
85	61·899	113·24	55·636	123·23	48·933	135·52

Two-sided tests for the χ^2-distribution (*cont.*)

	5%		1%		0·1%	
ν	$\underline{\chi}^2$	$\bar{\chi}^2$	$\underline{\chi}^2$	$\bar{\chi}^2$	$\underline{\chi}^2$	$\bar{\chi}^2$
86	62·750	114·39	56·440	124·43	49·684	136·77
87	63·601	115·54	57·245	125·62	50·436	138·02
88	64·454	116·68	58·052	126·81	51·189	139·26
89	65·307	117·83	58·859	128·01	51·944	140·51
90	66·161	118·98	59·668	129·20	52·700	141·75
91	67·016	120·12	60·477	130·39	53·457	142·99
92	67·871	121·26	61·288	131·58	54·216	144·23
93	68·728	122·41	62·100	132·76	54·976	145·47
94	69·585	123·55	62·912	133·95	55·738	146·71
95	70·443	124·69	63·726	135·14	56·500	147·95
96	71·302	125·83	64·540	136·32	57·264	149·19
97	72·161	126·97	65·356	137·51	58·029	150·42
98	73·021	128·11	66·172	138·69	58·795	151·66
99	73·882	129·25	66·990	139·87	59·562	152·89
100	74·744	130·39	67·808	141·05	60·331	154·12

The values of $\underline{\chi}^2$ and $\bar{\chi}^2$ satisfy the equations

$$\int_0^{\underline{\chi}^2} + \int_{\bar{\chi}^2}^{\infty} 2^{-\frac{1}{2}\nu}\{\Gamma(\tfrac{1}{2}\nu)\}^{-1} e^{-\frac{1}{2}x} x^{\frac{1}{2}\nu-1}\,dx = \alpha$$

and

$$(\underline{\chi}^2)^{\frac{1}{2}\nu} e^{-\frac{1}{2}\underline{\chi}^2} = (\bar{\chi}^2)^{\frac{1}{2}\nu} e^{-\frac{1}{2}\bar{\chi}^2}$$

for $\alpha = 0{\cdot}05$, 0·01 and 0·001.

Taken, with permission, from Lindley, D. V., East, D. A. and Hamilton, P. A. 'Tables for making inferences about the variance of a normal distribution.' *Biometrika*, **47**, 433–8.

BIBLIOGRAPHY

ALEXANDER, H. W. (1961). *Elements of Mathematical Statistics.* New York: John Wiley and Sons Inc.

BIRNBAUM, Z. W. (1962). *Introduction to Probability and Mathematical Statistics.* New York: Harper and Bros.

BLACKWELL, D. and GIRSHICK, M. A. (1954). *Theory of Games and Statistical Decisions.* New York: John Wiley and Sons Inc.

BRUNK, H. D. (1960). *An Introduction to Mathematical Statistics.* Boston: Ginn and Co. Ltd.

CHERNOFF, H. and MOSES, L. E. (1959). *Elementary Decision Theory.* New York: John Wiley and Sons Inc.

COCHRAN, W. G. (1963). *Sampling Techniques.* New York: John Wiley and Sons Inc.

COCHRAN, W. G. and COX, G. M. (1957). *Experimental Designs.* New York: John Wiley and Sons Inc.

COX, D. R. (1958). *Planning of Experiments.* New York: John Wiley and Sons Inc.

CRAMÉR, H. (1946). *Mathematical Methods of Statistics.* Princeton University Press.

DAVID, F. N. (1954). *Tables of the Ordinates and Probability Integral of the Distribution of the Correlation Coefficient in Small Samples.* Cambridge University Press.

DAVIES, O. L. (editor) (1957). *Statistical Methods in Research and Production,* 3rd edition. Edinburgh: Oliver and Boyd.

DAVIS, H. T. (1933). *Tables of the Higher Mathematical Functions,* volume I. Bloomington: Principia Press.

FISHER, R. A. (1958). *Statistical Methods for Research Workers,* 13th edition. Edinburgh: Oliver and Boyd.

FISHER, R. A. (1959). *Statistical Methods and Scientific Inference,* 2nd edition. Edinburgh: Oliver and Boyd.

FISHER, R. A. (1960). *The Design of Experiments,* 7th edition. Edinburgh: Oliver and Boyd.

FISHER, R. A. and YATES, F. (1963). *Statistical Tables for Biological, Agricultural and Medical Research,* 6th edition. Edinburgh: Oliver and Boyd.

FRASER, D. A. S. (1958). *Statistics: an Introduction.* New York: John Wiley and Sons Inc.

GREENWOOD, J. A. and HARTLEY, H. O. (1962). *Guide to Tables in Mathematical Statistics.* Princeton University Press.

HOEL, P. G. (1960). *Introduction to Mathematical Statistics,* 2nd edition. New York: John Wiley and Sons Inc.

HOGG, R. V. and CRAIG, A. T. (1959). *Introduction to Mathematical Statistics.* New York: Macmillan.

JEFFREYS, H. (1961). *Theory of Probability*, 3rd edition. Oxford: Clarendon Press.

KEMPTHORNE, O. (1952). *The Design and Analysis of Experiments*. New York: John Wiley and Sons Inc.

KENDALL, M. G. and STUART, A. (1958, 1961). *The Advanced Theory of Statistics*. Two volumes. London: Griffin and Co.

LINDLEY, D. V. and MILLER, J. C. P. (1961). *Cambridge Elementary Statistical Tables*. Cambridge University Press.

PEARSON, E. S. and HARTLEY, H. O. (1958). *Biometrika Tables for Statisticians*, volume I. Cambridge University Press.

PLACKETT, R. L. (1960). *Principles of Regression Analysis*. Oxford: Clarendon Press.

RAIFFA, H. and SCHLAIFER, R. (1961). *Applied Statistical Decision Theory*. Boston: Harvard University Graduate School of Business Administration.

SCHEFFÉ, H. (1959). *The Analysis of Variance*. New York: John Wiley and Sons Inc.

SCHLAIFER, R. (1959). *Probability and Statistics for Business Decisions*. New York: McGraw-Hill Book Co. Inc.

TUCKER, H. G. (1962). *An Introduction to Probability and Mathematical Statistics*. New York: Academic Press.

WEISS, L. (1961). *Statistical Decision Theory*. New York: McGraw-Hill Book Co. Inc.

WILKS, S. S. (1962). *Mathematical Statistics*. New York: John Wiley and Sons Inc.

SUBJECT INDEX

INDEX OF NOTATIONS

(much of the notation is defined in Part I)